Mechanical Vibration

Mechanical Vibration

Fundamentals with Solved Examples

Ivana Kovačić
University of Novi Sad
Serbia

Dragi Radomirović
University of Novi Sad
Serbia

WILEY

This edition first published 2018

Registered Offices
John Wiley & Sons, Inc., 111 River Street, Hoboken, NJ 07030, USA
John Wiley & Sons Ltd, The Atrium, Southern Gate, Chichester, West Sussex, PO19 8SQ, UK

Editorial Office
The Atrium, Southern Gate, Chichester, West Sussex, PO19 8SQ, UK

For details of our global editorial offices, customer services, and more information about Wiley products visit us at www.wiley.com.

Wiley also publishes its books in a variety of electronic formats and by print-on-demand. Some content that appears in standard print versions of this book may not be available in other formats.

Library of Congress Cataloging-in-Publication Data

Names: Kovačić, Ivana, author. | Radomirović, Dragi, author.
Title: Mechanical vibration : fundamentals with solved examples / by Ivana Kovačić, University of Novi Sad, Serbia, Dragi Radomirović, University of Novi Sad, Serbia.
Description: Singapore : Wiley, [2017] | Includes bibliographical references and index. |
Identifiers: LCCN 2017026938 (print) | LCCN 2017027118 (ebook) | ISBN 9781118927571 (pdf) | ISBN 9781118927588 (epub) | ISBN 9781118675151 (cloth)
Subjects: LCSH: Vibration.
Classification: LCC TA355 (ebook) | LCC TA355 .K689 2017 (print) | DDC 620.3–dc23
LC record available at https://lccn.loc.gov/2017026938

Cover design by Wiley
Cover image: Courtesy of the authors

Set in 10/12pt Warnock by SPi Global, Pondicherry, India
Printed and bound in Malaysia by Vivar Printing Sdn Bhd

10 9 8 7 6 5 4 3 2 1

Contents

About the Authors

Ivana Kovačić graduated in Mechanical Engineering from the Faculty of Technical Sciences (FTN), University of Novi Sad, Serbia. She obtained her MSc and PhD in the Theory of Nonlinear Vibrations at the FTN. She is currently a Full Professor of Mechanics at the FTN and the head of the Centre of Excellence for Vibro-Acoustic Systems and Signal Processing CEVAS at the same faculty. Kovačić is the Subject Editor of three academic journals: the Journal of Sound and Vibration, the European Journal of Mechanics A/Solids and Meccanica. Her research involves the use of quantitative and qualitative methods to study differential equations arising from nonlinear dynamics problems mainly in mechanical engineering, and recently also in biomechanics and tree vibrations.

Dragi Radomirović graduated in Mechanical Engineering from the Faculty of Technical Sciences (FTN), University of Novi Sad (UNS), Serbia. He obtained his MSc and PhD in Analytical Mechanics at the FTN. He is a Full Professor of Mechanics at the Faculty of Agriculture, UNS. His research interests are directed towards Mechanical Vibrations and Analytical Mechanics.

Preface

Mechanical Vibrations: Fundamentals with Solved Examples takes a logically organized, clear and thorough problem-solved approach to instructing the reader in the application of certain mechanical principles to derive mathematical models for oscillatory systems, while laying a foundation for vibration engineering analysis and design.

The methodology is mainly based on Analytical Mechanics and Lagrange's formalism, so that the approach, terminology and notation used are consistent, creating a coherent chain that links the chapters in the book. This enables the reader to learn gradually how to treat different oscillatory mechanical systems performing small oscillations.

Chapter 1 contains preliminaries, giving a brief overview of important facts and expressions which are needed as background knowledge from Statics, Kinematics, Kinetics and Strength of Materials. Chapter 2 provides theoretical basics related to Lagrangian Mechanics and the formation of equations of motion from Lagrange's equations of the second kind. Chapters 3–7 contain a variety of examples in which Lagrange's equations of the second kind are used to derive the corresponding equations of motion for small oscillations. Chapter 8 is concerned with oscillatory systems with infinite numbers of degrees of freedom, and Chapter 9 introduces some elements of equivalent stiffness and damping as well as Nonlinear Oscillations theory. There is also one Appendix with certain mathematical topics, based on which the book can be used on a stand-alone basis. There are a Chapter Outline and Objectives at the beginning of each chapter as well as a Theoretical Introduction. This theoretical portion is followed by a large number of fully solved examples, the solutions of which are presented in detail. The examples are arranged in order of increasing difficulty, so that the most difficult exercises are given near the end of the chapters. Some of these problems, inherent in the design and analysis of mechanical systems and engineering structures, are characterised by a complexity and originality that is rarely found in textbooks. Numerous pedagogical features, extensive explanations and absolutely unique techniques that stem from the authors' long-term teaching and research experience are included in the text in order to aid the reader's comprehension and retention. The book is visually rich and all the figures are original and produced by the authors themselves.

We, the authors, believe that the book can be used as a tutorial or supplementary material for self-study for senior students, graduate students, university professors and researchers in mechanics and engineering. We hope that they will find it helpful while indulging themselves in the magnificent world of mechanical oscillatory problems.

Due to unfortunate circumstances, during 2015 both authors lost beloved members of their families: Milorad Radomirović, Nikola Kovačić and Smiljana Radomirović. This book is dedicated to them.

Novi Sad, Serbia, 2016 *Ivana Kovačić and Dragi Radomirović*

1

Preliminaries

Chapter Outline

This chapter provides theoretical fundamentals for the considerations presented in the subsequent chapters. It is divided into four sections, containing some key concepts, facts and expressions from Statics, Kinematics, Kinetics and Strength of Materials.

Chapter Objectives

- To present preliminaries from Statics, Kinematics, Kinetics and Strength of Materials
- To focus only on the key concepts, facts and expressions of interest for the considerations presented in this book
- To provide help to readers to enable them to use this book on a stand-alone basis

1.1 From Statics

1.1.1 Mechanical Systems and Equilibrium Equations

Of interest here is the equilibrium of different systems of forces and torques lying in one plane. They are given separately in Table 1.1.1 together with the corresponding equilibrium equations and their descriptions. Note that before writing these equations, one must create a free-body diagram for the object under consideration if it is not free. The way to do this is described in Table 1.1.2.

1.1.2 Constraints and Free-Body Diagrams

When considering the equilibrium of an object or combinations of objects via equations of motion, it is essential to isolate them from all surrounding bodies. This isolation is accomplished by the free-body diagram, which shows all active forces and active torques acting on the object or combinations of objects as well as forces and torques that exist due to mechanical contacts with surrounding bodies, which represent the so-called *mechanical constraints*. There are several common types that can exist in a plane, and they are collected and described in Table 1.1.2. Note that these forces and torques are also called passive forces and passive torques.

Mechanical Vibration: Fundamentals with Solved Examples, First Edition.
Ivana Kovačić and Dragi Radomirović.

Table 1.1.1

Type	Mechanical model	Equilibrium equations
Concurrent forces	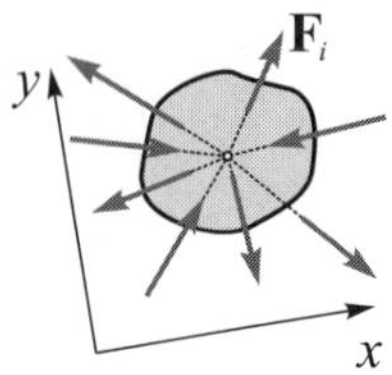	1. $\sum F_x = 0$ 2. $\sum F_y = 0$ Sum of the projections of all forces on two orthogonal axes is equal to zero.
Parallel forces and torques in the same plane	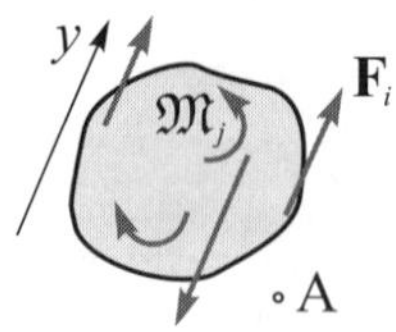	1. $\sum F_y = 0$ 2. $\sum M_{\mathrm{A}} = 0$ Sum of the projections of forces on the axis parallel to the direction of the forces is equal to zero. Sum of the moments about any point A on or off the body is equal to zero.
Arbitrary forces and torques in one plane	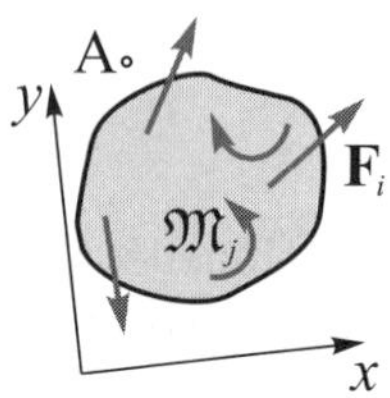	1. $\sum F_x = 0$ 2. $\sum F_y = 0$ 3. $\sum M_{\mathrm{A}} = 0$ Sum of the projections of all forces on two orthogonal axes is equal to zero. Sum of the moments about any point A on or off the body is equal to zero. The alternative equilibrium equations are: i) three equilibrium equations with the zero sum of moments for any three points that are not on the same straight line; ii) one force equilibrium equation in an arbitrary x-direction and two equilibrium equations with the zero sums of moments for any two points that must not lie on a line perpendicular to the x-direction.

1.1.3 Equilibrium Condition Via Virtual Work

Besides the approach based on the equilibrium equation, one can determine and investigate equilibria based on the principle of virtual work.

The virtual work of a force is the scalar product of the vector of the force and the virtual displacement of the point at which it acts. This can be further expressed in the rectangular/Cartesian coordinate system as follows:

$$\delta A^{\mathbf{F}} = \mathbf{F}\cdot\delta\mathbf{r}_{\mathrm{A}} = \left(F_x\mathbf{i} + F_y\mathbf{j}\right)\cdot\left(\delta x_{\mathrm{A}}\mathbf{i} + \delta y_{\mathrm{A}}\mathbf{j}\right) = F_x\delta x_{\mathrm{A}} + F_y\delta y_{\mathrm{A}}. \tag{1.1.1}$$

Table 1.1.2

Type	Schematics	Description
Smooth surface	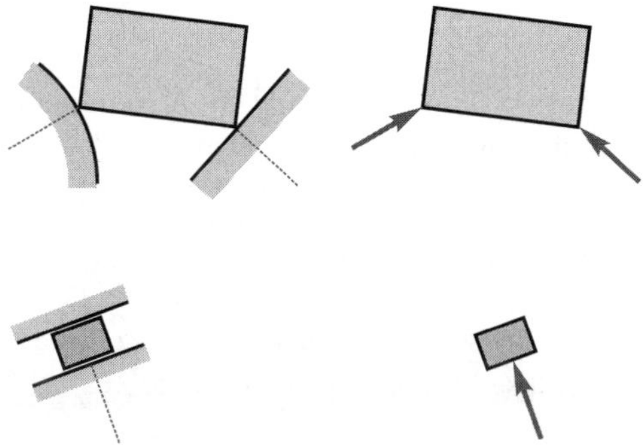	Contact force is compressive and is normal to the tangent at the point of contact.
Weightless straight rod	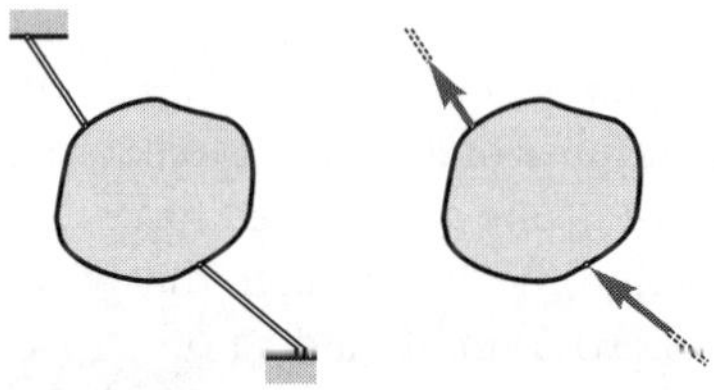	Force exerted is the direction of the straight rod, but it can be tensile or compressive.
Rope/ Cable	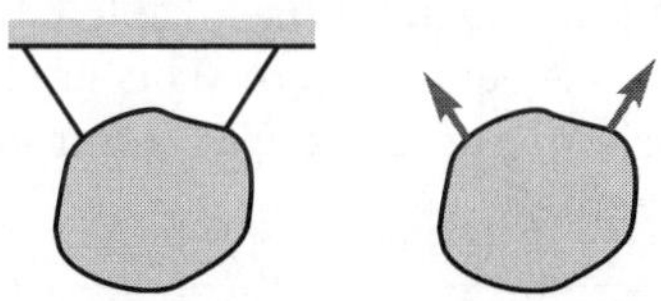	Force exerted is always a tension away from the body in the direction of the rope.
Pin connection	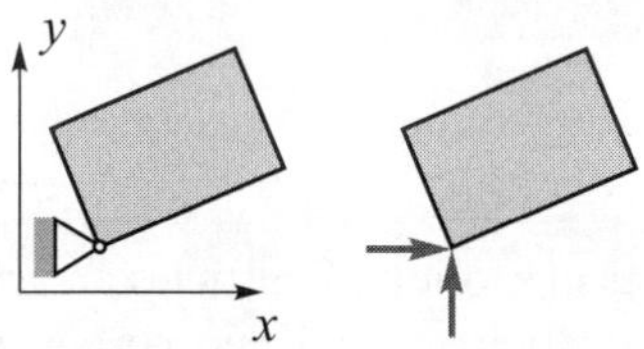	Force exerted lies in the plane normal to the pin axis; this force is usually shown in terms of two orthogonal components (for example, horizontal and vertical).
Roller/ Rocker/ Ball support	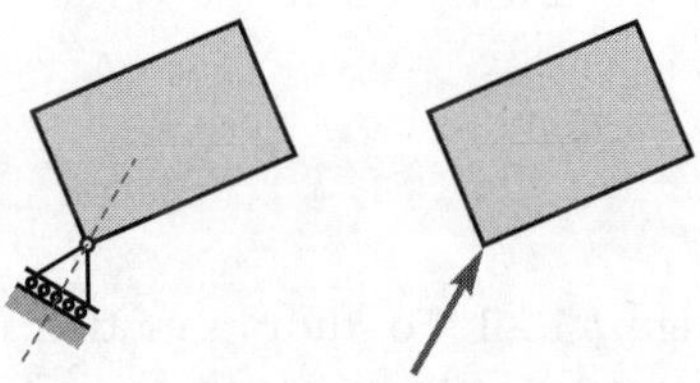	Compressive force normal to the supporting surface/guide.
Fixed or built-in support	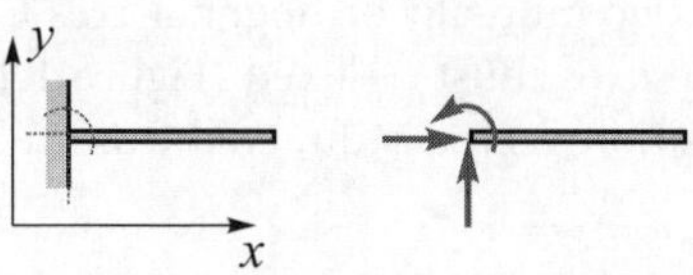	Two orthogonal components of a force and a torque.

The virtual work of a torque on the virtual rotation $\delta\varphi$ can be defined as

$$\delta A^{\mathfrak{M}} = \pm \mathfrak{M}\delta\varphi, \tag{1.1.2}$$

where the plus sign corresponds to the case when the torque helps to increase of the angle of rotation, while the minus sign holds in the opposite case.

For the case of a system of forces and torques, the overall virtual work is the sum of the virtual work of each of them:

$$\delta A = \sum_{j=1}^{u} \delta A^{\mathbf{F}_j} + \sum_{k=1}^{v} \delta A^{\mathfrak{M}_k}. \tag{1.1.3}$$

If the position of the mechanical system is depicted by N generalised coordinates q_i $(i = 1,\ldots, N)$, one can express Equation (1.1.3) in the form

$$\delta A = \sum_{i=1}^{N} Q_{q_i}\, \delta q_i, \tag{1.1.4}$$

where the coefficients Q_{q_i} represent the so-called *generalised forces* (see Section 2.4). In the equilibrium position, the generalised forces are equal to zero:

$$Q_{q_i} = 0. \tag{1.1.5}$$

Note that the number of homogeneous algebraic equations (1.1.5) is equal to the number of generalised coordinates, that is, the number of degrees of freedom. For example, if the system has one degree of freedom, there is one generalised force and one equation (1.1.5) to determine the equilibrium or any other parameter that yields it. For two-degree-of-freedom systems, two equations (1.1.5) exist, and so on.

It should also be noted that in the case of ideal constraints (see Table 1.1.2), the virtual work of the corresponding forces and torques is equal to zero. This implies that when solving the examples related to static equilibria, one does not need to introduce reactions of ideal constraints as their virtual work is zero.

1.2 From Kinematics

Kinematics deals with the geometrical aspects of motion of particles and rigid bodies, as well as with the mathematical description of their motion and certain velocities and accelerations they have over time.

1.2.1 Kinematics of Particles

A particle moving in a plane is shown in Figure 1.2.1. To study its motion, different coordinates can be used. For example, a set of two mutually orthogonal axes x and y with the origin O can be arbitrary chosen, but the axes must be fixed (Figure 1.2.1a). The unit vectors are, respectively, $\mathbf{i}$ and $\mathbf{j}$. The position vector is directed from the origin to the particle and is defined by:

$$\mathbf{r} = x\mathbf{i} + y\mathbf{j}, \tag{1.2.1}$$

where x and y, in general, change over time t, that is, $x = x(t)$ and $y = y(t)$.

The particle's velocity $\mathbf{v}$ and acceleration $\mathbf{a}$ are defined by:

$$\mathbf{v} = \frac{d\mathbf{r}}{dt} = \dot{x}\mathbf{i} + \dot{y}\mathbf{j}, \tag{1.2.2}$$

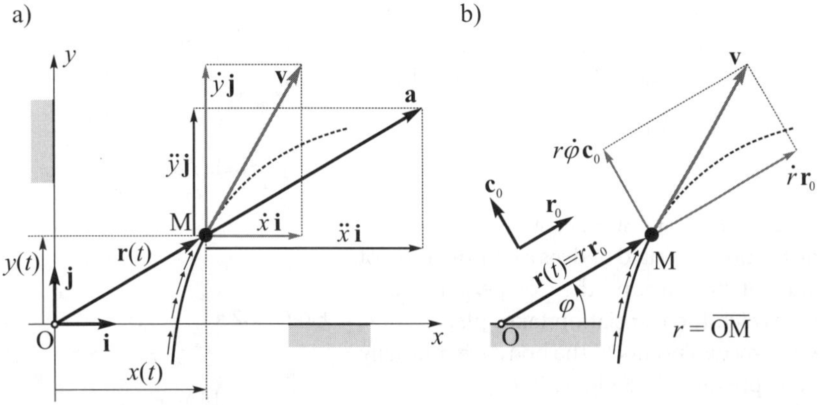

Figure 1.2.1

$$\mathbf{a} = \frac{d\mathbf{v}}{dt} = \frac{d^2\mathbf{r}}{dt^2} = \ddot{x}\mathbf{i} + \ddot{y}\mathbf{j}, \tag{1.2.3}$$

where the overdot indicates differentiation with respect to time.

Besides this, one can use polar coordinates (Figure 1.2.1b) $r = r(t)$ and $\varphi = \varphi(t)$ with the moveable unit vectors $\mathbf{r}_0$ and $\mathbf{c}_0$. The velocity of the particle has two components:

$$\mathbf{v} = \frac{d\mathbf{r}}{dt} = \dot{r}\mathbf{r}_0 + r\dot{\varphi}\mathbf{c}_0. \tag{1.2.4}$$

1.2.2 Kinematics of Rigid Bodies

1.2.2.1 Rigid Body in Translatory Motion

During translatory motion every line in the body remains parallel to its original position at all times. One can distinguish rectilinear translation (Figure 1.2.2), when all points move along parallel straight lines, and curvilinear translation (Figure 1.2.3), when all points move along congruent curves. During translatory motion all points have the same velocity as noted in Figures 1.2.2 and 1.2.3. Thus, specifying the motion of one point enables one to describe completely the translation of the whole body.

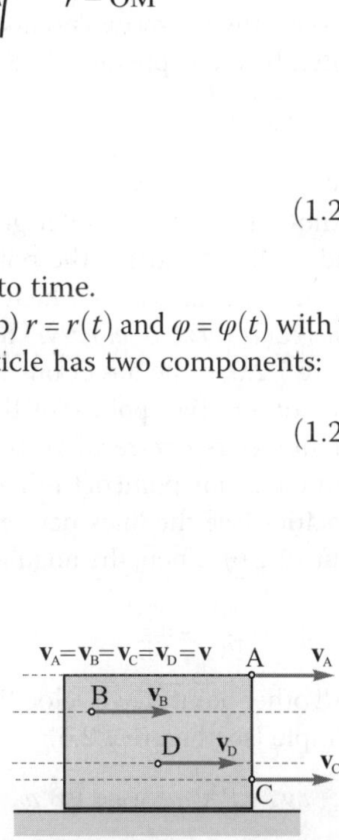

Figure 1.2.2

1.2.2.2 Rigid Body in Fixed-Axis Rotation

During rotation about a fixed axis (Figure 1.2.4) all points in a rigid body (other than those on the axis) move along concentric circular paths around the axis of rotation. Note that the axis of rotation in Figure 1.2.4 passes through the point O and is perpendicular to the plane of the figure. The position of the body is defined by the angle between the fixed line and a line attached to the body (this angle is labelled by φ in Figure 1.2.4) and the angular velocity of the body

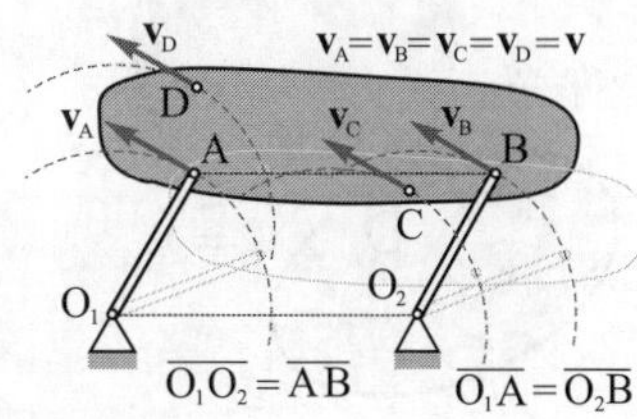

Figure 1.2.3

is $\omega = \dot{\varphi}$. The velocity of each point is proportional to it, as well as to its distance with respect to the axis of rotation. Referring again to Figure 1.2.4, one can write

$$v_{\mathrm{A}} = \overline{\mathrm{OA}}\ \omega, \quad v_{\mathrm{B}} = \overline{\mathrm{OB}}\ \omega, \quad v_{\mathrm{C}} = \overline{\mathrm{OC}}\ \omega. \tag{1.2.5}$$

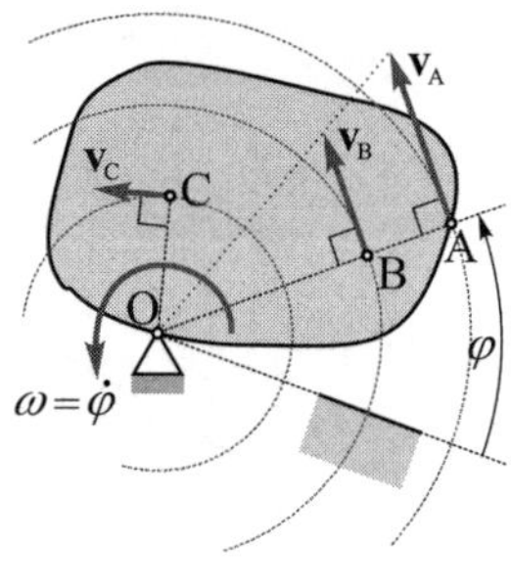

Figure 1.2.4

1.2.2.3 Rigid Body in General Plane Motion

General plane motion of a rigid body is a combination of translation and rotation around the axis perpendicular to the plane in which the translation takes place. Thus, the velocities of any two points of the body are mutually related by the expression (see Figure 1.2.5):

$$\mathbf{v}_{\mathrm{B}} = \mathbf{v}_{\mathrm{A}} + \mathbf{v}_{\mathrm{B/A}}, \tag{1.2.6}$$

where $v_{\mathrm{B/A}} = \overline{\mathrm{AB}}\ \omega$, $\mathbf{v}_{\mathrm{B/A}} \perp \mathrm{AB}$.

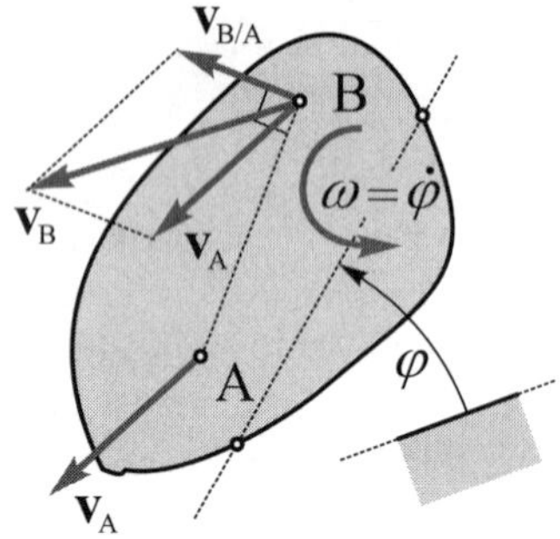

Figure 1.2.5

Another way to consider general plane motion of a rigid body is to locate the point whose instantaneous velocity is equal to zero, which is the so-called *instantaneous centre of zero velocity*. This point is labelled by P in Figure 1.2.6. If the directions of two non-parallel velocities of any two points of the body are known, the instantaneous centre of zero velocity corresponds to the intersection point of the lines perpendicular to these velocities (see the lines perpendicular to $\mathbf{v}_{\mathrm{C}}$ and $\mathbf{v}_{\mathrm{D}}$ in Figure 1.2.6). Then, the angular velocity of the body is

$$\omega = \frac{v_{\mathrm{C}}}{\overline{\overline{\mathrm{PC}}}} = \frac{v_{\mathrm{D}}}{\overline{\overline{\mathrm{PD}}}}. \tag{1.2.7}$$

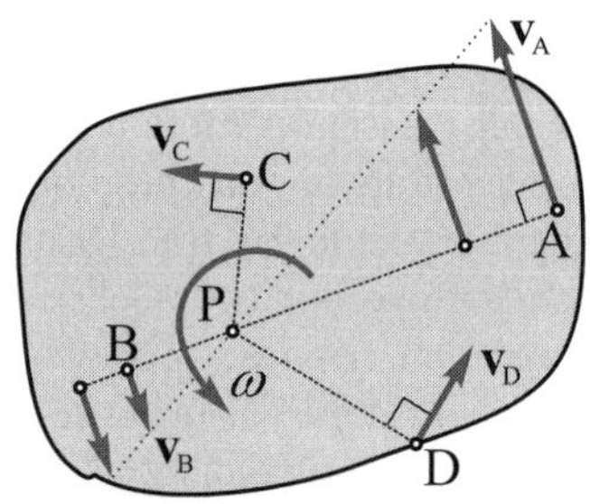

Figure 1.2.6

All other points have velocities proportional to it. For example (see Figure 1.2.6):

$$v_{\mathrm{A}} = \overline{\mathrm{PA}}\ \omega, \quad v_{\mathrm{B}} = \overline{\mathrm{PB}}\ \omega. \tag{1.2.8}$$

It is also of interest to note the case when two bodies roll along each other, but there is no slipping between

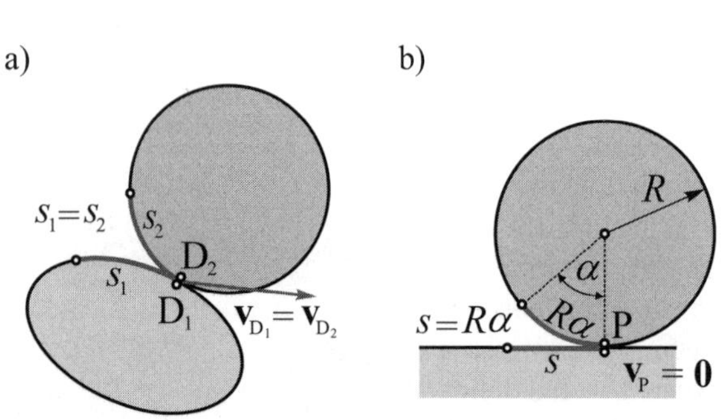

Figure 1.2.7

them, as shown in Figure 1.2.7a,b. Then, the equality of the arcs s_1 and s_2 depicted in Figure 1.2.7a must hold and the velocities of the contact points are equal ($\mathbf{v}_{D_1} = \mathbf{v}_{D_2}$). Analogously, for the disc rolling without slipping along a horizontal fixed plane shown in Figure 1.2.7b, the equality of the arc length $R\alpha$ and the distance s holds. The contact point is on the fixed plane, so its instantaneous velocity is zero and it represents the instantaneous centre of zero velocity ($\mathbf{v}_P = \mathbf{0}$).

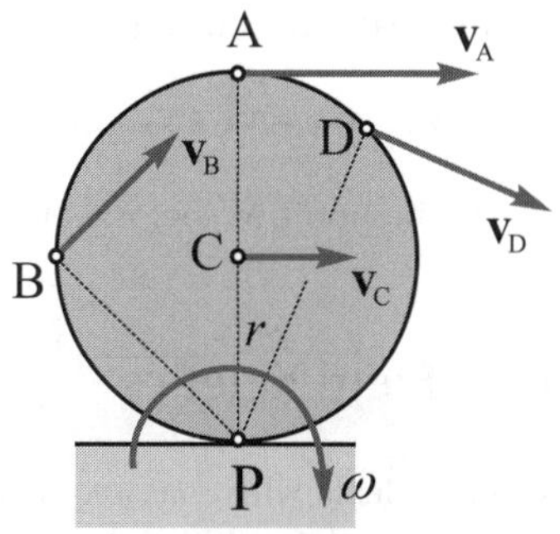

Figure 1.2.8

This case is shown in more detail in Figure 1.2.8. Relating the velocity of the centre of the disc to its angular velocity, one has

$$v_C = \overline{PC}\ \omega = r\omega \ \Rightarrow\ \omega = \frac{v_c}{r}. \tag{1.2.9}$$

The velocities of several other points are also shown.

1.2.3 Kinematics of Particles in Compound Motion

Let us consider the motion of a point M (Figure 1.2.9) relative to a rigid body that moves relatively to a fixed coordinate system depicted by the axes x and y. The motion (trajectory, velocity, acceleration) of the point M with respect to the fixed coordinate system x–y is called *absolute*. The motion of the point M with respect the body and the coordinate system ξ–η attached to the body is called *relative*. The motion of the body with respect to the fixed coordinate systems is called *transportation*.

The theorem on composition of velocities for a compound (composition/resultant) motion states that the absolute velocity of the point is equal to the vector sum of the relative and transportation velocities of the point:

$$\mathbf{v}_M = \mathbf{v}_{rel} + \mathbf{v}_{tr}. \tag{1.2.10}$$

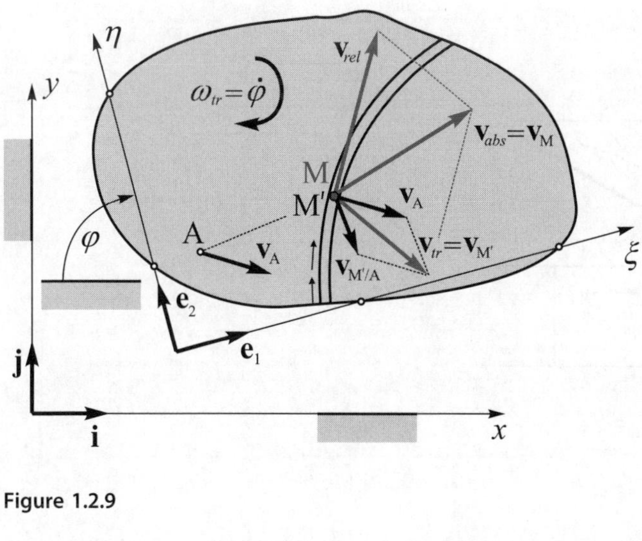

Figure 1.2.9

Note that if the body is in general plane motion, the transportation velocity is given by

$$\mathbf{v}_{tr} = \mathbf{v}_{\mathrm{M}'} = \mathbf{v}_{\mathrm{A}} + \mathbf{v}_{\mathrm{M}'/\mathrm{A}}, \tag{1.2.11}$$

where the point M′ belongs to the body and it is located right below the point M.

1.3 From Kinetics

Kinetics deals with massive objects in motion – particles and rigid bodies – where the former can be subjected to forces and the latter to forces and torques. This subsection contains the basic relationships from Newton's dynamics needed to form their equations of motion. They are all based on free-body diagrams. This implies that the object under consideration must be isolated, and the appropriate forces and torques that exist due to mechanical contacts with surrounding bodies should be introduced as described in Table 1.1.2.

1.3.1 Kinetics of Particles

Newton's second law gives the vector relationship between the force **F** and the acceleration **a** of the particle of mass m:

$$m\mathbf{a} = \mathbf{F}. \tag{1.3.1}$$

When the particle is subjected to the system of concurrent forces $\mathbf{F}_i$, this equation becomes

$$m\mathbf{a} = \sum \mathbf{F}_i. \tag{1.3.2}$$

This vector equation can be expressed in scalar form by using different coordinates (rectangular/Cartesian coordinates, polar coordinates, etc.). For example, in the rectangular coordinates (Figure 1.3.1), Equation (1.3.2) is transformed into two second-order ordinary differential equations

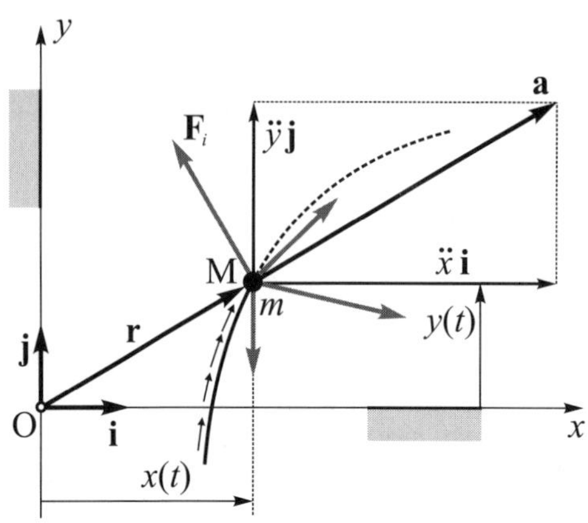

Figure 1.3.1

$$m\ddot{x} = \sum F_x, \tag{1.3.3}$$

$$m\ddot{y} = \sum F_y. \tag{1.3.4}$$

Their integration can give the velocity as well as the equations of motion $x(t)$ and $y(t)$. For the integration, the initial conditions for the particle initial position and initial velocity are needed.

Of interest for the discussions in this book is the form of the kinetic energy. A general expression for the kinetic energy of the particle of mass m is:

$$T = \frac{1}{2}m\mathbf{v}\cdot\mathbf{v} = \frac{1}{2}mv^2, \tag{1.3.5}$$

where $\mathbf{v}$ is the absolute velocity of the particle.

1.3.2 Kinetics of Rigid Bodies

1.3.2.1 Kinetics of Rigid Bodies in Translatory Motion

For a rigid body of mass m and centre of mass C, translating in plane (Figure 1.3.2), the scalar equations of motion in rectangular coordinates have the general form

$$m\ddot{x}_C = \sum F_x, \tag{1.3.6}$$

$$m\ddot{y}_C = \sum F_y, \tag{1.3.7}$$

where the projections of forces stem from external forces (active and passive). The condition for translatory motion is

$$\sum M_C = 0. \tag{1.3.8}$$

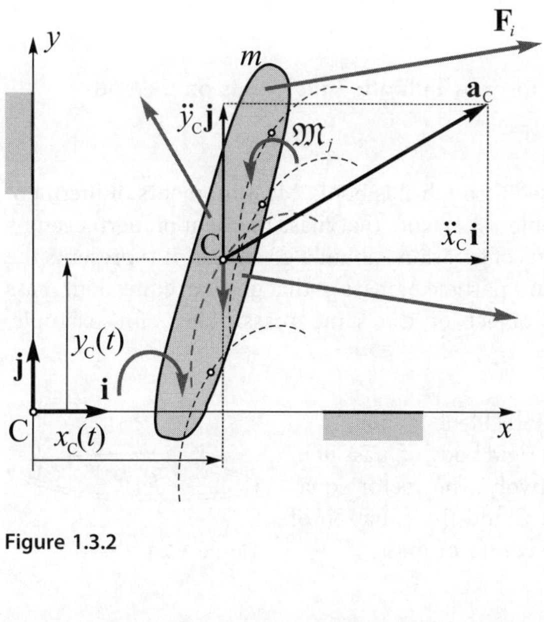

Figure 1.3.2

A general expression for the kinetic energy of the rigid body in translatory motion is:

$$T = \frac{1}{2} m\mathbf{v}_C \cdot \mathbf{v}_C = \frac{1}{2} m v_C^2. \tag{1.3.9}$$

Due to the characteristics of translatory motion (see Section 1.2.2.1), one can use the velocity of any other point of the rigid body instead of v_C.

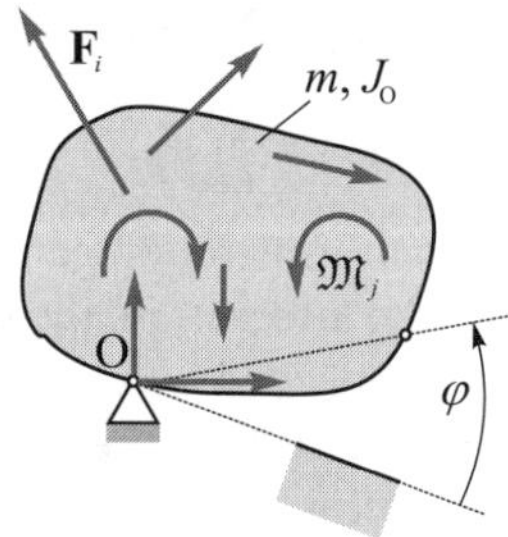

Figure 1.3.3

1.3.2.2 Kinetics of Rigid Bodies in Fixed-Axis Rotation

For a rigid body in fixed-axis rotation, the differential equation of motion is related to a moment equation about the rotation axis through O (Figure 1.3.3) and has the form

$$J_O \ddot{\varphi} = \sum M_O, \tag{1.3.10}$$

where φ is the angle of rotation, and J_O is the mass moment of inertia for the rotation axis through O. The positive sign of the moment corresponds to an increase in the angle of rotation.

A general expression for the kinetic energy of a rigid body in fixed-axis rotation is:

$$T = \frac{1}{2} J_O \omega^2. \tag{1.3.11}$$

The way to calculate the mass moment of inertia for different shapes of uniform bodies is presented below.

On the mass moment of inertia

According to the parallel-axis theorem, the relationship between the mass moment of inertia about the axis z through the centre of mass C and a parallel axis z_1 through another point O (Figure 1.3.4) is given by

$$J_{z_1} = J_z + mb^2. \tag{1.3.12}$$

This can be written in a simplified form by indicating the points on the body:

$$J_O = J_C + m\overline{OC}^2; \tag{1.3.13}$$

this is the form that will mainly be used through this book. Mass moments of inertia of some common shapes are given in Table 1.3.1. Note that mass moment of inertia can be calculated based on the so-called radius of gyration i about a given axis. It represents the perpendicular distance from the axis to a particle of mass m that gives an equivalent mass moment of inertia to the original object of the same mass. Thus, for example, $J_C \equiv J_z = m i_C^2 \equiv m i_z^2$ and $J_O \equiv J_{z_1} = m i_O^2 \equiv m i_{z_1}^2$.

1.3.2.3 Kinetics of Rigid Bodies in General Plane Motion

The general equations of motion for a rigid body of mass m in general plane motion (Figure 1.3.5) involve the vector equation of motion of the centre of mass C and the equation of rotation around the axis through the centre of mass.

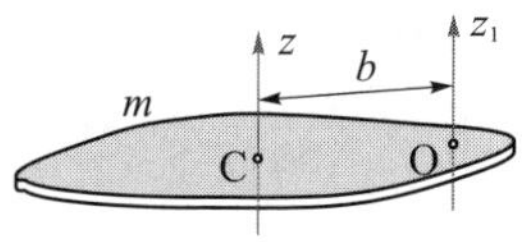

Figure 1.3.4

Table 1.3.1

Shape	Presentation	Mass moment of inertia
Disc	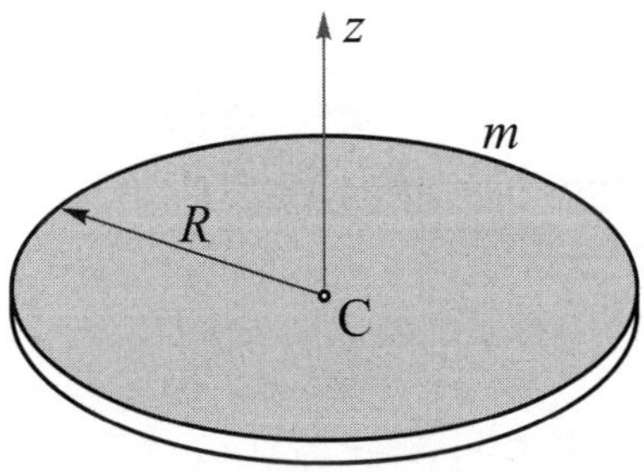	$J_C \equiv J_z = \frac{1}{2} m R^2$
Circular sector	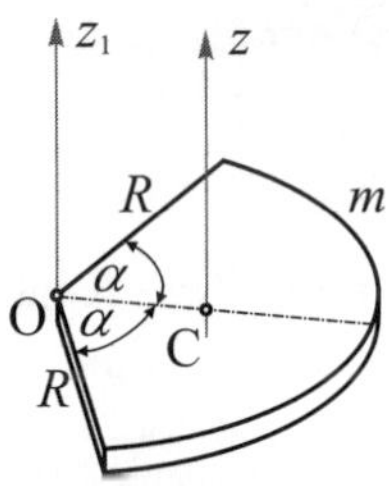	$J_O \equiv J_{z_1} = \frac{1}{2} m R^2$ $\overline{OC} = \frac{2}{3} R \frac{\sin \alpha}{\alpha}$
Ring	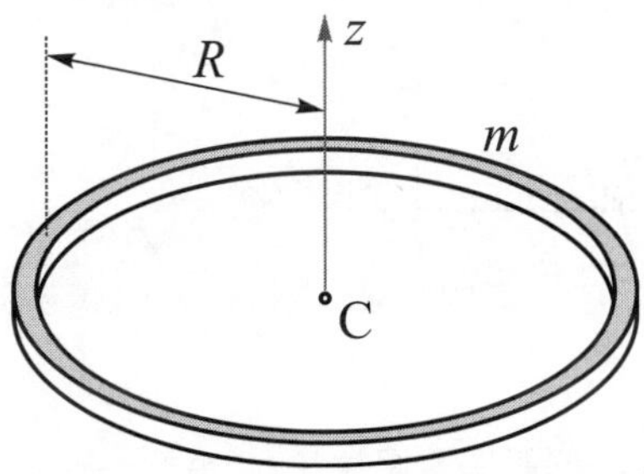	$J_C \equiv J_z = m R^2$
Circular arch	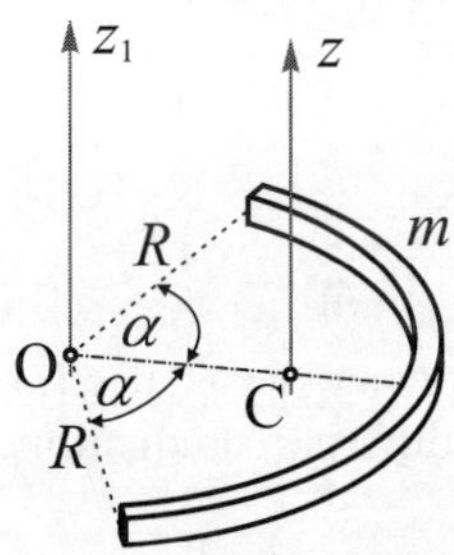	$J_O \equiv J_{z_1} = m R^2$ $\overline{OC} = R \frac{\sin \alpha}{\alpha}$
Rectangle	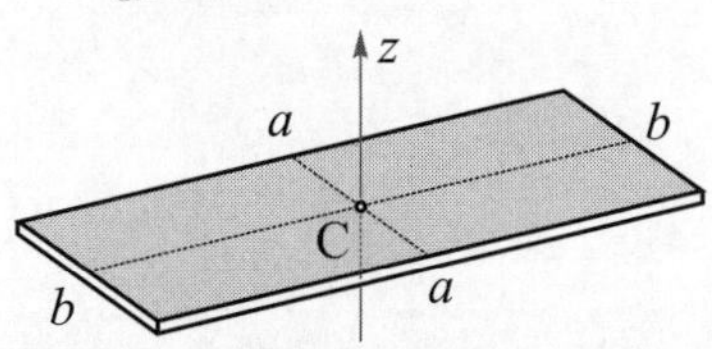	$J_C \equiv J_z = \frac{1}{12} m \left(a^2 + b^2\right)$

(Continued)

Table 1.3.1 (Continued)

Shape	Presentation	Mass moment of inertia
Ellipse	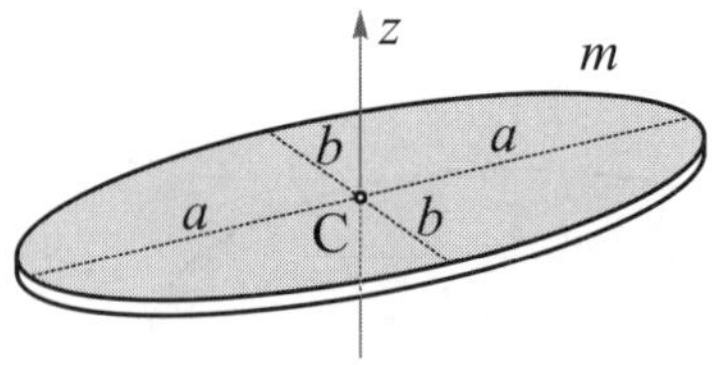	$J_C \equiv J_z = \frac{1}{4} m \left(a^2 + b^2\right)$
Slender rod	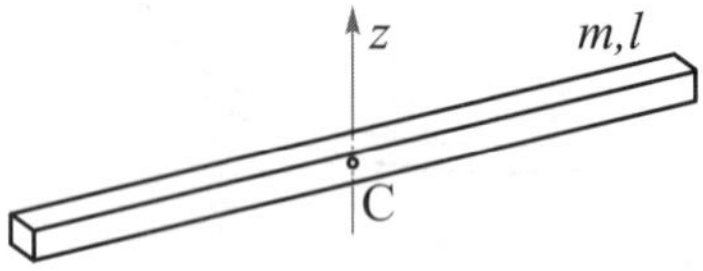	$J_C \equiv J_z = \frac{1}{12} m l^2$

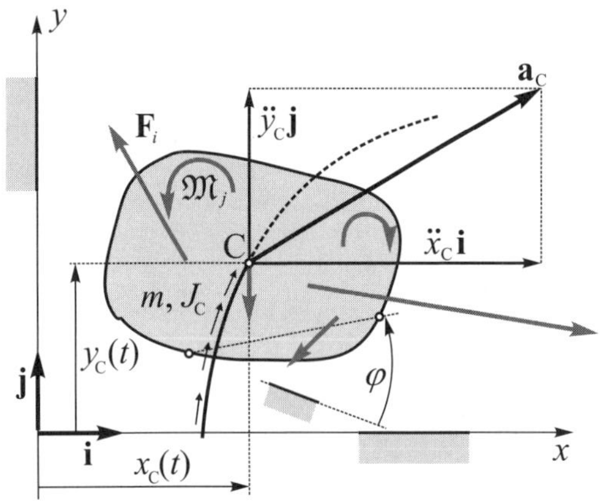

Figure 1.3.5

Using the rectangular coordinates for the former (see Figure 1.3.5), one can write them in the form of three second-order ordinary differential equations

$$m\ddot{x}_C = \sum F_x, \tag{1.3.14}$$

$$m\ddot{y}_C = \sum F_y. \tag{1.3.15}$$

$$J_C\ddot{\varphi} = \sum M_C. \tag{1.3.16}$$

Note that the mass moment of inertia corresponds to the axis through the centre of mass as well as that the moment on the right-hand side of Equation (1.3.16) should be calculated for the centre of mass. The positive sign of the moments corresponds to an increase in the angle of rotation.

A general expression for the kinetic energy of the rigid body in general plane motion is given by the sum of two terms – one corresponding to the kinetic energy of translatory motion and the other to rotation around the axis through the centre:

$$T = \frac{1}{2}mv_C^2 + \frac{1}{2}J_C\omega^2. \tag{1.3.17}$$

Alternatively, given the kinematics of this motion (see Section 1.2.2.3), the kinetic energy can also be determined based on:

$$T = \frac{1}{2}J_P\omega^2, \tag{1.3.18}$$

where P stands for the instantaneous centre of zero velocity.

1.4 From Strength of Materials

Chapter 8 of this book contains the investigations of oscillations of certain continuous systems: longitudinal (axial) vibration of bars, torsional vibration of shafts and transversal vibration of beams. These investigations require the use of certain facts and expressions from Strength of Materials and they are presented herein for three types of loading.

1.4.1 Axial Loading

A uniform elastic bar of cross-sectional area A is considered (Figure 1.4.1a). It is made of Hooke's material whose Young's modulus is E. Let us focus now on the infinitesimally small element of the bar (Figure 1.4.1b) placed at the location defined by x. Its length is dx, and the corresponding free-body diagram is presented in Figure 1.4.1c. If the bar is axially loaded, the internal force S is generated, and the displacement caused is labelled by $u = u(x)$.

To determine the expression for this internal force, Hooke's law can be used, relating the force to the axial stress σ and the axial strain $du(x)/dx$ as follows:

$$S = \sigma A = E\frac{du}{dx}A. \tag{1.4.1}$$

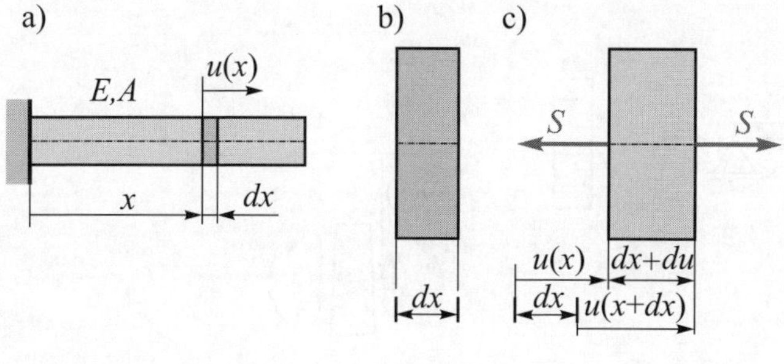

Figure 1.4.1

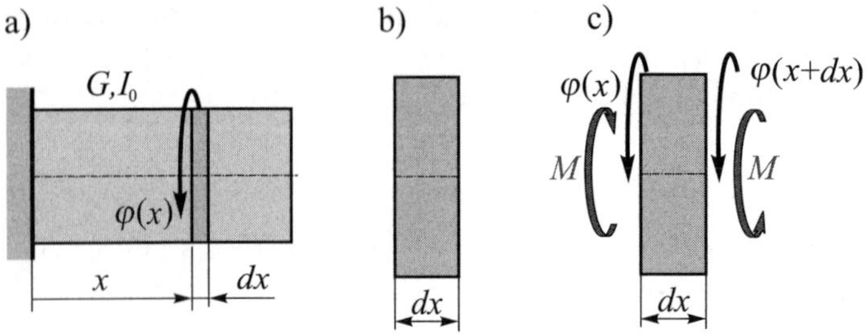

Figure 1.4.2

1.4.2 Torsion

The object of interest is a uniform elastic shaft of polar moment of inertia I_o (Figure 1.4.2a). The shaft is made of a material whose shear modulus is G. The infinitesimally small element of length dx, placed at location x is considered (Figure 1.4.2b), and the corresponding free-body diagram is presented in Figure 1.4.2c.

The internal torques (twisting torques) acting on this element are labelled by M, and the angle of rotation (twist) is labelled by $\varphi = \varphi(x)$. They are mutually related by the expression:

$$M = GI_0 \frac{d\varphi}{dx}. \tag{1.4.2}$$

1.4.3 Bending

Bending of beams is considered now (Figure 1.4.3a). The uniform beam is assumed to be made of Hooke's material whose Young's modulus is E. The moment of inertia of the beam cross-section about the neutral axis is labelled by I. The displacement u of each section depends on its location x, that is, $u = u(x)$. An infinitesimally small element of length dx is shown in Figure 1.4.3b. Its free-body diagram is presented in Figure 1.4.3c. This element is subjected to internal forces and bending moments M. The internal forces include shear forces F_s and axial forces, which are not shown for brevity.

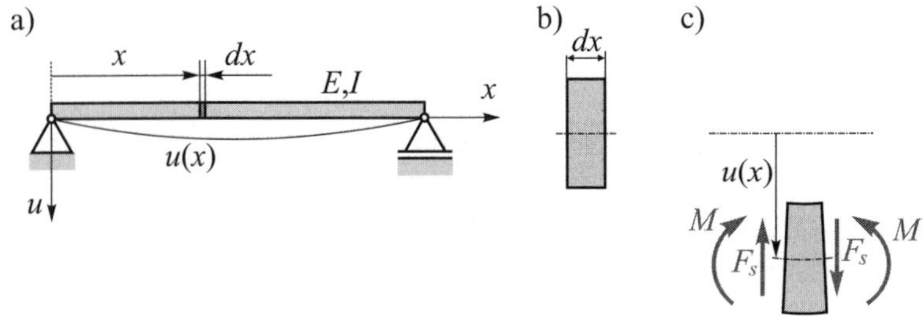

Figure 1.4.3

The relationship between the shear force F_s and the bending moment M is given by:

$$F_s = \frac{dM}{dx}, \tag{1.4.3}$$

while the expression relating the bending moment and the deflection has the form:

$$M = -EI\frac{d^2u}{dx^2}. \tag{1.4.4}$$

2

Lagrange's Equation for Mechanical Oscillatory Systems

Chapter Outline

This chapter provides theoretical basics related to Lagrangian mechanics and the formation of equations of motion from Lagrange's equation of the second kind. The quantities appearing in them (kinetic energy, potential energy, generalised forces and dissipation function) are explained and discussed. Their approximate forms that are valid for the systems performing small oscillations around a stable equilibrium position are derived. Special attention is given to the presentation of the original method for obtaining the potential energy of a linear spring by considering its deformations in two mutually orthogonal directions.

Chapter Objectives

- To present Lagrange's equation of the second kind and the advantages of its use
- To discuss the exact and approximate forms of the kinetic energy in mechanical oscillatory systems
- To discuss the exact and approximate forms of the potential energy in mechanical oscillatory systems
- To explain briefly how to obtain the gravitational potential energy
- To describe in detail how the potential energy of a spring can be determined
- To define non-conservative generalised forces
- To describe the dissipative function and its mathematical models

2.1 About Lagrange's Equation of the Second Kind

There are several ways which one can use to determine differential equations of motion of mechanical systems. For example, in Newtonian mechanics, the main roles are played by *vector* quantities: forces and accelerations expressed in terms of certain coordinates. Free-body diagrams need to be formed, and constraint and reaction forces appear. These forces are unknown in advance and some additional equations are often needed to make the number of the unknowns and the equations available equal. Consequently, this

Mechanical Vibration: Fundamentals with Solved Examples, First Edition.
Ivana Kovačić and Dragi Radomirović.

procedure can be cumbersome or less appealing. On the other side, in the basis of Analytical or Lagrangian mechanics, three *scalar* quantities are of particular importance: kinetic energy, potential energy and virtual work (or resulting generalised forces). Regardless of the number of degrees of freedom, the equations of motion are derived from these three quantities. In addition, there is no need to construct free-body diagrams as the system is considered as a whole, and constraints and reactions of ideal forces do not appear in the formulation. The number of Lagrange's equations coincides with the number of degrees of freedom, and no additional equations are needed as all the quantities need to be expressed in terms of generalised coordinates.

The method of deriving Lagrange's equation is omitted here as the focus is on its use for obtaining equations of motion of certain mechanical systems. Thus, only the basics needed for this purpose are presented. Its derivation can be found, for example, in [1,2]. Note also that the phrase 'Lagrange's equation' refers to 'Lagrange's equation of the second kind' only.

Let us start with a system that has N degrees of freedom. Then, it is described by N generalised coordinates q_i ($i = 1,\ldots, N$). These generalised coordinates are mutually independent and unconstrained, and uniquely define the configuration of the system. There are N associated Lagrange's equations, which have the following form:

$$\frac{d}{dt}\frac{\partial T}{\partial \dot{q}_i} - \frac{\partial T}{\partial q_i} + \frac{\partial D}{\partial \dot{q}_i} + \frac{\partial V}{\partial q_i} = Q_i, \tag{2.1.1}$$

where:

- the dot here stands for the derivative with respect to time t;
- $d(\ )/dt$ denotes the total time derivative;
- $\partial(\ )/\partial q_i$ and $\partial(\ )/\partial \dot{q}_i$ denote, respectively, the partial derivative with respect to the generalised coordinate q_i and generalised velocity $\dot{q}_i$;
- T is the kinetic energy of the system;
- D is the dissipation (Rayleigh's) function;
- V is the potential energy of the system;
- Q_i is the generalised non-conservative force, which is obtained from the consideration of the virtual work δA of these forces, with the exception of dissipative forces as they appear in the function D in Equation (2.1.1).

If the system performs small oscillations around the stable equilibrium position $q_i = 0$, these Lagrange's equations can be simplified to

$$\frac{d}{dt}\frac{\partial T}{\partial \dot{q}_i} + \frac{\partial D}{\partial \dot{q}_i} + \frac{\partial V}{\partial q_i} = Q_i, \tag{2.1.2}$$

while the forms of the kinetic energy, dissipation function, potential energy and generalised forces are discussed below.

If a system has one degree of freedom and is described by the generalised coordinate q, there is only one Lagrange's equation:

$$\frac{d}{dt}\frac{\partial T}{\partial \dot{q}} - \frac{\partial T}{\partial q} + \frac{\partial D}{\partial \dot{q}} + \frac{\partial V}{\partial q} = Q. \tag{2.1.3}$$

Its further simplification is concerned with the case when the system performs small oscillations around the stable equilibrium position $q = 0$. Lagrange's equation (2.1.3) now becomes

$$\frac{d}{dt}\frac{\partial T}{\partial \dot{q}} + \frac{\partial D}{\partial \dot{q}} + \frac{\partial V}{\partial q} = Q, \tag{2.1.4}$$

while the kinetic energy, dissipation function, potential energy and generalised forces can also be obtained in a simplified, approximate way with respect to their exact forms existing in Equations (2.1.1) and (2.1.3). The subsequent sections present theoretical basics for these simplifications, while a great number of examples in the following chapters demonstrate their suitability and advantages. Note that the term 'small oscillations' assumes that only linear terms exist in the equation of motion derived from Equations (2.1.2) and (2.1.4).

2.2 Kinetic Energy in Mechanical Oscillatory Systems

The kinetic energy of mechanical systems, in general, depends on time, the generalised coordinates and velocities

$$T = T(t, q_i, \dot{q}_i), \tag{2.2.1}$$

and in case of scleronomic constraints can be represented in the following quadratic form

$$T = \frac{1}{2}\sum_{i=1}^{N}\sum_{j=1}^{N}\widetilde{T}_{ij}\,\dot{q}_i\,\dot{q}_j, \tag{2.2.2}$$

where the so-called inertial coefficients $\widetilde{T}_{ij}$ depend on the generalised coordinates, that is, $\widetilde{T}_{ij} = \widetilde{T}_{ij}(q_i, q_j)$.

Without loss of generality, this form and its further transformations will be shown on the systems with two degrees of freedom ($N = 2$, and the generalised coordinates q_1 and q_2). The kinetic energy (2.2.2) is now given by

$$T = \frac{1}{2}\left(\widetilde{T}_{11}\dot{q}_1^2 + 2\widetilde{T}_{12}\dot{q}_1\dot{q}_2 + \widetilde{T}_{22}\dot{q}_2^2\right). \tag{2.2.3}$$

Of particular interest here is to show how this form changes if the system performs small oscillations about the stable equilibrium position $q_1 = 0$ and $q_2 = 0$. To that end, the inertial coefficients can be developed in series as follows

$$\widetilde{T}_{ij}(q_1, q_2) = \widetilde{T}_{ij}(0,0) + \left(\frac{\partial \widetilde{T}_{ij}}{\partial q_1}\right)_{(0,0)} q_1 + \left(\frac{\partial \widetilde{T}_{ij}}{\partial q_2}\right)_{(0,0)} q_2 + \ldots \tag{2.2.4}$$

Given the requirement that only linear terms appear in the equations of motion, one should actually simplify this expression to the first term only:

$$\widetilde{T}_{ij}(q_1, q_2) \approx \widetilde{T}_{ij}(0,0), \tag{2.2.5}$$

which brings us to a very important conclusion regarding the way how the kinetic energy can be calculated in a system performing small oscillations about a stable equilibrium position: there is no need to consider it in an arbitrary position, but only in the position when the system passes through the equilibrium position or it is assumed to be passing through this position.

If the system has only one degree of freedom, the analogous analysis holds. Starting from the following form of the kinetic energy:

$$T(q,\dot{q}) = \frac{1}{2}\tilde{T}\dot{q}^2, \tag{2.2.6}$$

one can develop the inertial coefficient into the series and truncate it to the first term only:

$$\tilde{T}(q) \approx \tilde{T}(0) = const., \tag{2.2.7}$$

leading to:

$$T = \frac{1}{2}\tilde{T}(0)\,\dot{q}^2, \tag{2.2.8}$$

which corresponds to the case when the system passes through the equilibrium position, so that the generalised coordinate is equal to zero and the generalised velocity is different from zero. The form given by Equation (2.2.8) is the reason why the term $\partial T/\partial q$ does not appear in Equation (2.1.4).

It is interesting to note that the form (2.2.8) corresponds completely to the kinetic energy of a block of mass m sliding along a guide (Figure 2.2.1a),

$$T = \frac{1}{2}m\dot{x}^2, \tag{2.2.9}$$

where the generalised coordinate is chosen to be the coordinate x.

The same form can be recognised in the kinetic energy of a pendulum (Figure 2.2.1b),

$$T = \frac{1}{2}ml^2\dot{\varphi}^2, \tag{2.2.10}$$

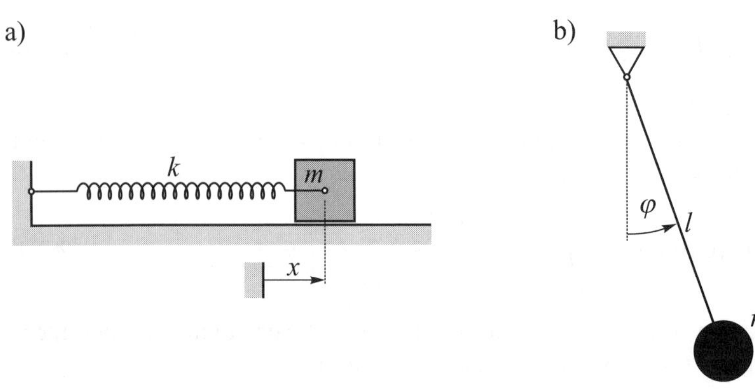

Figure 2.2.1

where the generalised coordinate is chosen to be the angle φ. Both inertial coefficient in (2.2.9) and (2.2.10) are obviously constant. These two examples are shown as they both represent paradigms for simple harmonic oscillators; also, in both cases the exact form of the kinetic energy has the same form as the one when the system passes through the equilibrium position, which is not always the case. Their potential energy will be discussed in Section 2.3.

2.3 Potential Energy in Mechanical Oscillatory Systems

The potential energy of mechanical systems, in general, depends on time and the generalised coordinates [1]

$$V = V(t, q_i), \tag{2.3.1}$$

and in case of scleronomic constraints can be represented in the following quadratic form

$$V = \frac{1}{2}\sum_{i=1}^{N}\sum_{j=1}^{N}\widetilde{V}_{ij}\, q_i\, q_j, \tag{2.3.2}$$

where $\widetilde{V}_{ij}$ represent the so-called elastic coefficients (equivalent stiffness) [3].

As in the case of the kinetic energy, the consideration of the potential energy of scleronomic systems and its approximations for small oscillations around the stable equilibrium position will be related to its form corresponding to systems with two degrees of freedom $V = V\left(q_i, q_j\right)$. Now, it can be developed in series as follows

$$\begin{aligned} V(q_1, q_2) &= V(0,0) + \left(\frac{\partial V}{\partial q_1}\right)_{(0,0)} q_1 + \left(\frac{\partial V}{\partial q_2}\right)_{(0,0)} q_2 \\ &+ \frac{1}{2}\left[\left(\frac{\partial^2 V}{\partial q_1^2}\right)_{(0,0)} q_1^2 + 2\left(\frac{\partial^2 V}{\partial q_1 \partial q_2}\right)_{(0,0)} q_1 q_2 + \left(\frac{\partial^2 V}{\partial q_2^2}\right)_{(0,0)} q_2^2\right] + \ldots \end{aligned} \tag{2.3.3}$$

Note that the series is truncated to the quadratic terms only, as this will result in the existence of linear terms in the equations of motion.

It is always possible to calibrate the potential energy to make it be equal to zero at the equilibrium position. Even if this has not been done, the term $V(0, 0)$ is constant and will be lost during the process of forming the equations for motion since it will be differentiated. Further, the system oscillates around the equilibrium position, which corresponds to the minimum of the potential energy, that is:

$$\left(\frac{\partial V}{\partial q_1}\right)_{(0,0)} = 0, \quad \left(\frac{\partial V}{\partial q_2}\right)_{(0,0)} = 0. \tag{2.3.4}$$

Thus, the potential energy (2.3.3) has the form:

$$V(q_1, q_2) \approx \frac{1}{2}\left[\left(\frac{\partial^2 V}{\partial q_1^2}\right)_{(0,0)} q_1^2 + 2\left(\frac{\partial^2 V}{\partial q_1 \partial q_2}\right)_{(0,0)} q_1 q_2 + \left(\frac{\partial^2 V}{\partial q_2^2}\right)_{(0,0)} q_2^2\right], \tag{2.3.5}$$

or, it can be expressed as

$$V(q_1,q_2) \approx \frac{1}{2}\left[\widetilde{V}_{11}q_1^2 + 2\widetilde{V}_{12}q_1q_2 + \widetilde{V}_{22}q_2^2\right]. \tag{2.3.6}$$

For systems with one degree of freedom, the form (2.3.5) becomes:

$$V(q) \approx \frac{1}{2}\widetilde{V}q^2, \tag{2.3.7}$$

where $\widetilde{V}$ is a constant.

2.3.1 Gravitational Potential Energy

The gravitational potential energy V_g is the energy that a massive object possesses because of its position in a gravitational field. The most common use of gravitational potential energy is for objects near the surface of the Earth, where the gravitational acceleration g can be assumed to be constant. There are several ways in which it can be obtained, but it is always proportional to the weight mg. The first method is presented in Figure 2.3.1a, when it depends on the coordinate z in a fixed coordinate system:

$$V_g = mgz_C. \tag{2.3.8}$$

If the axis has the opposite direction (Figure 2.3.1b), the potential energy is:

$$V_g = -mgy_C. \tag{2.3.9}$$

In addition, it is possible to define it with respect to the zero of gravitational potential energy ($V_g \equiv 0$), which can be chosen as a horizontal through any fixed point (like the choice of the zero of a coordinate system). Two of these are shown in Figure 2.3.1c;

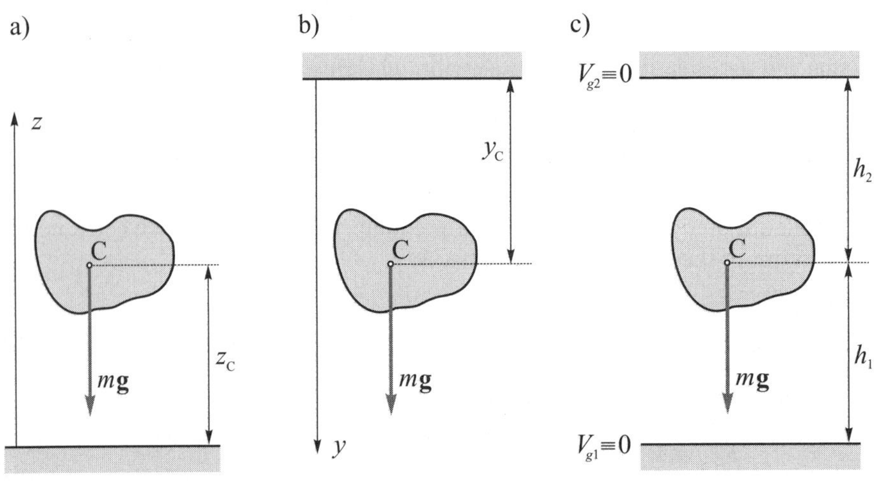

Figure 2.3.1

in this case, the potential energy depends on the vertical distance with respect to each of these zero lines:

$$V_g = +mgh_1, \tag{2.3.10}$$

or

$$V_g = -mgh_2. \tag{2.3.11}$$

The plus sign is used when the centre of gravity is above the zero of gravitational potential energy, and the minus sign when it is below this level. In the Lagrangian formalism, the height/vertical distance from the zero of gravitational potential energy is the function of generalised coordinates: $h = h(q_i)$. For the considerations in systems performing small oscillations around an equilibrium position, this should be developed into a series truncated to the quadratic terms of generalised coordinates. For example, for the pendulum shown in Figure 2.3.2a, the zero of gravitational potential energy corresponds to the horizontal through the equilibrium position. The mass is above this level for an increase in the generalised coordinate, and the potential energy takes the form

$$V_g = mgh = mgl(1-\cos\varphi) \approx \frac{1}{2}mgl\varphi^2. \tag{2.3.12}$$

This expression is in accordance with the general form (2.3.7), with the elastic coefficient (equivalent stiffness) being $\tilde{V} = mgl$.

Alternatively, one can choose the level to be at the point O (Figure 2.3.2b), which will result in the form

$$V_g = -mgl\cos\varphi \approx -mgl + \frac{1}{2}mgl\varphi^2. \tag{2.3.13}$$

It is seen that the constant term $-mgl$ appears in this expression. However, it can be ignored as it will be lost during the differentiation carried out in Lagrange's equation. Thus, this potential energy can be further used in the form

$$V_g \approx \frac{1}{2}mgl\varphi^2. \tag{2.3.14}$$

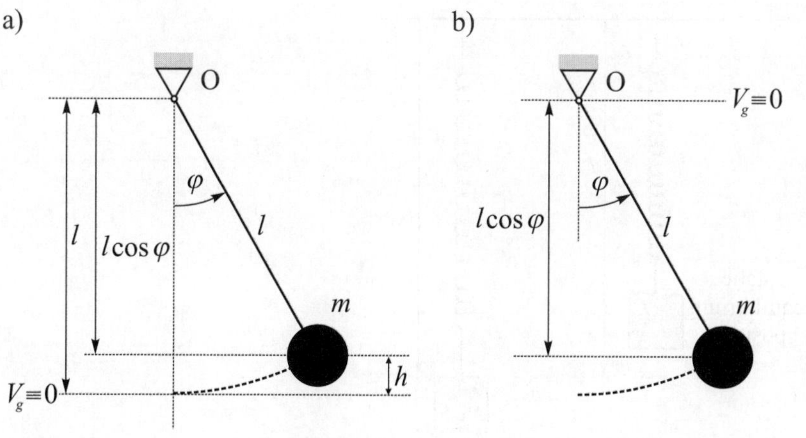

Figure 2.3.2

This fact about the possibility to ignore the constant term in the potential energy will be used in many examples in this book without further remark and is also seen in Equations (2.3.10) and (2.3.11): although these expressions represent the potential energy of the same object, they differ by a constant term.

2.3.2 Potential Energy of a Spring (Elastic Potential Energy)

A spring is a fundamental elastic component found in many mechanical systems. A spring is also used as a physical model for certain material elastic properties. Its basic characteristics are the stiffness/spring constant k and the deflection (total deflection) Δl. They are of particular importance as they define the corresponding restoring force F and the potential energy V. For linear springs, they are respectively given by

$$F = k\Delta l, \tag{2.3.15}$$

$$V = \frac{1}{2}k(\Delta l)^2. \tag{2.3.16}$$

The total deflection (Figure 2.3.3) is defined as

$$\Delta l = l - l_0, \tag{2.3.17}$$

where l is the length of the spring in an arbitrary position and l_0 is the length of the undeformed spring. This deflection can also be expressed as a sum of a static spring deflection f_{st} and the deflection measured from the static equilibrium position λ (additional deflection), that is,

$$\Delta l = f_{st} + \lambda. \tag{2.3.18}$$

As can be seen from Figure 2.3.3, the additional deflection λ is the difference between the length of the spring l and the length of the spring in the static equilibrium position l_{st}:

$$\lambda = l - l_{st}. \tag{2.3.19}$$

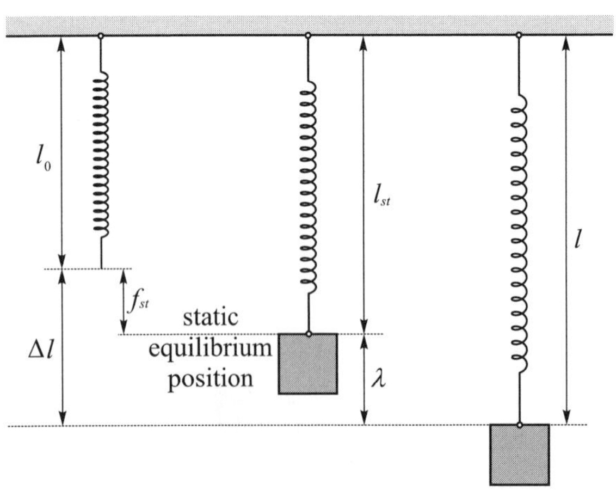

Figure 2.3.3

It is should be emphasised that the procedure for obtaining the spring deflection by using Equations (2.3.17) or (2.3.18) is easy and straightforward if the spring deforms axially, that is, collinearly with the direction of the spring in the static equilibrium position, as shown in Figure 2.3.3 or for the spring in Figure 2.2.1a with the potential energy $V = kx^2/2$. However, in many examples this is not the case as springs exhibit plane deformations. Then the length of the spring brings geometric nonlinearity into consideration. The corresponding expression can have a form that is unsuitable for analytical considerations and is developed directly into a polynomial series with respect to the generalised coordinate(s).

One question that naturally arises is: 'How can one determine the deflection and the potential energy alternatively, so as to avoid lengthy calculations related to the exact total deflection, but to take into account the static deflection (if it exists), and make the procedure easier?' This section aims at answering this question by presenting an original method to determine approximations for the deflection of a spring and its potential energy in a very convenient way. This method was developed originally for linear springs [4] by the second author of this publication, and it was extended to Duffing–type nonlinear springs jointly by both authors (this extension is presented in Chapter 9 and in [5]).

2.3.2.1 On the Approximations for Linear Spring Deflection

Let us consider a spring A_0B_0 in the static equilibrium position (Figure 2.3.4). In accordance with Figure 2.3.3, its length in this position is labelled by l_{st}. When further deformed, the spring takes the position $\overline{A_1B_1} = l$. To find the difference between the lengths l and l_{st}, the deformations in two characteristic directions are considered. The first one is the u-direction, which is collinear with the direction of the spring in the static equilibrium position. When the spring is extended, one holds $u>0$, and when it is compressed, $u<0$. The second direction of interest is the v-direction, which is orthogonal to the u-direction, as shown in Figure 2.3.4.

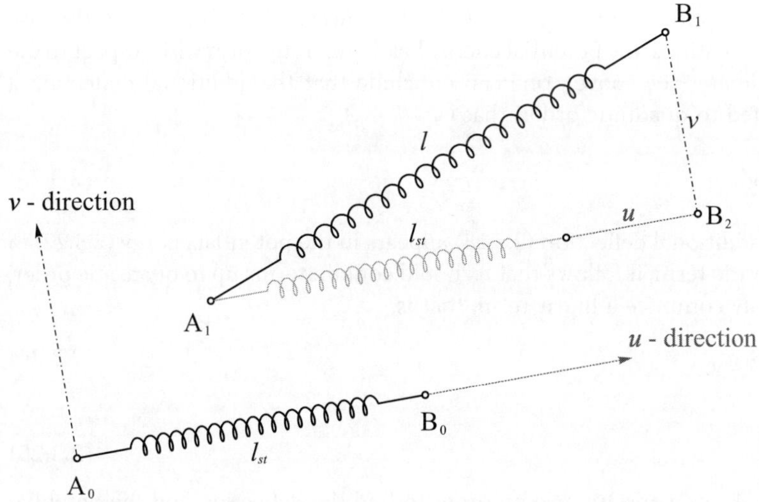

Figure 2.3.4

In accordance with (2.3.19) and based on Figure 2.3.4, the additional deflection λ can now be expressed as

$$\lambda \equiv \lambda(u,v) = \sqrt{(l_{st}+u)^2 + v^2} - l_{st}. \tag{2.3.20}$$

The additional deflection λ is the function of two variables u and v and can be developed into a Maclaurin series, which, truncated to fifth order takes the form:

$$\lambda(u,v) \approx u + \frac{1}{2l_{st}}v^2 - \frac{1}{2(l_{st})^2}uv^2 + \frac{1}{2(l_{st})^3}u^2v^2 - \frac{1}{8(l_{st})^3}v^4 - \frac{1}{2(l_{st})^4}u^3v^2 + \frac{3}{8(l_{st})^4}uv^4 + \ldots \tag{2.3.21}$$

This expression implies that one needs to determine the deflection in the u-direction and the v-direction, but the question of importance here is the order of truncation of the series (2.3.21) as well as the order of the expressions of u and v in terms of the generalised coordinate(s). This is discussed subsequently for linear springs that are deformed or undeformed in the static equilibrium position.

Linear Spring Deformed in the Static Equilibrium Position in a System That Performs Small Oscillations

Let us first consider a spring that is linear and prestressed (predeformed) in the static equilibrium position around which it oscillates, performing small oscillations. By using Equations (2.3.16) and (2.3.18), the potential energy is found to be

$$V = kf_{st}\lambda + \frac{1}{2}k\lambda^2 + const., \quad \text{or} \quad V = kf_{st}\lambda + \frac{1}{2}k\lambda^2, \tag{2.3.22}$$

where the third constant term in the first expression can be omitted as the potential energy does not depend on an additive constant, as discussed above. Now, the series (2.3.21) should be substituted into Equation (2.3.22). However, knowing that for the case of small (linear) oscillations, the potential energy has a quadratic form with respect to the generalised coordinate (see above), one can conclude that the additional deflection λ should be truncated to quadratic order, that is,

$$\lambda \approx u + \frac{1}{2l_{st}}v^2. \tag{2.3.23}$$

Further, as the additional deflection (2.3.23) appears in the potential energy (2.3.22) in a linear and quadratic term, it follows that u should contain terms up to quadratic order, while v should only comprise a linear term, that is,

$$u = A_1 q + B_1 q^2, \tag{2.3.24}$$

and

$$v = A_2 q, \tag{2.3.25}$$

where A_1, B_1 and A_2 are constants. So, in order to find the deflection and the potential energy of the prestressed linear spring, one should look for deflections (1) in the

u-direction, truncating it to the quadratic terms of the generalised coordinate, and (2) in the v-direction, expressing it as a linear function of the generalised coordinate.

Linear Spring Undeformed in the Static Equilibrium Position in a System That Performs Small Oscillations

If the linear spring considered is undeformed in the static equilibrium position, one holds $f_{st} = 0$. Given Equation (2.3.18), the potential energy (2.3.16) becomes

$$V = \frac{1}{2}k\lambda^2. \tag{2.3.26}$$

Using again the fact that for the system considered the potential energy is a quadratic function of the generalised coordinate, one concludes that

$$\lambda \equiv u = A_1 q. \tag{2.3.27}$$

Thus, in order to find the deflection and the potential energy of a linear spring that is not prestressed, one can ignore the deflection in the v-direction and should only determine the extension/compression in the u-direction and truncate it to a linear term with respect to the generalised coordinate.

Remark: Although these approximate expressions are derived and given in terms of one generalised coordinate, their generalisation to the systems with more than one degree of freedom is straightforward and will be illustrated in Chapter 6 and 7.

2.4 Generalised Forces in Mechanical Oscillatory Systems

The term 'generalised forces' refers to the coefficients in the expression for the virtual work δA (see the explanations after Equation (1.1.4)) next to the variations of the generalised coordinates

$$\delta A = \sum_{i=1}^{N} Q_{q_i}\, \delta q_i. \tag{2.4.1}$$

First, it should be noted that these generalised forces almost always have the units of a force or a torque. If, for example, the generalised coordinate is a Cartesian coordinate, the generalised force will be a force. However, if the generalised coordinate is an angle, the corresponding generalised force will be a torque.

In order to show how to determine these generalised forces, one can consider a system of n mass particles with their positions defined by the position vectors $\mathbf{r}_j = \mathbf{r}_j(q_1, q_2, \ldots, q_N)$, where $j = 1, \ldots, n$. The corresponding virtual displacements (variations of the position vector) can be expressed by using some rules of differential calculus as follows

$$\delta \mathbf{r}_j = \frac{\partial \mathbf{r}_j}{\partial q_1}\delta q_1 + \ldots + \frac{\partial \mathbf{r}_j}{\partial q_N}\delta q_N = \sum_{i=1}^{N}\frac{\partial \mathbf{r}_j}{\partial q_i}\delta q_i. \tag{2.4.2}$$

The virtual work of forces is defined by the scalar product

$$\delta A = \sum_{j=1}^{n} \mathbf{F}_j \cdot \delta \mathbf{r}_j, \tag{2.4.3}$$

which can be transformed into

$$\delta A = \sum_{j=1}^{n} \mathbf{F}_j \cdot \sum_{i=1}^{N} \frac{\partial \mathbf{r}_j}{\partial q_i} \delta q_i = \sum_{i=1}^{N} \left(\sum_{j=1}^{n} \mathbf{F}_j \cdot \frac{\partial \mathbf{r}_j}{\partial q_i} \right) \delta q_i = \sum_{i=1}^{N} Q_{q_i} \, \delta q_i, \tag{2.4.4}$$

with each generalised force Q_{q_i} being

$$Q_{q_i} = \sum_{j=1}^{n} \mathbf{F}_j \cdot \frac{\partial \mathbf{r}_j}{\partial q_i}. \tag{2.4.5}$$

If the system under consideration also contains, besides mass particles, a rigid body that rotates through an angle φ under the action of a torque M, the virtual work of this torque can easily be calculated from the expression:

$$\delta A^{\mathfrak{M}} = \pm \mathfrak{M} \delta\varphi, \tag{2.4.6}$$

where the torque and the rotation are assumed to be placed in the same plane (note that this whole book is concerned with plane problems and this is way the virtual work can be expressed in this way). The plus sign corresponds to the case when the torque helps to increase the angle, and the minus sign holds when it tries to reduce it.

It should be pointed out that while defining the virtual work of forces and torques, one does not need to write it down straight away in terms of generalised coordinates, but rather one can use the most appropriate coordinates and subsequently express the variations of these coordinates in terms of the chosen generalised coordinates. For example, let us consider again a pendulum under the action of a constant force F and a constant torque $\mathfrak{M}$ (Figure 2.4.1) and calculate the generalised force Q_φ.

Based on Figure 2.4.1, one can write down:

$$\delta A^{\mathbf{F}} = \mathbf{F} \cdot \delta \mathbf{r}_{\mathrm{A}} = F\mathbf{i} \cdot (\delta x_{\mathrm{A}} \mathbf{i} + \delta y_{\mathrm{A}} \mathbf{j}) = F \delta x_{\mathrm{A}} = F \delta (l \sin\varphi) = Fl\cos\varphi \delta\varphi, \tag{2.4.7}$$

$$\delta A^{\mathfrak{M}} = -\mathfrak{M} \delta\varphi. \tag{2.4.8}$$

The overall virtual work is

$$\delta A = \delta A^{\mathbf{F}} + \delta A^{\mathfrak{M}} = (Fl\cos\varphi - \mathfrak{M}) \delta\varphi, \tag{2.4.9}$$

which yields

$$Q_\varphi = Fl\cos\varphi - \mathfrak{M}. \tag{2.4.10}$$

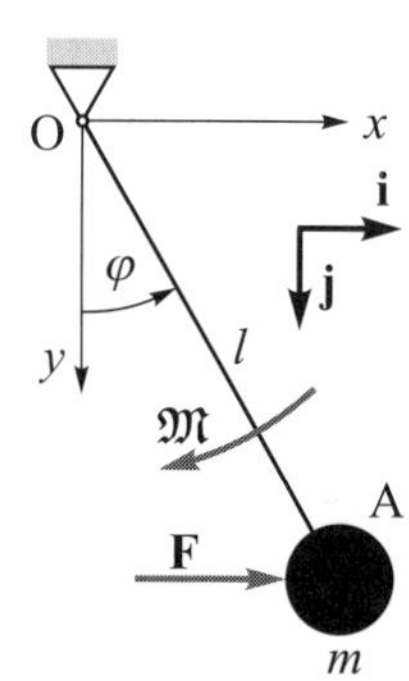

Figure 2.4.1

2.5 Dissipative Function in Mechanical Oscillatory Systems

Let us consider small oscillations of a system which is exposed to viscous damping forces, which can be modelled as

$$\mathbf{F}_j = -c_j \mathbf{v}_j, \tag{2.5.1}$$

where $\mathbf{v}_j$ denotes the velocity of the jth mass particle and c_j is a positive constant. In accordance with the definition of generalised forces, Equation (2.4.5), one has

$$Q_{q_i} = -\sum_{j=1}^{n} c_j \left(\mathbf{v}_j \cdot \frac{\partial \mathbf{r}_j}{\partial q_i} \right). \tag{2.5.2}$$

By using the identity

$$\frac{\partial \mathbf{r}_j}{\partial q_i} = \frac{\partial \mathbf{v}_j}{\partial \dot{q}_i}, \tag{2.5.3}$$

the expression (2.5.2) can be transformed into

$$Q_{q_i} = -\sum_{j=1}^{n} c_j \left(\mathbf{v}_j \cdot \frac{\partial \mathbf{v}_j}{\partial \dot{q}_i} \right) = -\frac{\partial}{\partial \dot{q}_i} \sum_{j=1}^{n} \left(\frac{1}{2} c_j v_j^2 \right). \tag{2.5.4}$$

Introducing

$$D = \frac{1}{2} \sum_{j=1}^{n} \left(c_j v_j^2 \right), \tag{2.5.5}$$

the generalised force stemming from linear viscous damping has the form

$$Q_{q_i} = -\frac{\partial D}{\partial \dot{q}_i}. \tag{2.5.6}$$

Let us focus again on the systems of n mass particles with two degrees of freedom and two generalised coordinates q_1 and q_2. The position vectors of each mass particle can now be expressed as $\mathbf{r}_j = \mathbf{r}_j(q_1, q_2)$ and the velocity of each of them is

$$\mathbf{v}_j = \sum_{i=1}^{2} \frac{\partial \mathbf{r}_j}{\partial q_i} \dot{q}_i. \tag{2.5.7}$$

Introducing this expression into Equation (2.5.5), one derives

$$D = \frac{1}{2} \sum_{i=1}^{2} \sum_{k=1}^{2} c_{ik} \dot{q}_i \dot{q}_k, \tag{2.5.8}$$

with the coefficients c_{ik} being

$$c_{ik} = \sum_{j=1}^{n} c_j \left(\frac{\partial \mathbf{r}_j}{\partial q_i} \cdot \frac{\partial \mathbf{r}_j}{\partial q_k} \right) = c_{ki}. \tag{2.5.9}$$

Note that the dissipation function, like the kinetic energy, represents a positive definite function in the generalised velocity. In general, the coefficients c_{ik} are functions of the generalised coordinates, but in the case of small oscillations around a stable equilibrium, they can be approximated by constants corresponding to the ones obtained when the system passes or is assumed to be passing through the equilibrium (as in the case of the kinetic energy).

References

1 Lurie, A. I. (2002) *Analytical Mechanics* (translated by A. Belyaev), Springer.
2 Meirovitch, L. (2001) *Fundamentals of Vibrations*, McGraw-Hill International Edition.
3 Balachandran, B. and Magrab, E. B. (2009) *Vibrations*, Second Edition, Cengage Learning.
4 Radomirovic, D. (1986) Forming of the potential energy in mechanical oscillatory systems, *Proceedings of the School of Engineering Sciences* Vol. 17, pp. 11–21.
5 Radomirovic, D. and Kovacic, I. (2013) Deflection and potential energy of linear and nonlinear springs: approximate expressions in terms of generalised coordinates, *European Journal of Physics*, Vol. 34, pp. 537–546.

3

Free Undamped Vibration of Single-Degree-of-Freedom Systems

Chapter Outline

This chapter starts with a brief preview of Lagrange's equation for free undamped vibration, both the exact one and the approximated one for small oscillations around a stable equilibrium. Fifteen examples are then solved in detail, showing how the equations of motion can be obtained in either of these approaches. Special attention is given to the potential energy (especially when the spring is predeformed) as well as to the way that the kinetic energy can be obtained more easily by considering the position when the system passes through the equilibrium position.

Chapter Objectives

- To present Lagrange's equation of the second kind for free undamped vibration
- To illustrate a variety of examples in which it yields the equation of motion
- To demonstrate the importance of the prestress of a spring for its potential energy
- To describe in detail how the potential energy of a spring can be determined by considering its deformations in two orthogonal directions
- To demonstrate how the kinetic energy can be obtained more easily by considering the position when the system passes through the equilibrium position

Theoretical Introduction

Free oscillations occur when a mechanical system is set in motion with an initial input (initial displacement and/or initial velocity). The corresponding equation of motion can be obtained by using Lagrange's equation of the second kind. Its form for a one-degree-of-freedom free conservative systems is

$$\frac{d}{dt}\frac{\partial T}{\partial \dot{q}} - \frac{\partial T}{\partial q} + \frac{\partial V}{\partial q} = 0. \tag{3.1}$$

When the system performs small oscillations around the equilibrium position $q = 0$, the previous equation can be simplified to

$$\frac{d}{dt}\frac{\partial T}{\partial \dot{q}} + \frac{\partial V}{\partial q} = 0, \tag{3.2}$$

Mechanical Vibration: Fundamentals with Solved Examples, First Edition.
Ivana Kovačić and Dragi Radomirović.

where

$$T = \frac{1}{2}\widetilde{T}\dot{q}^2, \tag{3.3}$$

with the coefficient $\widetilde{T}$ being $\widetilde{T}(q) = \widetilde{T}(0)$, while the potential energy in Equation (3.2) has the general form

$$V = \frac{1}{2}\widetilde{V}q^2, \tag{3.4}$$

where its coefficient corresponds to

$$\widetilde{V} = \left(\frac{\mathrm{d}^2 V}{\mathrm{d}q^2}\right)_{q=0}. \tag{3.5}$$

Substituting Equations (3.3) and (3.4) into Equation (3.2), one can derive the following general form of the equation of motion for free undamped small vibrations:

$$\ddot{q} + \omega_n{}^2 q = 0, \tag{3.6}$$

where ω_n represents the natural frequency and is defined by $\omega_n{}^2 = \widetilde{V}/\widetilde{T}$.

Knowing the natural frequency, one can determine the period of small vibrations T (without solving the equation of motion) from the expression

$$\mathrm{T} = \frac{2\pi}{\omega_n}, \tag{3.7}$$

which defines the time needed for the system to complete one full oscillation. The number of oscillations per second corresponds to the frequency of vibration $f = 1/\mathrm{T}$, and is expressed in Hertz (Hz).

A general solution of the equation of motion for small oscillations (3.6) has the form

$$q = A\cos(\omega_n t - \alpha), \tag{3.8}$$

where the constants A and α can be calculated from the initial conditions $q(0)$, $\dot{q}(0)$. Note that the constant A stands for the amplitude (a maximal displacement) and α for the phase shift, defining how much the oscillation leads/lags with respect to the pure cosine form (Figure 3.1). The related way to present this motion as the point M moves along a circle of the radius A is also shown in Figure 3.1.

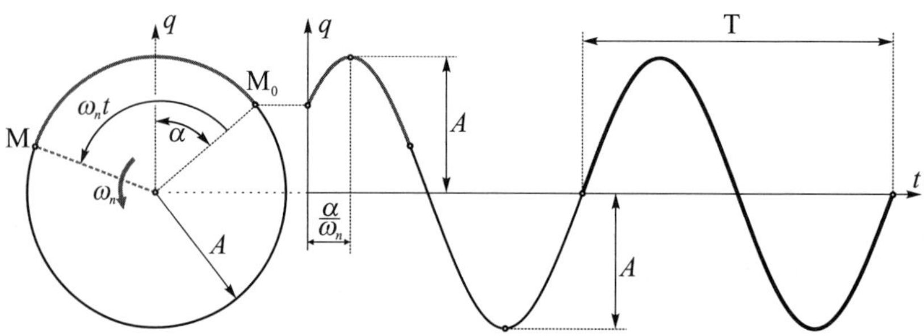

Figure 3.1

Example 3.1 A slender bar of mass m and length L can roll without slipping along a fixed circular surface of radius R (Figure 3.1.1). Determine the equation of motion for small vibrations as well as the natural frequency and the period of vibration.

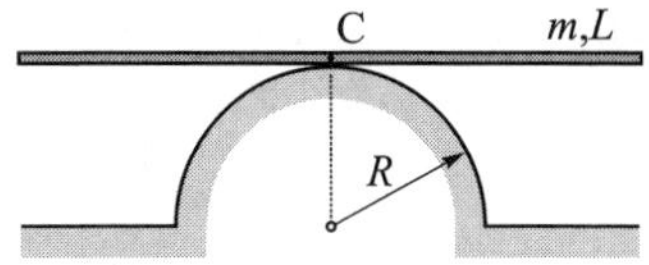

Figure 3.1.1

Solution

This problem will be approached in two ways. First, the expressions for the kinetic and potential energy will be written down in exact form, and the exact, nonlinear differential equation of motion will be derived. Then, by linearising it, the equation of motion for small vibrations will be obtained. The second approach involves obtaining the equation for motion for small vibrations based on the simplified expressions for the kinetic and potential energy.

The angle between the horizontal and the bar position is chosen as the generalised coordinate $q = \varphi$ (Figure 3.1.2a). The potential energy can be constructed as (Figure 3.1.2a):

$$V = mgy_C = mg(R\cos\varphi + R\varphi\sin\varphi) \tag{3.1.1}$$

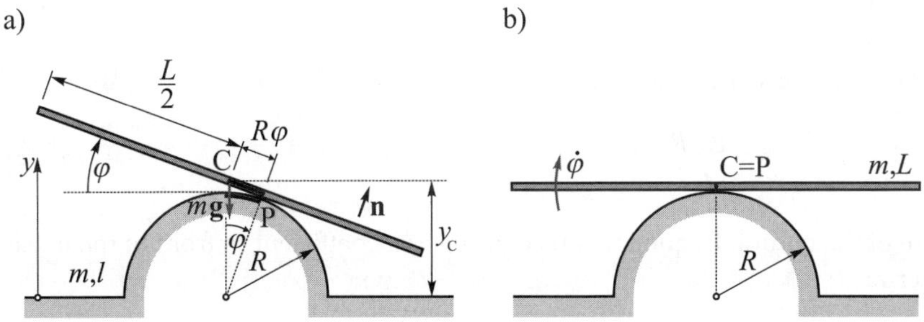

Figure 3.1.2

The bar executes general plane motion and its kinetic energy is defined by

$$T = \frac{m}{2}v_C^2 + \frac{1}{2}J_C\omega^2, \tag{3.1.2}$$

where v_C is the velocity of its centre of mass C, J_C is the mass moment of inertia of the bar about the axis perpendicular to the plane of motion through C ($J_C = mL^2/12$) and the angular velocity is $\omega = \dot{\varphi}$. The point on the bar in contact with the fixed circular surface has no velocity if the bar does not slip. It represents, therefore, the instantaneous centre of zero velocity P. The velocity of C with respect to P is defined by Figure 3.1.2a:

$$\mathbf{v}_C = R\varphi\omega\mathbf{n}, \quad v_C^2 = R^2\varphi^2\omega^2, \tag{3.1.3}$$

and the kinetic energy becomes

$$T = \frac{m}{2}\left(R^2\varphi^2 + \frac{1}{12}L^2\right)\dot{\varphi}^2. \tag{3.1.4}$$

To determine Lagrange's equation (3.1), one can find the following derivatives:

$$\begin{aligned} \frac{\partial T}{\partial \dot{\varphi}} = m\left(R^2\varphi^2 + \frac{1}{12}L^2\right)\dot{\varphi} \ \Rightarrow \ & \frac{d}{dt}\frac{\partial T}{\partial \dot{\varphi}} = 2mR^2\varphi\dot{\varphi}^2 + m\left(R^2\varphi^2 + \frac{1}{12}L^2\right)\ddot{\varphi}, \\ & \frac{\partial T}{\partial \varphi} = mR^2\varphi\dot{\varphi}^2, \ \frac{\partial V}{\partial \varphi} = mgR\varphi\cos\varphi. \end{aligned} \tag{3.1.5}$$

Transforming Equation (3.1) into

$$\left(R^2\varphi^2 + \frac{1}{12}L^2\right)\ddot{\varphi} + R^2\varphi\dot{\varphi}^2 + gR\varphi\cos\varphi = 0. \tag{3.1.6}$$

Linearising this, that is, approximating $\cos\varphi \approx 1$ and omitting the nonlinear terms, leads to

$$\frac{1}{12}L^2\ddot{\varphi} + gR\varphi = 0. \tag{3.1.7}$$

This equation can be written as

$$\ddot{\varphi} + \omega_n{}^2\varphi = 0, \ \ \omega_n{}^2 = \frac{12gR}{L^2}. \tag{3.1.8}$$

The square of the natural frequency is recognised as a coefficient in front of the linear geometric term:

$$\omega_n = \frac{2}{L}\sqrt{3gR}. \tag{3.1.9}$$

The period of vibration is

$$\mathrm{T} = \frac{L\pi}{\sqrt{3gR}}. \tag{3.1.10}$$

Let us demonstrate the second approach. It requires first the approximations of the sine and cosine function in Equation (3.1.1) as follows

$$\sin q \approx q, \ \ \cos q \approx 1 - q^2/2, \tag{3.1.11}$$

so that the approximation for the potential energy is obtained as follows

$$V \approx \frac{mgR\varphi^2}{2}. \tag{3.1.12}$$

When the bar passes through the equilibrium position, the following holds

$$\varphi = 0, \quad \dot{\varphi} \neq 0. \tag{3.1.13}$$

Given Figure 3.1.2b, which shows this position, one has

$$v_C = 0, \tag{3.1.14}$$

and the approximate expression for the kinetic energy is

$$T \approx \frac{m}{24} L^2 \dot{\varphi}^2. \tag{3.1.15}$$

Substituting Equations (3.1.12) and (3.1.15) into Equation (3.2), one can derive Equation (3.1.8), which confirms that both approaches give the same result.

Example 3.2 A uniform slender rod of length $2a$ and mass m is pivoted at O and supported at end A by a spring of stiffness k. If the system in Figure 3.2.1 is in its equilibrium position and performs small oscillations around it, determine the natural frequency and the period of these oscillations.

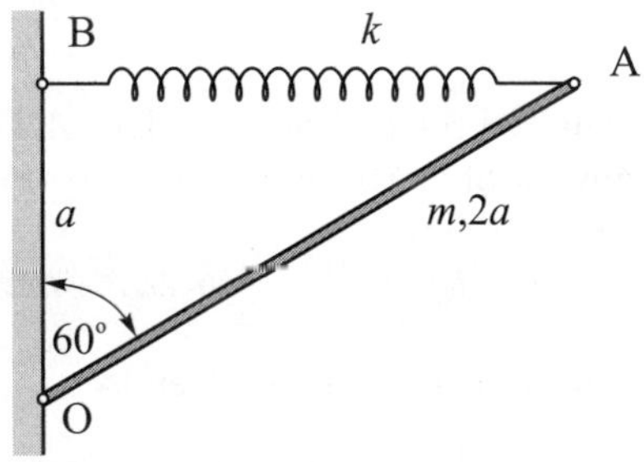

Figure 3.2.1

Solution

This problem will be solved by using two approaches, as for the previous one. First, it should be noted that the spring is predeformed in the equilibrium position. To determine its static deflection f_{st}, a free-body diagram is formed (Figure 3.2.2a) and a zero moment sum about O is considered:

$$\sum M_O = 0 \;\Rightarrow\; kf_{st}a - mga\sin 60^o = 0 \;\Rightarrow\; f_{st} = \frac{mg\sqrt{3}}{2k}. \tag{3.2.1}$$

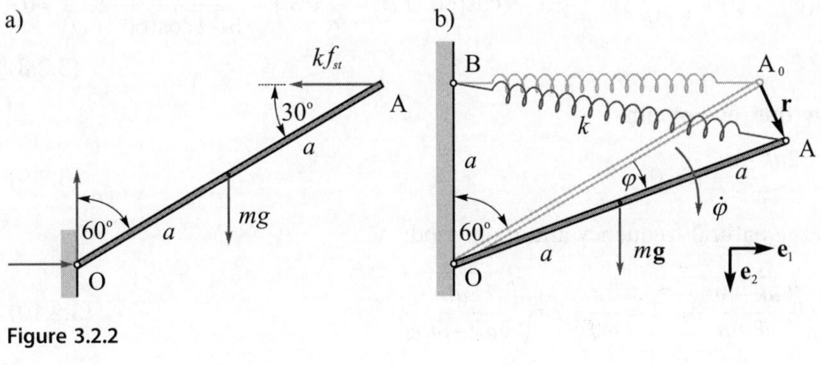

Figure 3.2.2

The generalised coordinate q is chosen to be the angle φ measured from the equilibrium position of the rod (Figure 3.2.2b). The potential energy of the rod in the displaced position becomes

$$V = mga\cos(60^\circ + \varphi) + \frac{k}{2}(f_{st} + \lambda)^2. \tag{3.2.2}$$

The additional deflection of the spring λ measured from the static equilibrium position depends on the angle φ and can be presented as

$$\begin{aligned} \lambda(\varphi) &= \overline{\mathrm{BA}}(\varphi) - \overline{\mathrm{BA}}(0) = \sqrt{a^2 + (2a)^2 - 4a^2\cos(60^\circ + \varphi)} - a\sqrt{3} \\ &\Rightarrow \lambda(\varphi) = a\sqrt{5 - 4\cos(60^\circ + \varphi)} - a\sqrt{3}. \end{aligned} \tag{3.2.3}$$

Thus, the exact expression for the potential energy is

$$V = mga\cos(60^\circ + \varphi) + \frac{k}{2}\left(\frac{mg\sqrt{3}}{2k} + a\sqrt{5 - 4\cos(60^\circ + \varphi)} - a\sqrt{3}\right)^2. \tag{3.2.4}$$

The rod rotates with angular velocity $\omega = \dot{\varphi}$ about the fixed axes through O, and the moment of inertia about O can be found from the parallel-axis theorem as follows

$$J_O = J_C + ma^2 = \frac{1}{12}m(2a)^2 + ma^2 = \frac{4}{3}ma^2. \tag{3.2.5}$$

The kinetic energy is given by

$$T = \frac{1}{2}J_O\omega^2 = \frac{2}{3}ma^2\dot{\varphi}^2. \tag{3.2.6}$$

Differentiating Eqs. (3.2.4) and (3.2.6) lead to

$$\begin{aligned} &\frac{\partial T}{\partial \dot{\varphi}} = \frac{4}{3}ma^2\dot{\varphi} \Rightarrow \frac{d}{dt}\frac{\partial T}{\partial \dot{\varphi}} = \frac{4}{3}ma^2\ddot{\varphi}, \quad \frac{\partial T}{\partial \varphi} = 0, \\ &\frac{\partial V}{\partial \varphi} = -mga\sin(60^\circ + \varphi) + k\left(\frac{mg\sqrt{3}}{2k} + a\sqrt{5 - 4\cos(60^\circ + \varphi)} - a\sqrt{3}\right)\frac{2a\sin(60^\circ + \varphi)}{\sqrt{5 - 4\cos(60^\circ + \varphi)}}. \end{aligned} \tag{3.2.7}$$

Thus, Equation (3.1) gives the exact equation of motion

$$\frac{4}{3}a\ddot{\varphi} - g\sin(60^\circ + \varphi) + \left(\frac{g\sqrt{3}}{2k} + \frac{a}{m}\sqrt{5 - 4\cos(60^\circ + \varphi)} - \frac{a}{m}\sqrt{3}\right)\frac{2k\sin(60^\circ + \varphi)}{\sqrt{5 - 4\cos(60^\circ + \varphi)}} = 0. \tag{3.2.8}$$

Linearising this, one obtains

$$\frac{4}{3}a\ddot{\varphi} + \frac{2ak - mg}{2m}\varphi = 0, \tag{3.2.9}$$

which gives the natural frequency and the period:

$$\omega_n = \sqrt{\frac{6ak - 3mg}{8am}}, \quad \mathrm{T} = \frac{2\pi}{\omega_n} = 2\pi\sqrt{\frac{8am}{6ak - 3mg}}. \tag{3.2.10}$$

In the second approach, the equation of motion will be obtained from Equation (3.2) and by using approximate forms of the kinetic and potential energy. It should be noted that the kinetic energy (3.2.6) has the identical form to the approximate one. Therefore, only the approximation of the potential energy needs to be obtained. To that end, one can represent Equation (3.2.4) as a Maclaurin series by a second-degree polynomial in the generalised coordinate. Alternatively, one can use the technique presented in Section 2.3.2 and look for the deflection λ of the spring as follows

$$\lambda = u + \frac{1}{2l_{st}}v^2, \tag{3.2.11}$$

where u stands for the deformation in the horizontal direction, v is the deformation in the vertical direction, while the length of the spring at equilibrium is

$$l_{st} = a\sqrt{3}. \tag{3.2.12}$$

Using Figure 3.2.2b, one can write

$$\begin{gathered} u = \mathbf{r}\cdot\mathbf{e}_1 = 2a\sin(60^\circ + \varphi) - a\sqrt{3} = 2a(\sin 60^\circ \cos\varphi + \cos 60^\circ \sin\varphi) - a\sqrt{3} \;\Rightarrow \\ u \approx 2a\left[\frac{\sqrt{3}}{2}\left(1 - \frac{\varphi^2}{2}\right) + \frac{1}{2}\varphi\right] - a\sqrt{3} = a\varphi - \frac{a\sqrt{3}}{2}\varphi^2, \end{gathered} \tag{3.2.13}$$

as well as

$$\begin{gathered} v = \mathbf{r}\cdot\mathbf{e}_2 = a - 2a\cos(60^\circ + \varphi) = a - 2a(\cos 60^\circ \cos\varphi - \sin 60^\circ \sin\varphi) \;\Rightarrow \\ v \approx a - 2a\left(\frac{1}{2}1 - \frac{\sqrt{3}}{2}\varphi\right) = a\sqrt{3}\varphi. \end{gathered} \tag{3.2.14}$$

Substituting Equations (3.2.13) and (3.2.14) into Equation (3.2.11), one derives

$$\lambda = a\varphi - \frac{a\sqrt{3}}{2}\varphi^2 + \frac{1}{2\sqrt{3}a}\left(a\sqrt{3}\varphi\right)^2 = a\varphi. \tag{3.2.15}$$

Finally, Equations (3.2.1), (3.2.2) and (3.2.15) give the expression for the approximate potential energy:

$$V \approx mga\left[\frac{1}{2}\left(1 - \frac{\varphi^2}{2}\right) - \frac{\sqrt{3}}{2}\varphi\right] + \frac{k}{2}\left(\frac{mg\sqrt{3}a\varphi}{k} + a^2\varphi^2\right) \;\Rightarrow\; V \approx \frac{1}{2}\left(ka^2 - \frac{mga}{2}\right)\varphi^2. \tag{3.2.16}$$

Introducing Equations (3.2.6) and (3.2.16) into Equation (3.2), the equation of motion identical to Equation (3.2.9) is obtained.

Example 3.3 The ends of a light bar of length L move freely in horizontal and vertical slots (Figure 3.3.1). The masses of the sliders at A and B are, respectively, M and m. The slider A is attached to a horizontal spring of stiffness k. If the system is shown in its equilibrium

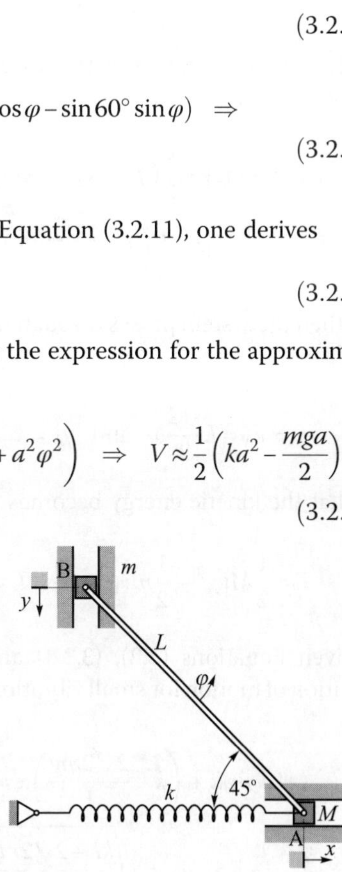

Figure 3.3.1

position, demonstrate that the natural frequency ω_n has the same value regardless of the choice of the generalised coordinates x, y or φ.

Solution

Figure 3.3.2 a and b depict the free-body diagrams for the sliders in the equilibrium position. Summing forces in the vertical position (Figure 3.3.2a) and in the horizontal position (Figure 3.3.2b) yields the static deflection of the spring as follows:

$$\sum F_y = 0 \;\Rightarrow\; -mg + F_s \sin 45^\circ = 0 \;\Rightarrow\; F_s = \sqrt{2}mg, \tag{3.3.1}$$

$$\sum F_x = 0 \;\Rightarrow\; -kf_{st} + F_s \cos 45^\circ = 0 \;\Rightarrow\; f_{st} = \frac{mg}{k}. \tag{3.3.2}$$

First, we choose x as the generalised coordinate. In order to determine the potential energy of the upper slider, one needs to define how y changes with x (Figure 3.3.3b). The exact and the approximate expressions for y are given by

$$y(x) = L\frac{\sqrt{2}}{2} - \sqrt{L^2 - \left(L\frac{\sqrt{2}}{2} + x\right)^2} \;\Rightarrow\; y(x) \approx x + \frac{\sqrt{2}}{L}x^2. \tag{3.3.3}$$

The potential energy of the lower slider is zero, and the overall potential energy is constructed as

$$\begin{aligned} V &= -mgy + \frac{k}{2}(f_{st} + x)^2 \approx -mg\left(x + \frac{\sqrt{2}}{L}x^2\right) + \frac{k}{2}\left(\frac{mg}{k} + x\right)^2 \\ &\Rightarrow\; V \approx \frac{1}{2}\left(k - \frac{2\sqrt{2}mg}{L}\right)x^2. \end{aligned} \tag{3.3.4}$$

When the system passes through the equilibrium position (Figure 3.3.3a), the following holds:

$$v_B = v_A = L\frac{\sqrt{2}}{2}\omega \;\text{ and }\; v_A = \dot{x}, \tag{3.3.5}$$

so that the kinetic energy becomes

$$T = \frac{1}{2}Mv_A^2 + \frac{1}{2}mv_B^2 \;\Rightarrow\; T \approx \frac{1}{2}(M+m)\dot{x}^2. \tag{3.3.6}$$

Given Equations (3.2), (3.3.4) and (3.3.6), the equation of motion for small vibrations is obtained:

$$\begin{aligned} &(M+m)\ddot{x} + \left(k - \frac{2\sqrt{2}mg}{L}\right)x = 0 \\ &\Rightarrow\; \omega_n = \sqrt{\frac{kL - 2\sqrt{2}mg}{(M+m)L}}. \end{aligned} \tag{3.3.7}$$

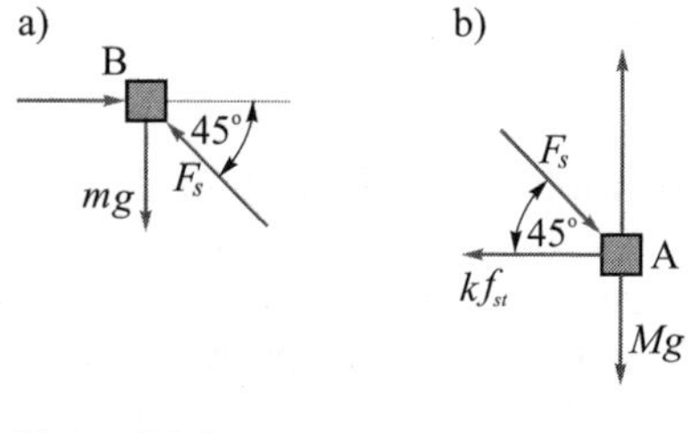

Figure 3.3.2

Let us now choose the displacement y as the generalised coordinate. Now, one needs to define how x changes with y in an arbitrary position. Based on Figure 3.3.3b, one finds

$$x(y)=\sqrt{L^2-\left(L\frac{\sqrt{2}}{2}-y\right)^2}-L\frac{\sqrt{2}}{2} \;\Rightarrow\; x(y)\approx y-\frac{\sqrt{2}}{L}y^2. \tag{3.3.8}$$

The overall potential energy now reads as

$$V=-mgy+\frac{k}{2}(f_{st}+x)^2\approx -mgy+\frac{k}{2}\left(\frac{mg}{k}+y-\frac{\sqrt{2}}{L}y^2\right)^2$$
$$\Rightarrow V\approx\frac{1}{2}\left(k-\frac{2\sqrt{2}mg}{L}\right)y^2. \tag{3.3.9}$$

When the system passes through the equilibrium position, the velocities are given by

$$v_B=v_A=L\frac{\sqrt{2}}{2}\omega,\quad v_B=\dot{y}. \tag{3.3.10}$$

The overall kinetic energy takes the form

$$T=\frac{1}{2}Mv_A{}^2+\frac{1}{2}mv_B{}^2 \;\Rightarrow\; T\approx\frac{1}{2}(M+m)\dot{y}^2. \tag{3.3.11}$$

Using Equations (3.3.9) and (3.3.11), the natural frequency given in Equation (3.3.7) can be obtained.

Finally, let us demonstrate the case when $q=\varphi$. Based on Figure 3.3.3b, one can write:

$$x(\varphi)=L\sin(45^\circ+\varphi)-L\frac{\sqrt{2}}{2} \;\Rightarrow\; x(\varphi)\approx L\frac{\sqrt{2}}{2}\varphi-L\frac{\sqrt{2}}{4}\varphi^2, \tag{3.3.12}$$

$$y(\varphi)=L\frac{\sqrt{2}}{2}-L\cos(45^\circ+\varphi) \;\Rightarrow\; y(\varphi)\approx L\frac{\sqrt{2}}{2}\varphi+L\frac{\sqrt{2}}{4}\varphi^2. \tag{3.3.13}$$

The potential energy now becomes

$$V=-mgy+\frac{k}{2}(f_{st}+x)^2\approx -mg\left(L\frac{\sqrt{2}}{2}\varphi+L\frac{\sqrt{2}}{4}\varphi^2\right)+\frac{k}{2}\left(\frac{mg}{k}+L\frac{\sqrt{2}}{2}\varphi-L\frac{\sqrt{2}}{4}\varphi^2\right)^2$$
$$\Rightarrow\; V\approx\frac{1}{2}\left(\frac{kL^2}{2}-\sqrt{2}mgL\right)\varphi^2. \tag{3.3.14}$$

The velocities corresponding to the case when the system passes through the equilibrium position are given by (Figure 3.3.3a):

$$v_B=v_A=L\frac{\sqrt{2}}{2}\omega \text{ and } \omega=\dot{\varphi}, \tag{3.3.15}$$

so that the kinetic energy of the system is

$$T=\frac{1}{2}Mv_A{}^2+\frac{1}{2}mv_B{}^2 \;\Rightarrow\; T\approx\frac{1}{2}\frac{(M+m)L^2}{2}\dot{\varphi}^2. \tag{3.3.16}$$

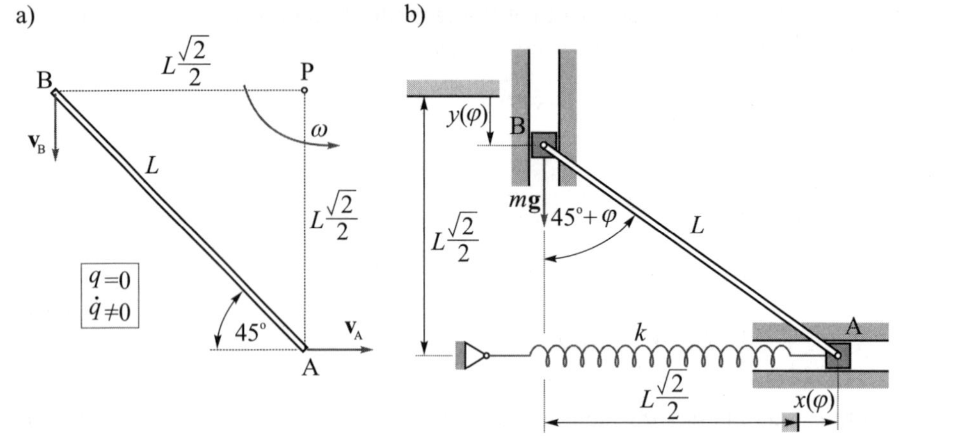

Figure 3.3.3

By using Equations (3.3.14) and (3.3.16), one derives

$$\frac{(M+m)L^2}{2}\ddot{\varphi}+\left(k\frac{L^2}{2}-\sqrt{2}mgL\right)\varphi=0 \;\Rightarrow\; \omega_n=\sqrt{\frac{kL-2\sqrt{2}mg}{(M+m)L}}, \tag{3.3.17}$$

which agrees with the result derived previously.

Example 3.4 A circular ring of mass m and radius R rolls without slipping on the surface of the fixed cylinder of radius r (Figure 3.4.1). Determine the natural frequency and the period of small oscillations about the equilibrium position shown.

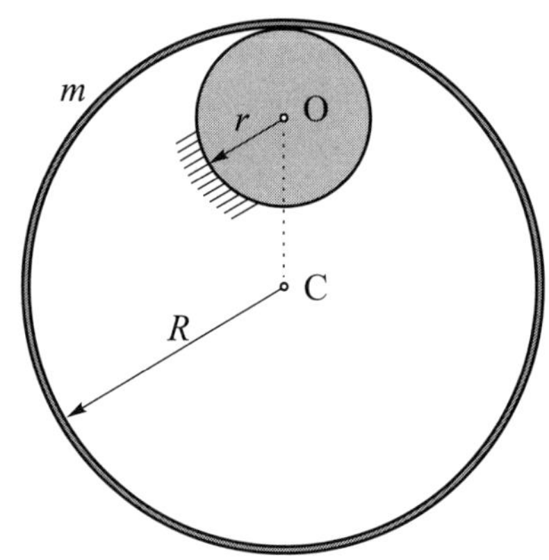

Figure 3.4.1

Solution

The generalised coordinate is chosen to be the angle φ (Figure 3.4.2a) between the vertical line and the OC-direction. Note that one can treat OC as a 'virtual' rod – then the angle φ would define its rotation.

The exact expression for the potential energy is (see Figure 3.4.2a):

$$V=-mgy_C=-mg(R-r)\cos\varphi. \tag{3.4.1}$$

Approximating this, one can derive

$$V\approx mg(R-r)\frac{\varphi^2}{2}. \tag{3.4.2}$$

The angle ψ (Figure 3.4.2a) is the angle of rotation of the ring and its relationship with the generalised coordinate φ stems from the equality related to the arc lengths

$$r\varphi = R(\varphi - \psi) \;\Rightarrow\; \psi = \frac{R-r}{R}\varphi. \tag{3.4.3}$$

On the other hand, the 'virtual' rod OC performs rotational motion around O, and this fact defines the velocity of point C (Figure 3.4.2b) when $\varphi = 0$,

$$\mathbf{v}_C = (R-r)\dot{\varphi}\mathbf{i} \;\Rightarrow\; v_C{}^2 = (R-r)^2\dot{\varphi}^2. \tag{3.4.4}$$

The angular velocity of the ring corresponding to the position when the system passes through the equilibrium is given by (note the location of the instantaneous centre P of zero velocity in Figure 3.4.2b):

$$\omega = \dot{\psi} = \frac{v_C}{R} = \frac{R-r}{R}\dot{\varphi}. \tag{3.4.5}$$

Note also that this expression can be obtained by differentiating Equation (3.4.3) with respect to time. Thus, the kinetic energy takes the form

$$T = \frac{m}{2}v_C{}^2 + \frac{1}{2}J_C\omega^2 = \frac{m}{2}(R-r)^2\dot{\varphi}^2 + \frac{1}{2}mR^2\frac{(R-r)^2}{R^2}\dot{\varphi}^2 \;\Rightarrow\; T = m(R-r)^2\dot{\varphi}^2. \tag{3.4.6}$$

The following expression has been used for the mass moment of inertia of the ring with respect to the axis through C:

$$J_C = mR^2. \tag{3.4.7}$$

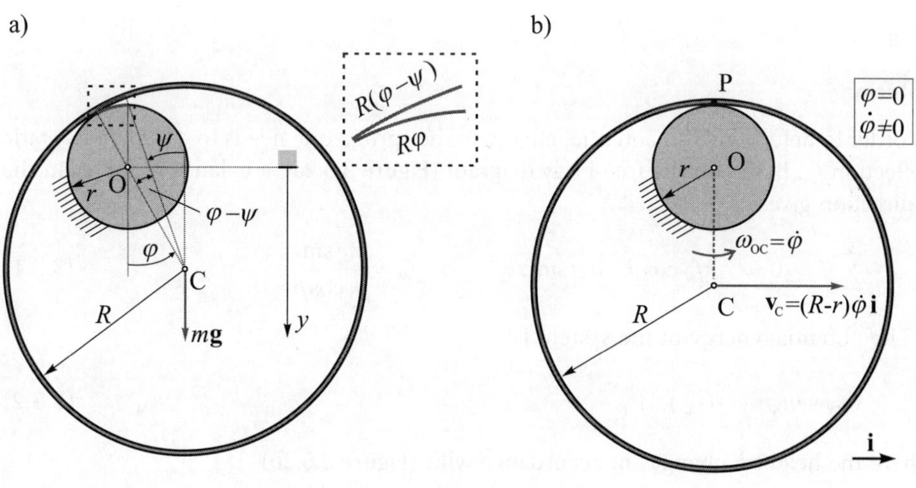

Figure 3.4.2

Lagrange's equation (3.2) yields

$$2(R-r)\ddot{\varphi}+g\varphi=0, \tag{3.4.8}$$

with the natural frequency and the period of small vibrations being

$$\omega_n=\sqrt{\frac{g}{2(R-r)}},\quad \mathrm{T}=2\pi\sqrt{\frac{2(R-r)}{g}}. \tag{3.4.9}$$

Example 3.5 A slider of mass m can slide along a frictionless guide and is attached to a spring that has stiffness k. The system performs small oscillations around the position $x=0$ shown in Figure 3.5.1. The length of the spring is then l, and the angle between the guide and the spring is β. Determine the natural frequency of small oscillations and plot the solution for motion $x(t)$, assuming: $l=1$ m, $m=6$ kg, $k=900$ N/m, $\beta=60^\circ$, $x(0)=0.1\text{m}$, $\dot{x}(0)=0$.

Consider first a horizontal guide ($\alpha=\alpha_0=0$); then assume that it is inclined ($\alpha=\alpha_1=30°$).

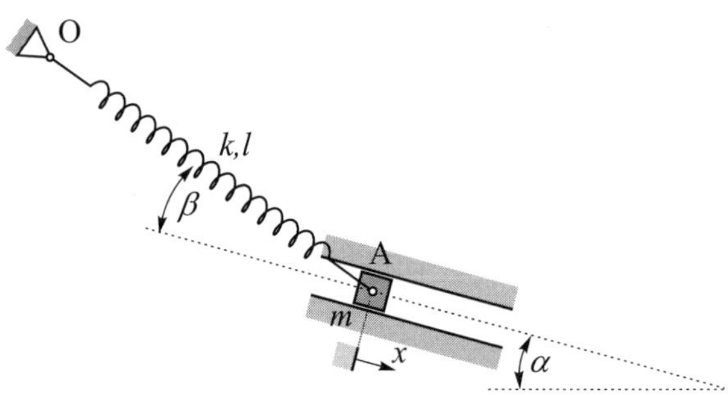

Figure 3.5.1

Solution

In order to determine the potential energy of the spring, one needs to calculate its static deflection f_{st}. Based on the free-body diagram (Figure 3.5.2a), a balance of forces in the x-direction gives:

$$\sum F_x=0\Rightarrow -kf_{st}\cos\beta+mg\sin\alpha=0\ \Rightarrow\ f_{st}=\frac{mg\sin\alpha}{k\cos\beta}. \tag{3.5.1}$$

The potential energy of the system is:

$$V\approx -mgh+\frac{k}{2}(f_{st}+\lambda)^2, \tag{3.5.2}$$

where the height h changes in accordance with (Figure 3.5.2b)

$$h=x\sin\alpha. \tag{3.5.3}$$

Two characteristic deformations of the spring are

$$u = \mathbf{r}\cdot\mathbf{e}_1 = x\cos\beta, \quad v = \mathbf{r}\cdot\mathbf{e}_2 = x\sin\beta, \tag{3.5.4}$$

which lead to

$$\lambda \approx u + \frac{1}{2l}v^2 \Rightarrow \lambda \approx \cos\beta x + \frac{1}{2l}\sin^2\beta x^2. \tag{3.5.5}$$

Substituting Equations (3.5.3) and (3.5.5) into Equation (3.5.2), the potential energy becomes

$$V \approx \frac{1}{2}\left(k\cos^2\beta + \frac{mg\sin\alpha\sin^2\beta}{l\cos\beta}\right)x^2. \tag{3.5.6}$$

The kinetic energy of the slider is

$$T = \frac{1}{2}m\dot{x}^2. \tag{3.5.7}$$

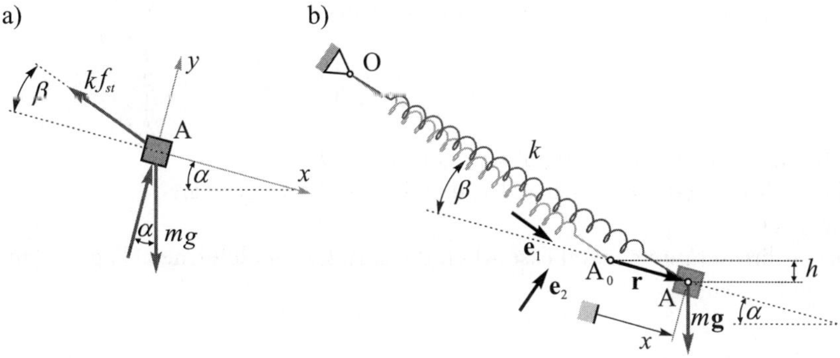

Figure 3.5.2

Lagrange's equation of motion (3.2) becomes

$$m\ddot{x} + \left(k\cos^2\beta + \frac{mg\sin\alpha\sin^2\beta}{l\cos\beta}\right)x = 0, \tag{3.5.8}$$

enabling one to recognise the expression for the natural frequency:

$$\omega_n = \sqrt{\frac{k}{m}\cos^2\beta + \frac{g\sin\alpha\sin^2\beta}{l\cos\beta}}. \tag{3.5.9}$$

The solutions for motion in both cases required are:

$$x(t)|_{\alpha=0} = 0.1\cos(8.660t), \quad x(t)|_{\alpha=30^\circ} = 0.1\cos(9.075t). \tag{3.5.10}$$

They are both plotted in Figure 3.5.3. It is seen that the one corresponding to the horizontal guide has a lower natural frequency, and, therefore, a longer period of vibration.

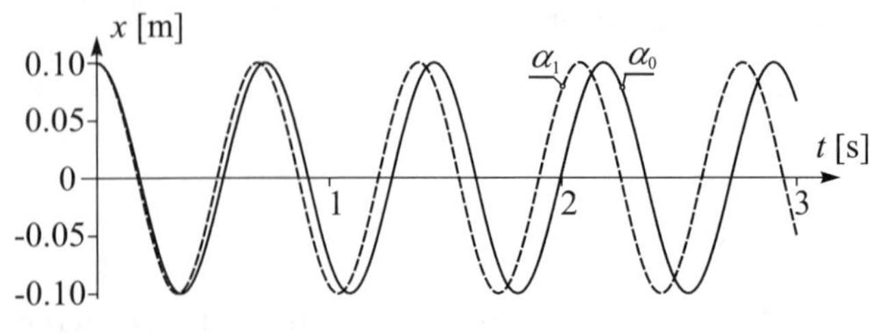

Figure 3.5.3

Example 3.6 A disc of mass m and radius r rolls without slipping inside a fixed cylinder of radius R. The disc is supported in its equilibrium position shown in Figure 3.6.1 by a vertical spring of stiffness k and length l. Choosing the angle φ as the generalised coordinate, determine the natural frequency of small vibrations in two cases: (a) the spring is predeformed by a force F, and (b) the spring is not predeformed ($F = 0$).

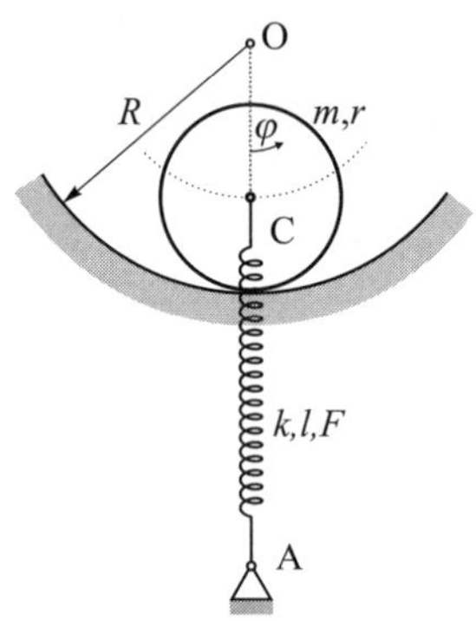

Figure 3.6.1

Solution

It should be pointed out that the generalised coordinate φ represents the angle of rotation of a 'virtual' rod OC around the axis through O.

Let us consider first a more general case when the spring is predeformed. The system potential energy is

$$V \approx -mgh + \frac{k}{2}(f_{st} + \lambda)^2, \tag{3.6.1}$$

where the height h is seen in Figure 3.6.2a:

$$h = (R - r)\cos\varphi. \tag{3.6.2}$$

Given the magnitude of the force F, the static deflection of the spring is

$$f_{st} = \frac{F}{k}, \tag{3.6.3}$$

and the additional deflection of the spring when the disc oscillates is defined by

$$\lambda \approx u + \frac{1}{2l}v^2, \tag{3.6.4}$$

where

$$u = \mathbf{r}\cdot\mathbf{e}_1 = (R - r)(1 - \cos\varphi) \;\Rightarrow\; u \approx (R - r)\frac{\varphi^2}{2}, \tag{3.6.5}$$

$$v = \mathbf{r}\cdot\mathbf{e}_2 = (R - r)\sin\varphi \;\Rightarrow\; v \approx (R - r)\varphi, \tag{3.6.6}$$

so that

$$\lambda \approx \frac{R-r}{2}\left(1+\frac{R-r}{l}\right)\varphi^2. \tag{3.6.7}$$

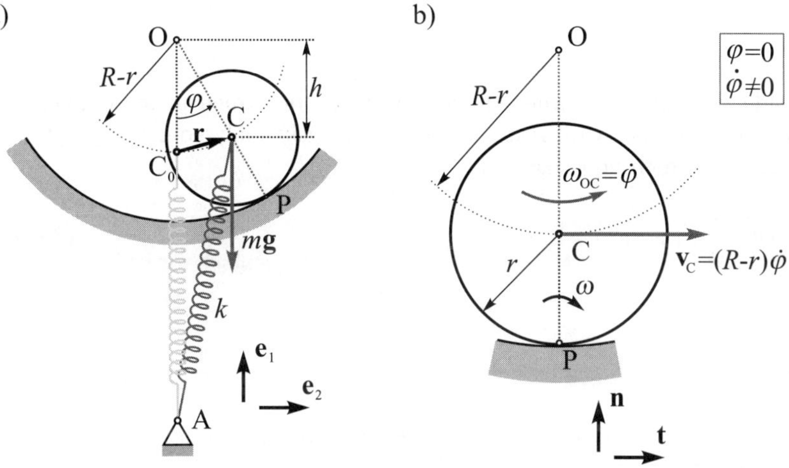

Figure 3.6.2

The potential energy now becomes

$$V \approx \frac{R-r}{2}\left[mg+F\left(1+\frac{R-r}{l}\right)\right]\varphi^2. \tag{3.6.8}$$

As the disc executes general plane motion, its kinetic energy is defined by

$$T = \frac{m}{2}v_C{}^2 + \frac{1}{2}J_C\omega^2, \tag{3.6.9}$$

where the velocity of the centre C is

$$\mathbf{v}_C = \overline{OC}\,\omega_{OC}\,\mathbf{t} = (R-r)\dot{\varphi}\mathbf{t}, \quad v_C = (R-r)\dot{\varphi}. \tag{3.6.10}$$

The position of the instantaneous centre of zero velocity P is shown in Figure 3.6.2b, and it can be used to find the angular velocity of the disc:

$$\omega = \frac{v_C}{\overline{CP}} = \frac{R-r}{r}\dot{\varphi}. \tag{3.6.11}$$

Knowing that the mass moment of inertia of the disc about the axis perpendicular to the plane of motion through C is

$$J_C = \frac{1}{2}mr^2, \tag{3.6.12}$$

the expression for the kinetic energy takes the form

$$T = \frac{3}{4}m(R-r)^2\dot{\varphi}^2. \tag{3.6.13}$$

Equations (3.6.8) and (3.6.13) yield the following differential equation of motion for small vibrations and associated natural frequency:

$$\frac{3}{2}m(R-r)\ddot{\varphi}+\left[mg+F\left(1+\frac{R-r}{l}\right)\right]\varphi=0 \;\Rightarrow\; \omega_n=\sqrt{\frac{2mgl+2F(l+R-r)}{3m(R-r)l}}. \tag{3.6.14}$$

Let us consider now the case when the spring in not deformed in the equilibrium position. The only difference with respect to the previously analysed case is related to the potential energy, which now reads as:

$$V\approx -mgh+\frac{k}{2}\lambda^2,\ \lambda\approx u,\ u\approx 0, \tag{3.6.15}$$

that is,

$$V\approx\frac{1}{2}mg(R-r)\varphi^2. \tag{3.6.16}$$

The following equation of motion for small vibrations and associated natural frequency are derived:

$$\frac{3}{2}(R-r)\ddot{\varphi}+g\varphi=0 \;\Rightarrow\; \omega_n=\sqrt{\frac{2g}{3(R-r)}}. \tag{3.6.17}$$

The same results can be obtained from the natural frequency in Equation (3.6.14) by setting $F=0$.

Example 3.7 A block of mass m, which can move in the horizontal direction without friction, is attached to two springs of stiffness k_1 and k_2, and lengths l_1 and l_2 in the equilibrium position (Figure 3.7.1), one of them being horizontal and the other inclined at an angle α. The horizontal spring is extended by a force F. Determine the equation of motion of small vibrations and the natural frequency.

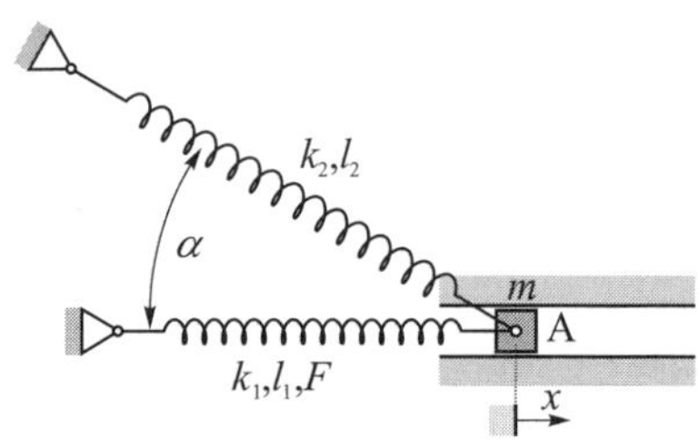

Figure 3.7.1

Solution

The static deflection of the horizontal spring can be calculated based on the force given

$$f_{st1}=\frac{F}{k_1}, \tag{3.7.1}$$

while a balance of forces acting on the block in the x-direction (see the free-body diagram presented in Figure 3.7.2a) gives the static deflection of the other spring:

$$\sum F_x=k_2 f_{st2}\cos\alpha-F=0 \;\Rightarrow\; f_{st2}=\frac{F}{k_2\cos\alpha}. \tag{3.7.2}$$

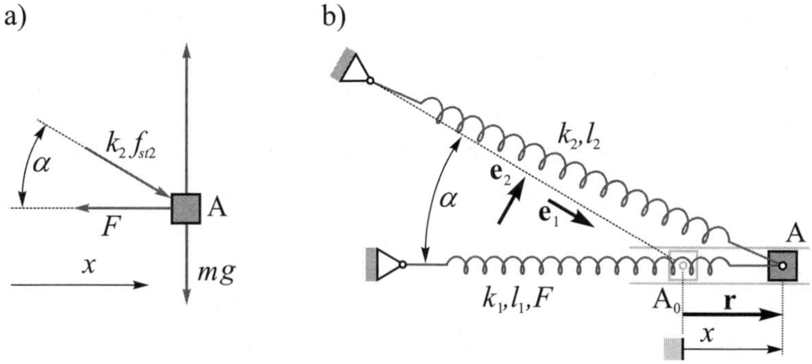

Figure 3.7.2

The potential energy of the system stems from the two springs, and is defined by

$$V \approx \frac{1}{2}k_1(f_{st1}+x)^2+\frac{1}{2}k_2(-f_{st2}+\lambda)^2. \tag{3.7.3}$$

The additional deformation of the inclined spring is

$$\lambda \approx u+\frac{1}{2l_2}v^2, \tag{3.7.4}$$

where (see Figure 3.7.2b):

$$u=\mathbf{r}\cdot\mathbf{e}_1=x\cos\alpha, \quad v=\mathbf{r}\cdot\mathbf{e}_2=x\sin\alpha. \tag{3.7.5}$$

Thus, this additional deformation becomes

$$\lambda \approx x\cos\alpha+\frac{1}{2l_2}x^2\sin^2\alpha. \tag{3.7.6}$$

The overall potential energy is found to be

$$V \approx \frac{1}{2}\left(k_1+k_2\cos^2\alpha-\frac{F\sin^2\alpha}{l_2\cos\alpha}\right)x^2. \tag{3.7.7}$$

The kinetic energy is easily constructed as

$$T=\frac{1}{2}m\dot{x}^2, \tag{3.7.8}$$

so that the equation of small vibrations and the corresponding natural frequency are:

$$m\ddot{x}+\left(k_1+k_2\cos^2\alpha-\frac{F\sin^2\alpha}{l_2\cos\alpha}\right)x=0, \tag{3.7.9}$$

$$\omega_n=\sqrt{\frac{(k_1+k_2\cos^2\alpha)l_2\cos\alpha-F\sin^2\alpha}{ml_2\cos\alpha}}. \tag{3.7.10}$$

Example 3.8 A weightless bar of length l is hinged to two sliders A and B, the masses of which are, respectively, M and m, and which move without friction along the guides shown in Figure 3.8.1. Each of them is attached to the spring, so that the system oscillates around the position presented in Figure 3.8.1. Choosing the angle φ for the generalised coordinate, determine the natural frequency when the spring of stiffness k_1 is not pre-deformed while the one of stiffness k_2 is prestressed, and vice versa.

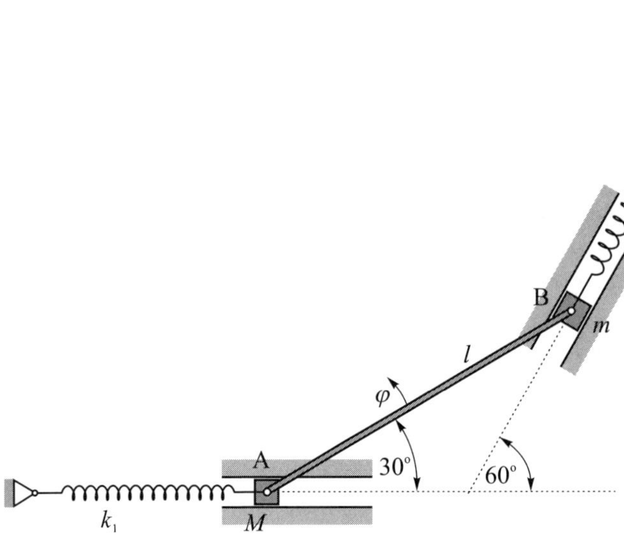

Figure 3.8.1

Solution

Creating the free-body diagram shown in Figure 3.8.2a and equating to zero the moment sum about P, the static deflection of the second spring can be obtained:

$$\sum M_P = 0 \Rightarrow k_2 f_{st2} l - mgl\cos 30^\circ = 0 \;\Rightarrow\; f_{st2} = \frac{mg\sqrt{3}}{2k_2}. \tag{3.8.1}$$

In order to find the displacements of points A and B in terms of the generalised coordinate φ, it is suitable to sketch how the system looks like in an arbitrary position and use the Law of Sines (Sine Rule) for the triangle OAB (Figure 3.8.3a):

$$\frac{l\sqrt{3}/3 - x(\varphi)}{\sin(30^\circ - \varphi)} = \frac{l}{\sin 120^\circ} \Rightarrow x(\varphi) = \frac{l\sqrt{3}}{3} - \frac{2l\sqrt{3}}{3}\sin(30^\circ - \varphi)$$
$$\Rightarrow\; x(\varphi) \approx l\varphi + (\sqrt{3}l/6)\varphi^2, \tag{3.8.2}$$

as well as

$$\frac{l\sqrt{3}/3 + y(\varphi)}{\sin(30^\circ + \varphi)} = \frac{l}{\sin 120^\circ} \;\Rightarrow\; y(\varphi) \approx l\varphi - \left(\sqrt{3}l/6\right)\varphi^2. \tag{3.8.3}$$

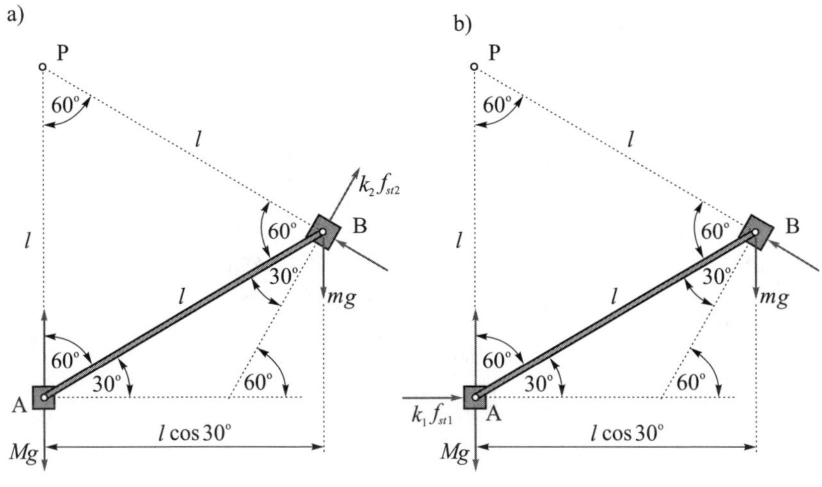

Figure 3.8.2

The overall potential energy is given by

$$V = \frac{k_1}{2}\lambda_1{}^2 + \frac{k_2}{2}(f_{st2} + \lambda_2)^2 + mgh, \tag{3.8.4}$$

where

$$\lambda_1 = x(\varphi) \approx l\varphi, \quad \lambda_2 = -y(\varphi) \approx -l\varphi + \frac{\sqrt{3}l}{6}\varphi^2, \quad h = y(\varphi)\sin 60^o \approx \frac{\sqrt{3}l}{2}\varphi - \frac{l}{4}\varphi^2. \tag{3.8.5}$$

This leads to the following expression for the potential energy:

$$V \approx \frac{1}{2}(k_1 + k_2)l^2\varphi^2. \tag{3.8.6}$$

Given the equal distance of the points A and B with respect to the instantaneous centre of zero velocity P, their velocities are the same (Figure 3.8.3b):

$$v_B = v_A = l\dot{\varphi} \tag{3.8.7}$$

The kinetic energy is defined by

$$T = \frac{1}{2}Mv_A{}^2 + \frac{1}{2}mv_B{}^2 \;\Rightarrow\; T \approx \frac{1}{2}(M + m)l^2\dot{\varphi}^2. \tag{3.8.8}$$

The equation of motion for small vibrations and the associated natural frequency are

$$(M + m)\ddot{\varphi} + (k_1 + k_2)\varphi = 0, \quad \omega_n = \sqrt{\frac{k_1 + k_2}{M + m}}. \tag{3.8.9}$$

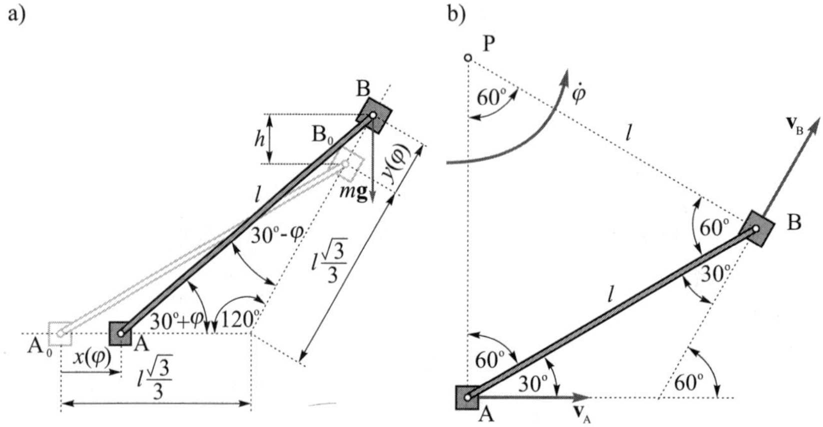

Figure 3.8.3

The free-body diagram corresponding to the second case is presented in Figure 3.8.2b. The zero sum of the moment about point P yields the static deflection of the first spring:

$$\sum M_P = 0 \Rightarrow k_1 f_{st1} l - mgl\cos 30^\circ = 0 \ \Rightarrow \ f_{st1} = \frac{mg\sqrt{3}}{2k_1}. \tag{3.8.10}$$

The potential energy of the system has the form

$$V = \frac{k_1}{2}(-f_{st1} + \lambda_1)^2 + \frac{k_2}{2}\lambda_2{}^2 + mgh, \tag{3.8.11}$$

where:

$$\lambda_1 = x(\varphi) \approx l\varphi + \frac{\sqrt{3}l}{6}\varphi^2, \ \ \lambda_2 = -y(\varphi) \approx -l\varphi, \quad h = y(\varphi)\sin 60^\circ \approx \frac{\sqrt{3}l}{2}\varphi - \frac{l}{4}\varphi^2. \tag{3.8.12}$$

Using these expressions for the additional deflections of the spring, the potential energy becomes

$$V \approx \frac{1}{2}\left[(k_1 + k_2)l^2 - mgl\right]\varphi^2. \tag{3.8.13}$$

The equation of motion for small vibrations and the associated natural frequency are now obtained as

$$(M+m)l\ddot{\varphi} + [(k_1 + k_2)l - mg]\varphi = 0, \ \ \omega_n = \sqrt{\frac{(k_1 + k_2)l - mg}{(M+m)l}}. \tag{3.8.14}$$

Example 3.9 An elliptical disc of mass m and half-axes a and b rolls without slipping along a horizontal surface. Its equilibrium position is shown in Figure 3.9.1. Determine its natural frequency for small oscillations and the period of these oscillations.

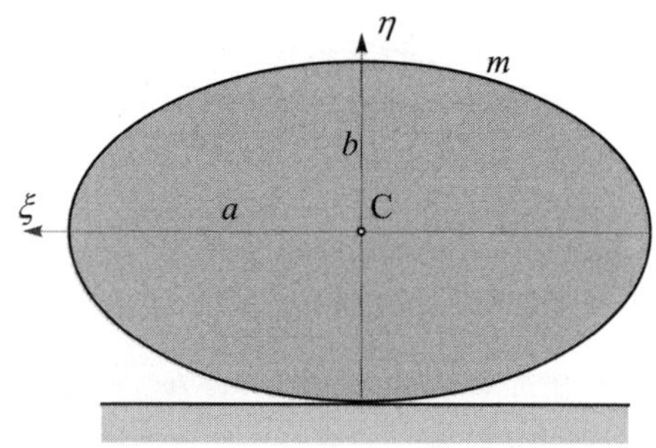

Figure 3.9.1

Solution

The generalised coordinate is chosen to be the angle of the rotation φ of the disc (Figure 3.9.2a). The potential energy is

$$V = mgy_C(\varphi), \tag{3.9.1}$$

where y_C can be obtained from the expression

$$y_C(\varphi) = \xi(\varphi)\sin\varphi + |\eta(\varphi)|\cos\varphi, \tag{3.9.2}$$

with ξ and η representing the coordinates of the contact point D (see Figure 3.9.2a).

To obtain these coordinates, the equation of the ellipse will be used,

$$\frac{\xi^2}{a^2} + \frac{\eta^2}{b^2} = 1, \tag{3.9.3}$$

together with the fact that the tangent of the angle φ corresponds to the first derivative of the ellipse $\eta(\varphi)$ at D:

$$\frac{d\eta}{d\xi} = \tan\varphi \;\Rightarrow\; -\frac{\xi b^2}{\eta a^2} = \tan\varphi. \tag{3.9.4}$$

Solving Equations (3.9.3) and (3.9.4), one obtains

$$\xi(\varphi) = \frac{a^2\tan\varphi}{\sqrt{b^2 + a^2\tan^2\varphi}}, \quad |\eta(\varphi)| = \frac{b^2}{\sqrt{b^2 + a^2\tan^2\varphi}}. \tag{3.9.5}$$

Thus, the potential energy becomes

$$V(\varphi) = mg\sqrt{b^2 + a^2\tan^2\varphi}\cos\varphi. \tag{3.9.6}$$

Developing this expression into a Maclaurin series, one obtains

$$V(\varphi) \approx mg\left(b + \frac{1}{2}\frac{a^2}{b}\varphi^2\right)\left(1 - \frac{1}{2}\varphi^2\right) \;\Rightarrow\; V(\varphi) \approx \frac{1}{2}mg\left(\frac{a^2}{b} - b\right)\varphi^2. \tag{3.9.7}$$

The kinetic energy of the disc is

$$T = \frac{m}{2}{v_C}^2 + \frac{1}{2}J_C\omega^2, \tag{3.9.8}$$

where v_C can easily be found based on Figure 3.9.2b by considering the case when the system passes through the equilibrium position and writing

$$v_C = \overline{PC}\omega \;\Rightarrow\; v_C = b\omega = b\dot{\varphi}. \tag{3.9.9}$$

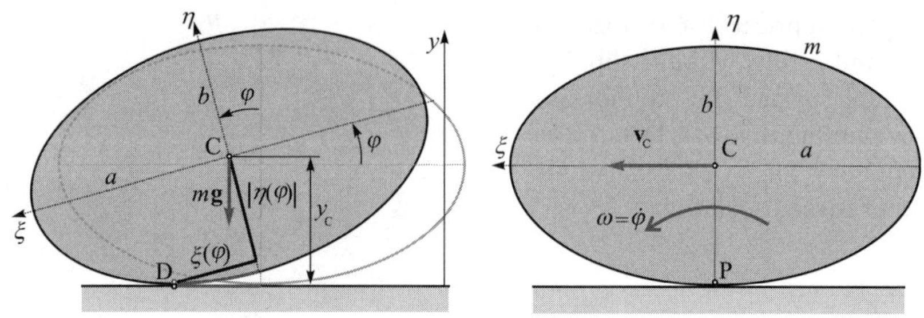

Figure 3.9.2

The mass moment of inertia of the elliptical disc about the axis perpendicular to the plane of motion through C is

$$J_C = \frac{1}{4}m\left(a^2+b^2\right), \tag{3.9.10}$$

and the kinetic energy is found to be

$$T = \frac{m}{8}\left(a^2+5b^2\right)\dot{\varphi}^2. \tag{3.9.11}$$

Finally, the equation of motion for small oscillations and the associated natural frequency are found to be:

$$b\left(a^2+5b^2\right)\ddot{\varphi}+4g\left(a^2-b^2\right)\varphi=0,\quad \omega_n=2\sqrt{\frac{g(a^2-b^2)}{b(a^2+5b^2)}}. \tag{3.9.12}$$

Example 3.10 Four equal circular discs of mass m and radius r are connected by three weightless links, which are pinned to the discs at their centres A, C, S and T. The discs roll without slipping along a horizontal surface. There are two more discs, of mass M and radius R, placed on the top of them, as well as a bar of mass M_1 on the top of these two discs, as shown in Figure 3.10.1. There is no slipping between the discs and the bar. Point A is supported by a spring of stiffness k. The lengths of the links are labelled in Figure 3.10.1. Taking the displacement of point x as the generalised coordinate, determine the natural frequency.

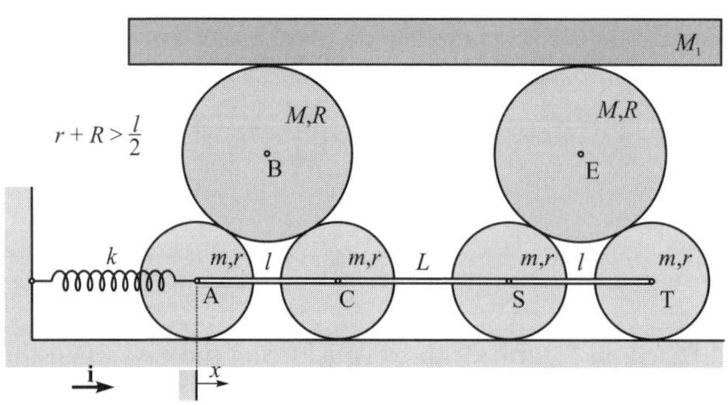

Figure 3.10.1

Solution

As the centres of all massive elements in this system move horizontally, their potential energy is constant and can be omitted, while the potential energy stemming from the spring is given by

$$V = \frac{1}{2}kx^2. \tag{3.10.1}$$

The four lower discs and two upper ones have mutually equal kinetic energies, so that the overall kinetic energy can be written down as

$$T = 4\left(\frac{1}{2}mv_A{}^2 + \frac{1}{2}J_A\omega_1{}^2\right) + 2\left(\frac{1}{2}Mv_B{}^2 + \frac{1}{2}J_B\omega_2{}^2\right) + \frac{1}{2}M_1v^2, \tag{3.10.2}$$

where ω_1 and ω_2 stand for the angular velocities (Figure 3.10.2), J_A and J_B are the mass moment of inertia of the discs with respect to the axes through the points noted, and v represents the velocity of the bar.

The centres of all discs have the same velocity

$$\mathbf{v}_A = \mathbf{v}_B = \dot{x}\mathbf{i}, \quad v_A = v_B = \dot{x}. \tag{3.10.3}$$

The lower discs execute general plane motion, and given the position of the instantaneous centre of zero velocity P_1 (see Figure 3.10.2), their angular velocity is

$$\omega_1 = \frac{v_A}{\overline{AP_1}} = \frac{\dot{x}}{r}. \tag{3.10.4}$$

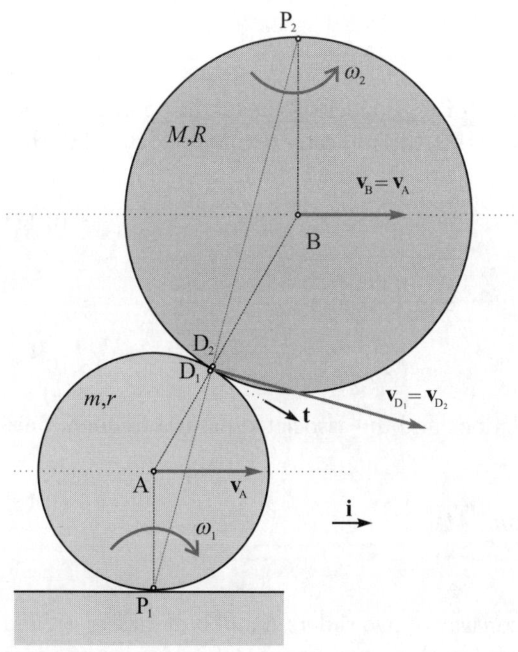

Figure 3.10.2

The velocity of point D_1 with respect to A is

$$\mathbf{v}_{D_1/A} = \overline{AD_1}\,\omega_1\,\mathbf{t} = \dot{x}\,\mathbf{t}. \tag{3.10.5}$$

The overall velocity of this point can now be presented as

$$\mathbf{v}_{D_1} = \mathbf{v}_A + \mathbf{v}_{D_1/A} = \mathbf{v}_A + \dot{x}\,\mathbf{t}. \tag{3.10.6}$$

The velocity of point D_2, which belongs to the upper disc, is defined by

$$\mathbf{v}_{D_2} = \mathbf{v}_B + \mathbf{v}_{D_2/B}, \tag{3.10.7}$$

where

$$\mathbf{v}_{D_2/B} = \overline{BD_2}\,\omega_2\,\mathbf{t} = R\omega_2\,\mathbf{t}. \tag{3.10.8}$$

As there is no slipping between the discs, the following holds

$$\mathbf{v}_{D_1} = \mathbf{v}_{D_2},\ \mathbf{v}_{D_1/A} = \mathbf{v}_{D_2/B} \quad \Rightarrow \quad \omega_2 = \frac{\dot{x}}{R}. \tag{3.10.9}$$

In order to prove the location of the instantaneous centre of zero velocity for the upper disc (P_2), one can write down the following relationship

$$\mathbf{v}_{P_2} = \mathbf{v}_B + \mathbf{v}_{P_2/B}, \tag{3.10.10}$$

where

$$\mathbf{v}_{P_2/B} = -\overline{P_2B}\,\omega_2\mathbf{i} = -R\omega_2\mathbf{i} = -\dot{x}\mathbf{i}, \tag{3.10.11}$$

which, given Equation (3.10.3), yields

$$v_{P_2} = 0. \tag{3.10.12}$$

This implies that the instantaneous centre of zero velocity P_2 lies at the point of contact between the upper disc and the bar. Besides this, this indicates that the velocity of the bar is zero, that is, it does not move

$$v = v_{P_2} = 0. \tag{3.10.13}$$

Thus, its kinetic energy is zero.

Equation (3.10.2) for the overall kinetic energy becomes

$$T = \frac{1}{2}(6m + 3M)\dot{x}^2. \tag{3.10.14}$$

The equation of motion for small oscillations and the associated natural frequency are

$$(6m + 3M)\ddot{x} + kx = 0, \quad \omega_n = \sqrt{\frac{k}{3(2m + M)}}. \tag{3.10.15}$$

Example 3.11 An oscillatory system consists of two sliders A and B of masses m_1 and m_2, respectively, which move without friction along two horizontal guides, and a slider C of mass M, which moves without friction along a vertical guide. Two weightless links are pinned to the sliders (their lengths are shown in Figure 3.11.1), and there is a spring of stiffness k between A and B. The system oscillates about the position shown in Figure 3.11.1. Taking the displacement x as the generalised coordinate, determine the natural frequency.

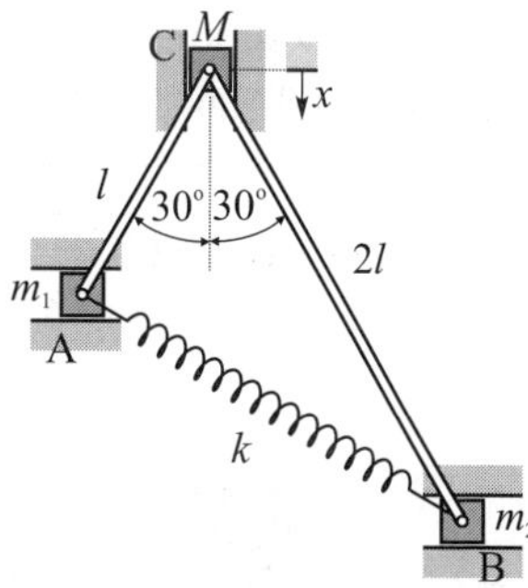

Figure 3.11.1

Solution

The first objective is to determine the static deflection of the spring. Based on the free-body diagrams shown in Figure 3.11.2a, b and c, one can write down the following equilibrium equations: first, for the system as a whole

$$\sum F_x = 0 \quad \Rightarrow F_{n3} = 0, \tag{3.11.1}$$

then for the slider C

$$\sum F_x = 0 \quad \Rightarrow \quad F_1 \sin 30^\circ - F_2 \sin 30^\circ = 0 \quad \Rightarrow \quad F_1 = F_2 = F, \tag{3.11.2}$$

$$\sum F_y = 0 \quad \Rightarrow \quad 2F \cos 30^\circ - Mg = 0 \quad \Rightarrow \quad F = \frac{Mg\sqrt{3}}{3}, \tag{3.11.3}$$

and finally for the slider A

$$\sum F_x = 0 \quad \Rightarrow \quad -F \sin 30^\circ + kf_{st} \cos 30^\circ = 0 \quad \Rightarrow \quad f_{st} = \frac{Mg}{3k}. \tag{3.11.4}$$

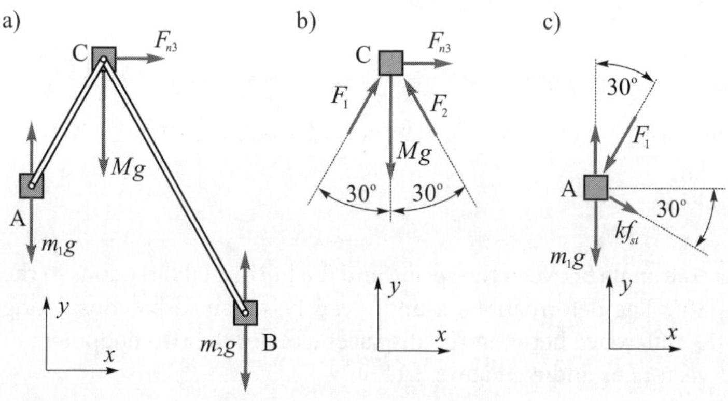

Figure 3.11.2

The system potential energy comprises the gravitational potential energy of the slider C and the potential energy of the spring

$$V = -Mgx + \frac{1}{2}k(f_{st} + \lambda)^2, \tag{3.11.5}$$

where the additional deflection λ needs to be found. Based on Figure 3.11.3a, one can express the displacements y and z as follows

$$y(x)=\sqrt{l^2-\left(l\frac{\sqrt{3}}{2}-x\right)^2}-\frac{l}{2},\quad z(x)=\sqrt{4l^2-\left(l\sqrt{3}-x\right)^2}-l. \tag{3.11.6}$$

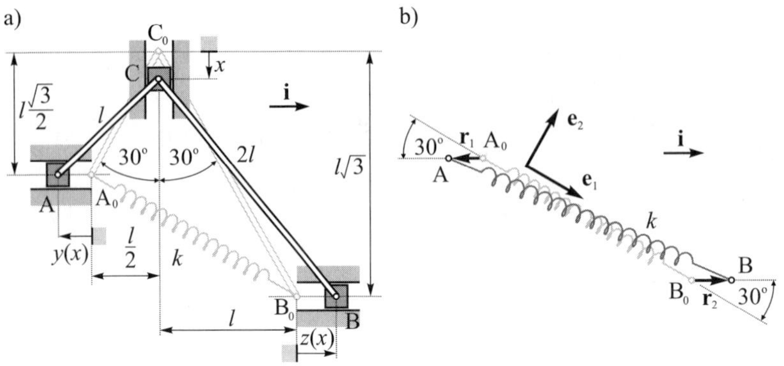

Figure 3.11.3

Developing these into series, one has

$$y(x)\approx\sqrt{3}x-\frac{4}{l}x^2,\quad z(x)\approx\sqrt{3}x-\frac{2}{l}x^2. \tag{3.11.7}$$

Now, to determine λ, one can use the expression

$$\lambda\approx u+\frac{1}{2l_{st}}v^2. \tag{3.11.8}$$

The static length of the spring l_{st} can be obtained by considering the triangle ABC from Figure 3.11.1. as

$$\overline{AB}\Big|_{x=0}=l_{st}=l\sqrt{3}. \tag{3.11.9}$$

One can also note that the angle between the spring and the horizontal directions in the equilibrium position is 30°. The deformations u and v can be obtained by considering the scalar products of the difference between the displacements of the two endpoints of the spring and the unit vectors $\mathbf{e}_1$ and $\mathbf{e}_2$ (Figure 3.11.3b):

$$\mathbf{r}_1=-y(x)\mathbf{i},\quad \mathbf{r}_2=z(x)\mathbf{i}, \tag{3.11.10}$$

$$u=(\mathbf{r}_2-\mathbf{r}_1)\cdot\mathbf{e}_1=[y(x)+z(x)]\cos 30^\circ,\quad v=(\mathbf{r}_2-\mathbf{r}_1)\cdot\mathbf{e}_2=[y(x)+z(x)]\sin 30^\circ. \tag{3.11.11}$$

Substituting Equation (3.11.7) into Equation (3.11.11), one derives

$$u\approx 3x-\frac{3\sqrt{3}}{l}x^2,\quad v\approx\sqrt{3}x. \tag{3.11.12}$$

Introducing these expressions, and substituting Equation (3.11.9) into Equation (3.11.8), one obtains

$$\lambda \approx 3x - \frac{5\sqrt{3}}{2l}x^2. \tag{3.11.13}$$

Thus, the potential energy (3.11.5) takes the form

$$V \approx \frac{1}{2}\left(9k - \frac{5\sqrt{3}}{3}\frac{Mg}{l}\right)x^2. \tag{3.11.14}$$

By considering the velocities corresponding to the position when the system passes through the equilibrium (Figure 3.11.4), one can conclude

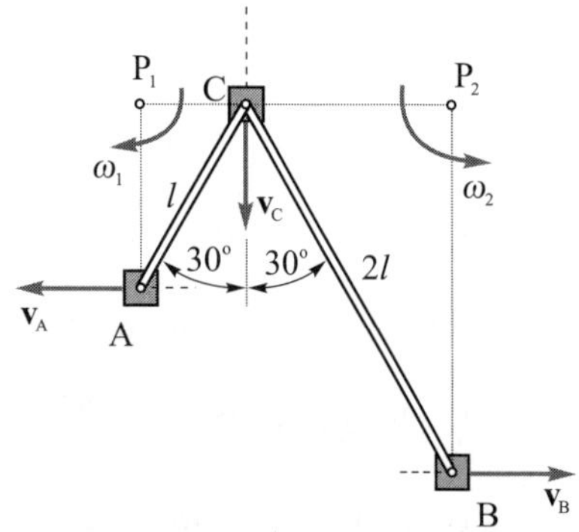

Figure 3.11.4

$$v_C = \dot{x},\ \omega_1 = \frac{v_C}{\overline{CP_1}} = \frac{2\dot{x}}{l},\ \ v_A = \overline{AP_1}\omega_1 = \sqrt{3}\dot{x},\ \ \omega_2 = \frac{v_C}{\overline{CP_2}} = \frac{\dot{x}}{l},\ \ v_B = \overline{BP_2}\omega_2 = \sqrt{3}\dot{x}, \tag{3.11.15}$$

so that the kinetic energy becomes

$$T = \frac{1}{2}Mv_C{}^2 + \frac{1}{2}m_1v_A{}^2 + \frac{1}{2}m_2v_B{}^2 \ \Rightarrow\ T \approx \frac{1}{2}[M + 3(m_1 + m_2)]\dot{x}^2. \tag{3.11.16}$$

Given Equations (3.1.14) and (3.11.16), one can derive the following equation of motion for small oscillations and expression for the corresponding natural frequency:

$$[M + 3(m_1 + m_2)]\ddot{x} + \left(9k - \frac{5\sqrt{3}}{3}\frac{Mg}{l}\right)x = 0,\ \ \omega_n = \sqrt{\frac{27kl - 5\sqrt{3}Mg}{3[M + 3(m_1 + m_2)]l}}. \tag{3.11.17}$$

Example 3.12 A half-disc oscillates around the equilibrium position shown in Figure 3.12.1. Its mass is m and its radius is R. It can roll without slipping along a horizontal surface. There are two identical springs attached to it at points A and B: their stiffness is k, the length in the equilibrium position is l, and they are extended by the force F. Using the angle φ as the generalised coordinate, determine the natural frequency in the case when the springs are predeformed ($F \neq 0$) as well as in the case when they are not predeformed ($F = 0$). Plot the resulting solution for motion for the following values of the system parameters: $l = 1$ m, $m = 10$ kg, $k = 1000$ N/m, $R = 0.5$ m, $\varphi(0) = 0.1$, $\dot{\varphi}(0) = 0$, using three possible values for the magnitude of the extension force: $F = F_0 = 0$, $F = F_1 = 50$ N and $F = F_2 = 100$ N.

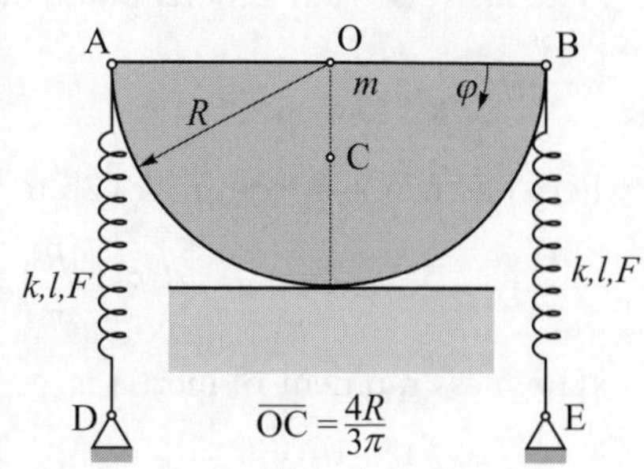

Figure 3.12.1

Solution

Let us first analyse the case when the springs are predeformed. The overall potential energy can be expressed as

$$V \approx -mgh + \frac{k}{2}(f_{st} + \lambda_1)^2 + \frac{k}{2}(f_{st} + \lambda_2)^2, \tag{3.12.1}$$

where the height h (Figure 3.12.2a) is

$$h = \overline{OC}\cos\varphi \approx \overline{OC}\left(1 - \frac{\varphi^2}{2}\right), \quad \overline{OC} = \frac{4R}{3\pi}. \tag{3.12.2}$$

The static deflection of the springs is equal to

$$f_{st} = \frac{F}{k}. \tag{3.12.3}$$

The additional deflections can be represented as

$$\lambda_1 \approx u_1 + \frac{1}{2l}{v_1}^2, \quad \lambda_2 \approx u_2 + \frac{1}{2l}{v_2}^2, \tag{3.12.4}$$

where u_1 and v_1 (see Figure 3.12.2a) are:

$$u_1 = \mathbf{r}_1 \cdot \mathbf{e}_1 = R\sin\varphi \;\Rightarrow\; u_1 \approx R\varphi, \tag{3.12.5}$$

$$v_1 = \mathbf{r}_1 \cdot \mathbf{e}_2 = R\varphi + R(1 - \cos\varphi) \;\Rightarrow\; v_1 \approx R\varphi, \tag{3.12.6}$$

while u_2 and v_2 are

$$u_2 = \mathbf{r}_2 \cdot \mathbf{e}_1 = -R\sin\varphi \;\Rightarrow\; u_2 \approx -R\varphi, \tag{3.12.7}$$

$$v_2 = \mathbf{r}_2 \cdot \mathbf{e}_2 = R\varphi - R(1 - \cos\varphi) \;\Rightarrow\; v_2 \approx R\varphi. \tag{3.12.8}$$

Finally, one can derive

$$\lambda_1 \approx R\varphi + \frac{1}{2l}R^2\varphi^2, \quad \lambda_2 \approx -R\varphi + \frac{1}{2l}R^2\varphi^2. \tag{3.12.9}$$

The potential energy now becomes

$$V = \frac{1}{2}\left[2\left(k + \frac{F}{l}\right)R^2 + mg\frac{4R}{3\pi}\right]\varphi^2. \tag{3.12.10}$$

The half-disc is in general plane motion and its kinetic energy has the form

$$T = \frac{1}{2}m{v_C}^2 + \frac{1}{2}J_C\omega^2, \tag{3.12.11}$$

where one can use Figure 3.12.2b to express v_C:

$$v_C = \overline{PC}\omega \;\Rightarrow\; v_C = \left(R - \frac{4R}{3\pi}\right)\dot{\varphi}. \tag{3.12.12}$$

The mass moment of inertia is

$$J_C = \left(\frac{1}{2} - \frac{16}{9\pi^2}\right)mR^2, \tag{3.12.13}$$

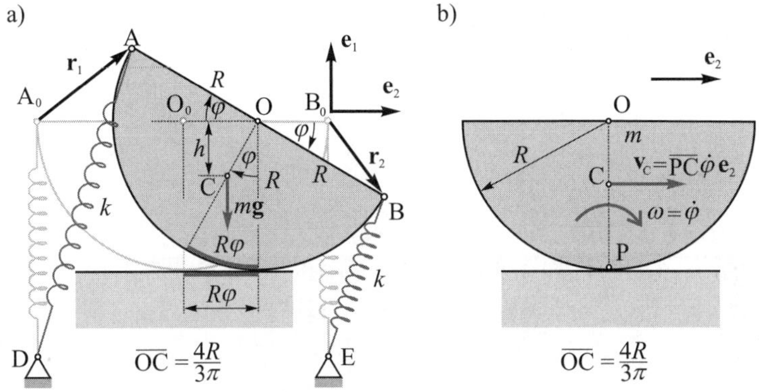

Figure 3.12.2

and the kinetic energy while the system passes through the equilibrium position takes the form

$$T \approx \frac{1}{2}\left(\frac{3}{2} - \frac{8}{3\pi}\right) mR^2\dot{\varphi}^2. \tag{3.12.14}$$

Equations (3.12.10) and (3.12.14) yield the following equation for small oscillations and the expression for the natural frequency

$$\left(\frac{3}{2} - \frac{8}{3\pi}\right) mR\ddot{\varphi} + \left[2\left(k + \frac{F}{l}\right)R + \frac{4mg}{3\pi}\right]\varphi = 0, \quad \omega_n = \sqrt{\frac{12\pi(kl + F)R + 8mgl}{(9\pi - 16)mRl}}. \tag{3.12.15}$$

Let us now analyse the second case when the springs are not predeformed in the equilibrium position. The potential energy (3.12.1) now simplifies to

$$V \approx -mgh + \frac{k}{2}\lambda_1{}^2 + \frac{k}{2}\lambda_2{}^2, \tag{3.12.16}$$

$$\lambda_1 \approx u_1 \approx R\varphi, \quad \lambda_2 \approx u_2 \approx -R\varphi. \tag{3.12.17}$$

which can be further transformed to

$$V \approx \frac{1}{2}\left(2kR + \frac{4mg}{3\pi}\right) R\varphi^2. \tag{3.12.18}$$

Using the expressions for the kinetic energy (3.2.14), and potential energy (3.2.18), the following equation for small oscillations and the expression for the associated natural frequency are obtained:

$$\left(\frac{3}{2} - \frac{8}{3\pi}\right) mR\ddot{\varphi} + \left(2kR + \frac{4mg}{3\pi}\right)\varphi = 0, \quad \omega_n = \sqrt{\frac{12\pi kR + 8mg}{(9\pi - 16)mR}}. \tag{3.12.19}$$

The same results can be obtained by putting F =0 into Equation (3.12.15).

The motion corresponding to three values of F is plotted in Figure 3.12.3. The higher the magnitude of the force, the higher the value of ω_n and the lower the period of vibration.

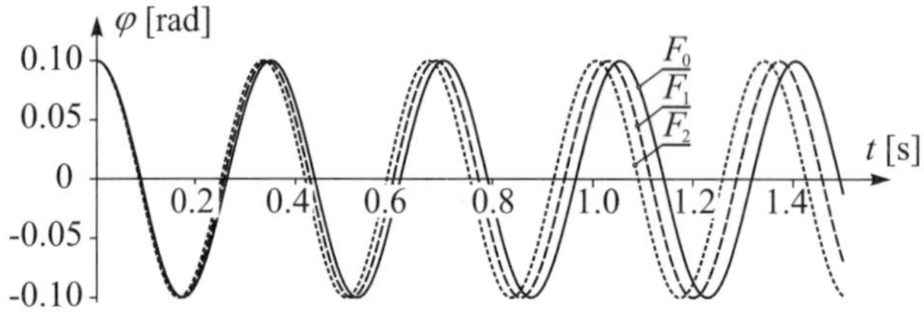

Figure 3.12.3

Example 3.13 A right-angle link consists of two rigidly connected bars OA of mass M and length L and AB of mass m and length l. The link is pivoted at O and also attached to it via a torsional spring of stiffness k. The system can oscillate freely in the vertical plane around the equilibrium position shown in Figure 3.13.1. Determine the natural frequency of small oscillations. The generalised coordinate is the angle of rotation φ.

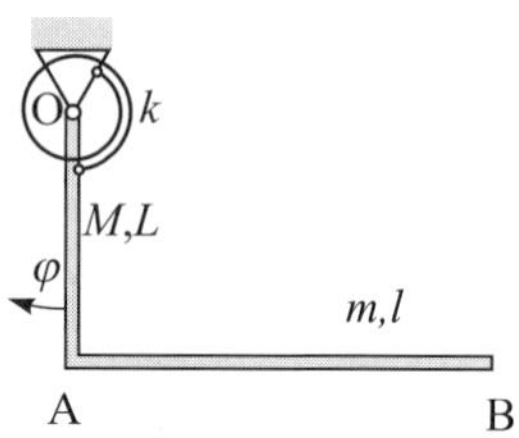

Figure 3.13.1

Solution

To assure this equilibrium position, the torsional spring must be predeformed. Its static deflection can be obtained by considering the free-body diagram (Figure 3.13.2a) and writing the following equilibrium equation

$$\sum M_O = 0 \;\Rightarrow\; k\alpha_{st} - mg\frac{l}{2} = 0 \;\Rightarrow\; \alpha_{st} = \frac{mgl}{2k}. \tag{3.13.1}$$

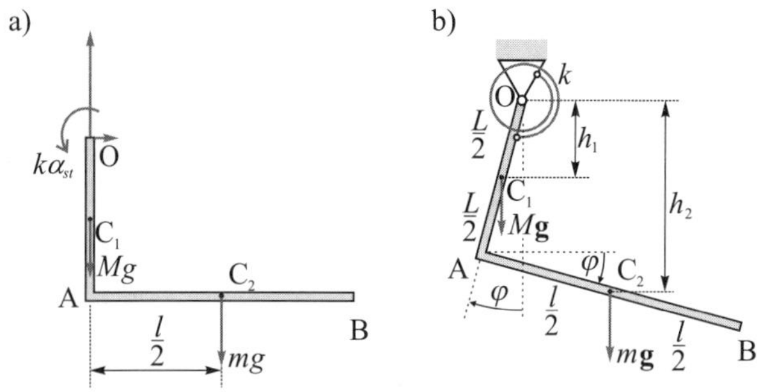

Figure 3.13.2

The potential energy consists of the gravitational energy of the link (two bars) and the spring:

$$V \approx -Mgh_1 - mgh_2 + \frac{1}{2}k(\alpha_{st} + \varphi)^2, \tag{3.13.2}$$

where the heights h_1 and h_2 are (see Figure 3.13.2b)

$$h_1 = \frac{L}{2}\cos\varphi \approx \frac{L}{2}\left(1-\frac{\varphi^2}{2}\right),\quad h_2 = L\cos\varphi + \frac{l}{2}\sin\varphi \approx L\left(1-\frac{\varphi^2}{2}\right) + \frac{l}{2}\varphi. \tag{3.13.3}$$

It is easy to see that the additional deflection of the spring is equal to φ, so that the potential energy becomes

$$V \approx \frac{1}{2}\left[\left(\frac{M}{2}+m\right)gL + k\right]\varphi^2. \tag{3.13.4}$$

The link rotates around the axis through O, so that the kinetic energy has the form

$$T = \frac{1}{2}J_O\dot{\varphi}^2, \tag{3.13.5}$$

where the mass moment of inertia J_O can be obtained from the parallel-axis theorem as follows

$$\begin{aligned} &J_O = J_O^{\,OA} + J_O^{\,AB},\quad J_O^{\,OA} = J_{C_1}^{\,OA} + M\cdot\overline{OC_1}^2,\quad J_O^{\,AB} = J_{C_2}^{\,AB} + m\cdot\overline{OC_2}^2, \\ &J_{C_1}^{\,OA} = \frac{1}{12}ML^2,\quad \overline{OC_1} = \frac{L}{2},\quad J_{C_2}^{\,AB} = \frac{1}{12}ml^2,\quad \overline{OC_2}^2 = L^2 + \left(\frac{l}{2}\right)^2. \end{aligned} \tag{3.13.6}$$

Thus, the mass moment of inertia J_O is

$$J_O = \left(\frac{M}{3}+m\right)L^2 + \frac{1}{3}ml^2. \tag{3.13.7}$$

This leads to the following expression for the kinetic energy:

$$T = \frac{1}{2}\left[\left(\frac{M}{3}+m\right)L^2 + \frac{1}{3}ml^2\right]\dot{\varphi}^2. \tag{3.13.8}$$

Using the expressions for the potential and kinetic energy (3.13.4) and (3.13.8), one can derive the following equation of motion for small oscillations and the associated natural frequency

$$\left[\left(\frac{M}{3}+m\right)L^2 + \frac{1}{3}ml^2\right]\ddot{\varphi} + \left[\left(\frac{M}{2}+m\right)gL + k\right]\varphi = 0,\quad \omega_n = \sqrt{\frac{3(M+2m)gL+6k}{2(M+3m)L^2+2ml^2}}. \tag{3.13.9}$$

Example 3.14 A linkage consists of two slender bars: OA of mass m and length l, and AB of mass M and length $L = 2\,l$ (Figure 3.14.1). The sliding block at B has negligible mass. There are two torsional springs: the first one has stiffness k_1 and is attached to

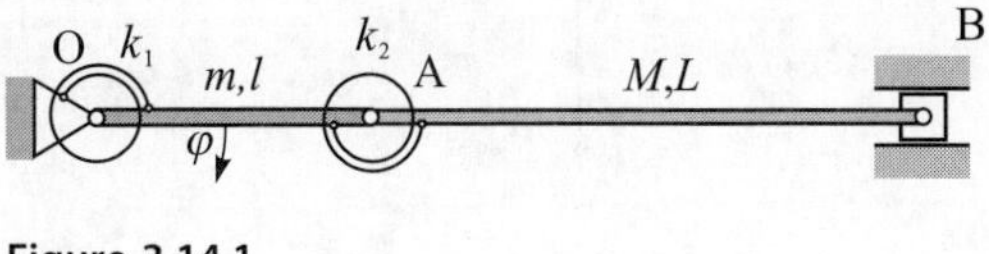

Figure 3.14.1

the bar OA and the fixed base at O; the second one has stiffness k_2 and connects the two bars. The system oscillates around the equilibrium position when the bars are horizontal. Determine the expression for the natural frequency when only one of the springs is predeformed. The generalised coordinate is the angle φ.

Solution

The free-body diagram for the system as a whole (Figure 3.14.2a) and for the case when the left spring is not predeformed can be used to write down the following equilibrium equation:

$$\sum M_B = 0 \;\Rightarrow\; -F_O 3l + mg\frac{5}{2}l + Mgl = 0 \;\Rightarrow\; F_O = \frac{5m+2M}{6}g, \tag{3.14.1}$$

while the free-body diagram for the bar OA (Figure 3.14.2b) and the moment equilibrium equation can be written down to obtain the static deflection of the right spring:

$$\sum M_A = 0 \;\Rightarrow\; -F_O l + mg\frac{l}{2} + k_2\alpha_{st2} = 0 \;\Rightarrow\; \alpha_{st2} = \frac{m+M}{3k_2}gl. \tag{3.14.2}$$

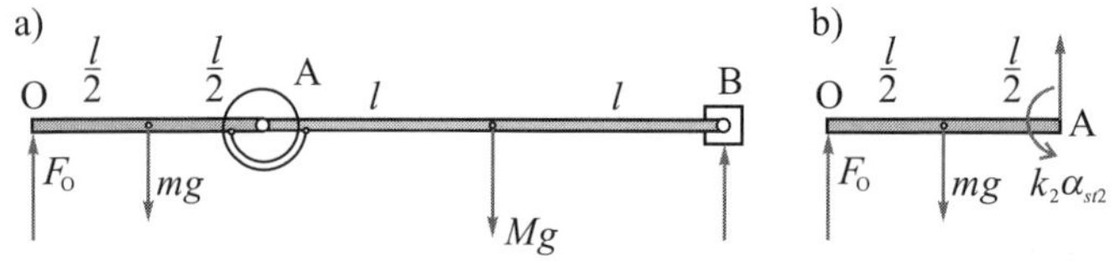

Figure 3.14.2

The system potential energy has the form

$$V = -mgh_1 - Mgh_2 + \frac{1}{2}k_1(0+\varphi)^2 + \frac{1}{2}k_2[\alpha_{st2}+\varphi+\psi(\varphi)]^2, \tag{3.14.3}$$

where the following geometric relationships and Figure 3.14.3a can be utilised to define $\psi(\varphi)$, h_1 and h_2:

$$h_A = l\sin\varphi = 2l\sin\psi, \tag{3.14.4}$$

$$\Rightarrow \; \psi(\varphi) = \arcsin\left(\frac{1}{2}\sin\varphi\right) \approx \frac{1}{2}\varphi + \ldots, \tag{3.14.5}$$

$$h_1 = h_2 = \frac{h_A}{2} = \frac{l}{2}\sin\varphi \approx \frac{l}{2}\varphi + \ldots \tag{3.14.6}$$

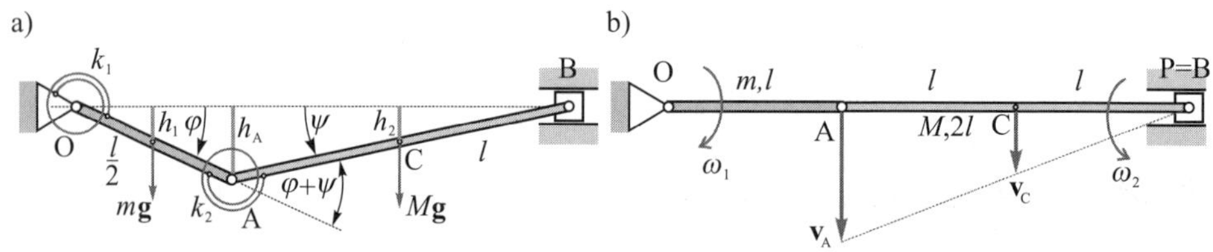

Figure 3.14.3

The expression for the potential energy is found to be

$$V = \frac{1}{2}\left(k_1 + \frac{9}{4}k_2\right)\varphi^2. \tag{3.14.7}$$

To determine the kinetic energy, one can notice that the bar OA rotates around the axis at O and the bar AB performs general plane motion, so that the following expression defines the overall kinetic energy:

$$T = \frac{1}{2}J_O\omega_1{}^2 + \frac{1}{2}Mv_C{}^2 + \frac{1}{2}J_C\omega_2{}^2, \tag{3.14.8}$$

where the mass moments of inertia are

$$J_C = \frac{1}{12}ML^2 = \frac{1}{12}M(2l)^2 = \frac{1}{3}Ml^2, \tag{3.14.9}$$

$$J_O = \frac{1}{12}ml^2 + m\left(\frac{l}{2}\right)^2 = \frac{1}{3}ml^2. \tag{3.14.10}$$

Considering the position when the system passes through the equilibrium position (Figure 3.14.3b), one can obtain:

$$\omega_1 = \dot{\varphi}, \quad v_A = l\omega_1 = l\dot{\varphi}, \quad \omega_2 = \frac{v_A}{\overline{AP}} = \frac{\dot{\varphi}}{2}, \quad v_C = \overline{PC}\,\omega_2 = \frac{l\dot{\varphi}}{2}. \tag{3.14.11}$$

Thus, the kinetic energy now takes the form

$$T = \frac{1}{6}(m+M)l^2\dot{\varphi}^2. \tag{3.14.12}$$

Equations (3.14.7) and (3.14.12) result in the following equation of motion for small oscillations and the corresponding natural frequency:

$$\frac{1}{3}(m+M)l^2\ddot{\varphi} + \left(k_1 + \frac{9}{4}k_2\right)\varphi = 0, \quad \omega_n = \frac{1}{2l}\sqrt{\frac{12k_1 + 27k_2}{m+M}}. \tag{3.14.13}$$

Let us now analyse the second case, when the left spring is predeformed. The free-body diagram corresponding to the bar AB is shown in Figure 3.14.4a and the following equilibrium equation holds:

$$\sum M_B = 0 \;\Rightarrow\; -F_A 2l + Mgl = 0 \;\Rightarrow\; F_A = \frac{Mg}{2}, \tag{3.14.14}$$

while the free-body diagram corresponding to the bar OA shown in Figure 3.14.4b and the zero sum of the moments about the point O give:

$$\sum M_O = 0 \;\Rightarrow\; -F_A l - mg\frac{l}{2} + k_1\alpha_{st1} = 0 \;\Rightarrow\; \alpha_{st1} = \frac{m+M}{2k_1}gl. \tag{3.14.15}$$

The potential energy now reads as

$$V = -mgh_1 - Mgh_2 + \frac{1}{2}k_1(\alpha_{st1} + \varphi)^2 + \frac{1}{2}k_2[0 + \varphi + \psi(\varphi)]^2, \tag{3.14.16}$$

and given Equations (3.14.4)–(3.14.6) and (3.14.15), it becomes

$$V = \frac{1}{2}\left(k_1 + \frac{9}{4}k_2\right)\varphi^2. \tag{3.14.17}$$

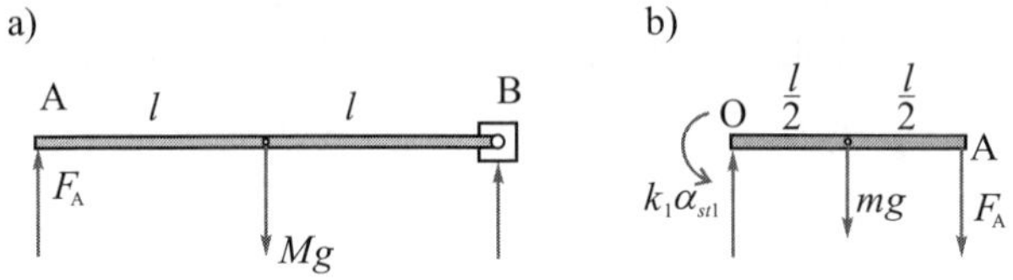

Figure 3.14.4

Thus, given the same expressions for the kinetic and potential energy as in the previous case, the expressions for the natural frequency are identical as well.

Example 3.15 A disc of mass m and radius R oscillates in the vertical plane (Figure 3.15.1). The disc is suspended by a cable and a spring that has a stiffness k. The generalised coordinate is the displacement x of the centre C. Determine the solution for small oscillations $x(t)$, assuming that the initial conditions are given by $x(0)=a$ and $\dot{x}(0) = 0$. In addition, find the critical value of the initial amplitude a leading to the cable being extended during the motion.

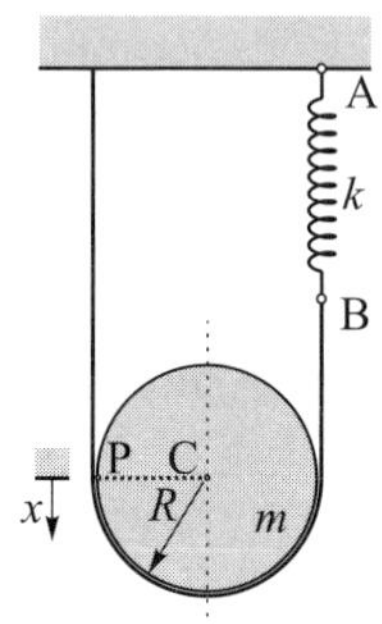

Figure 3.15.1

Solution

It should be noted that the spring is predeformed in the equilibrium position. The free-body diagram is presented in Figure 3.15.2a and can be used to obtain the static deflection from the following equilibrium:

$$\sum M_P = 0 \;\Rightarrow\; kf_{st}2R - mgR = 0 \;\Rightarrow\; f_{st} = \frac{mg}{2k}. \tag{3.15.1}$$

The system potential energy can be written down as

$$V = -mgx + \frac{1}{2}k(f_{st} + z)^2, \tag{3.15.2}$$

where the displacement z needs to be expressed in terms of the generalised coordinate (Figure 3.15.2b). To that end, the system passing through the arbitrary position is considered (Figure 3.15.2c) with the following findings:

$$\omega = \dot{\varphi}, \quad \mathbf{v}_C = \dot{x}\mathbf{j}, \quad \mathbf{v}_B = \dot{z}\mathbf{j}. \tag{3.15.3}$$

Given the position of the instantaneous centre P_1 of zero velocity, one has:

$$\omega = \frac{v_c}{\overline{P_1C}} \;\Rightarrow\; \dot{\varphi} = \frac{\dot{x}}{R}, \quad v_{B_1} = v_B = \overline{P_1B_1}\,\omega \;\Rightarrow\; \dot{z} = 2\dot{x}. \tag{3.15.4}$$

Integrating these kinematic relationships, one obtains

$$\varphi = \frac{x}{R}, \quad z = 2x. \tag{3.15.5}$$

On the other hand, finding their time derivatives gives the relationships between the following accelerations:

$$\ddot{\varphi} = \frac{\ddot{x}}{R}, \quad \ddot{z} = 2\ddot{x}. \tag{3.15.6}$$

Thus, the potential energy becomes:

$$V \approx 2kx^2. \tag{3.15.7}$$

The kinetic energy is defined by

$$T = \frac{1}{2}mv_C{}^2 + \frac{1}{2}J_C\dot{\varphi}^2, \quad J_C = \frac{1}{2}mR^2, \tag{3.15.8}$$

which can be combined with Equations (3.15.3) and (3.15.4) to derive

$$T = \frac{3}{4}m\dot{x}^2. \tag{3.15.9}$$

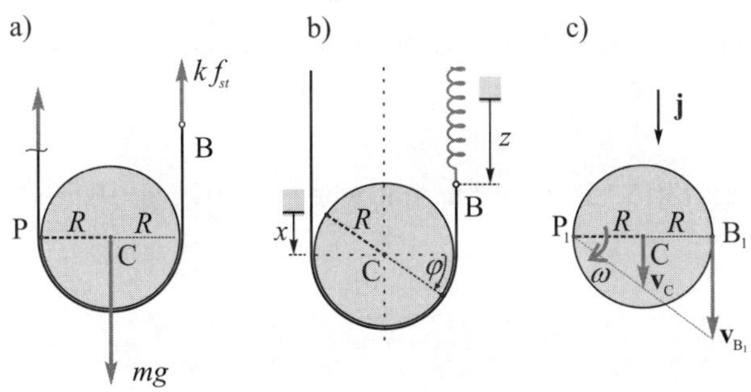

Figure 3.15.2

The differential equation of motion for small oscillations and the expression for the corresponding natural frequency are found to be

$$\frac{3}{2}m\ddot{x} + 4kx = 0, \quad \omega_n = \sqrt{\frac{8k}{3m}}. \tag{3.15.10}$$

The solution for motion has the form

$$x(t) = a\cos\omega_n t, \tag{3.15.11}$$

which also yields

$$z(t) = 2a\cos\omega_n t, \quad \varphi(t) = \frac{a}{R}\cos\omega_n t, \quad \ddot{x}(t) = -\frac{8k}{3m}a\cos\omega_n t, \quad \ddot{\varphi}(t) = -\frac{8k}{3m}\frac{a}{R}\cos\omega_n t. \tag{3.15.12}$$

The dynamic equations of motion have the form (see Figure 3.15.3)

$$m\ddot{x} = mg - F_1 - F_2, \quad J_C\ddot{\varphi} = F_1R - F_2R, \tag{3.15.13}$$

which can be combined with Equation (3.15.12) to derive the following system of algebraic equations with respect to F_1 and F_2:

$$F_1 + F_2 = mg + \frac{8ka}{3}\cos\omega_n t, \quad F_1 - F_2 = -\frac{4ka}{3}\cos\omega_n t. \tag{3.15.14}$$

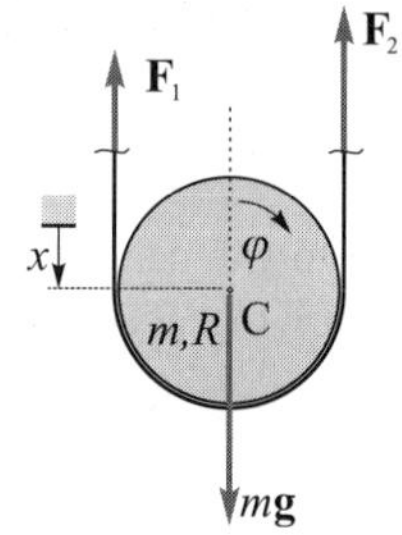

Figure 3.15.3

Solving these equations, one obtains

$$F_1 = \frac{mg}{2} + \frac{2}{3}ka\cos\omega_n t, \quad F_2 = \frac{mg}{2} + 2ka\cos\omega_n t. \tag{3.15.15}$$

The critical value of the initial amplitude leading to the cable being extended during the motion can be obtained by considering the minimal value of the forces and the condition that this cannot be smaller than zero. It is seen that the lowest value of the force is achieved in F_2 and is given by

$$F_{2min} = \frac{mg}{2} - 2ka. \tag{3.15.16}$$

Making it equal to zero, one finds the critical value of the initial amplitude a_c:

$$a_c = \frac{mg}{4k}. \tag{3.15.17}$$

Figure 3.15.4 shows how the forces (3.15.15) change in time when m=10 kg, g=10 m/s^2, k=1000 N/m, and a= a_c=0.025 m.

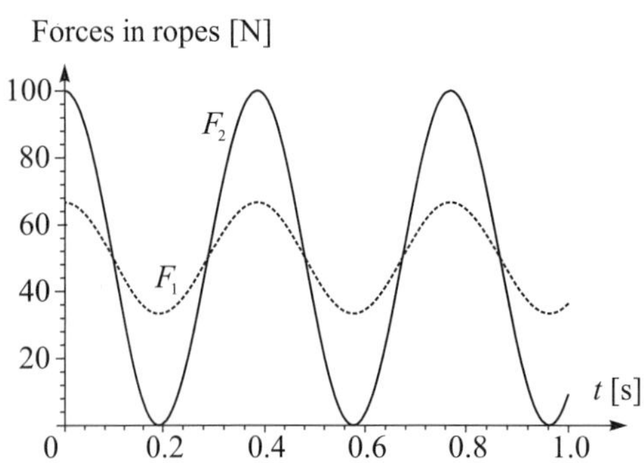

Figure 3.15.4

4

Free Damped Vibration of Single-Degree-of-Freedom Systems

Chapter Outline

This chapter first presents Lagrange's equation for free damped vibration, the quantities that exist in it and the resulting equation of motion for the case when the system performs small oscillations around an equilibrium position. Eleven examples are solved subsequently. Special attention is again given to the potential energy when the spring is predeformed, and the kinetic energy obtained when the system passes through the equilibrium position. It is also demonstrated that the dissipation function can be obtained by considering the velocities when the system passes through the equilibrium position.

Chapter Objectives

- To present Lagrange's equation of the second kind for free damped vibration
- To use it to get the equation of motion of a variety of damped mechanical systems that perform small oscillations around their equilibrium position
- To demonstrate how the dissipation function can be easily obtained by considering the velocities while the system passes through the equilibrium position

Theoretical Introduction

Lagrange's equation for one-degree-of-freedom free damped systems has the form

$$\frac{d}{dt}\frac{\partial T}{\partial \dot{q}} - \frac{\partial T}{\partial q} + \frac{\partial D}{\partial \dot{q}} + \frac{\partial V}{\partial q} = 0. \tag{4.1}$$

When the system performs small oscillations around the equilibrium position $q = 0$, this Lagrange's equation can be simplified to

$$\frac{d}{dt}\frac{\partial T}{\partial \dot{q}} + \frac{\partial D}{\partial \dot{q}} + \frac{\partial V}{\partial q} = 0, \tag{4.2}$$

Mechanical Vibration: Fundamentals with Solved Examples, First Edition.
Ivana Kovačić and Dragi Radomirović.

where the kinetic energy takes the form

$$T = \frac{1}{2}\widetilde{T}\dot{q}^2, \tag{4.3}$$

with the coefficient $\widetilde{T}$ being constant and calculated while the system passes through the equilibrium position: $\widetilde{T}(q) = \widetilde{T}(0)$.

The potential energy in Equation (4.2) has the general form

$$V = \frac{1}{2}\widetilde{V}q^2, \tag{4.4}$$

where its coefficient $\widetilde{V}$ corresponds to

$$\widetilde{V} = \left(\frac{\mathrm{d}^2 V}{\mathrm{d}q^2}\right)_{q=0}. \tag{4.5}$$

The dissipative function D in Equation (4.2) has the general form

$$D = \frac{1}{2}\widetilde{D}\dot{q}^2, \tag{4.6}$$

where $\widetilde{D}$ is a constant that can be calculated by considering the system passing through the equilibrium position.

Substituting (4.3), (4.4) and (4.6) into Equation (4.2), one can derive

$$\ddot{q} + 2\delta\dot{q} + \omega_n{}^2 q = 0. \tag{4.7}$$

where ω_n represents the natural frequency and is defined by $\omega_n{}^2 = \widetilde{V}/\widetilde{T}$, while the damping ratio is $2\delta = \widetilde{D}/\widetilde{T}$.

Knowing these two parameters, one can determine the quasiperiod of small vibrations T^* from the expression

$$T^* = \frac{2\pi}{\widetilde{\omega}} = \frac{2\pi}{\sqrt{\omega_n^2 - \delta^2}}. \tag{4.8}$$

A general solution of the differential equation of motion for small oscillations (4.7) has the form

$$q = Ae^{-\delta t}\cos(\widetilde{\omega}t + \alpha), \tag{4.9}$$

where the constants A and α can be calculated from the initial conditions $q(0)$, $\dot{q}(0)$. The parameter

$$\widetilde{\omega} = \sqrt{\omega_n^2 - \delta^2} \tag{4.10}$$

is called the damped natural frequency. If $\omega_n > \delta$, the system is under-damped, which means that it performs quasiperiodic oscillations (the amplitude decrease exponentially, while the quasiperiod is constant); when $\omega_n < \delta$, the system is over-damped and the system returns to its equilibrium position without oscillations; the limiting case between these two cases corresponds to $\omega_n = \delta$, and it is said that the system is critically damped – there are no oscillations, and the system returns to equilibrium in the shortest time.

Example 4.1 A uniform slender rod of length $2a$ and mass m is pivoted at O and supported at its end A by a spring of stiffness k. A linear viscous damper of damping coefficient c is also attached at A. The system in shown in Figure 4.1.1 in its equilibrium position and performs small oscillations around it. The spring is horizontal and the damper is vertical. Determine the equation of motion for small vibrations and the value of the damping coefficient c_c for which the system is critically damped. Assume that the system parameters have the values: $a = 1\,\text{m}$, $m = 6\,\text{kg}$, $k = 2000\,\text{N/m}$, $\varphi(0) = 0.1$, $\dot{\varphi}(0) = 0$. Plot the time-history diagrams corresponding to these values as well as to $c = 2c_c$ and $c = c_c/10$.

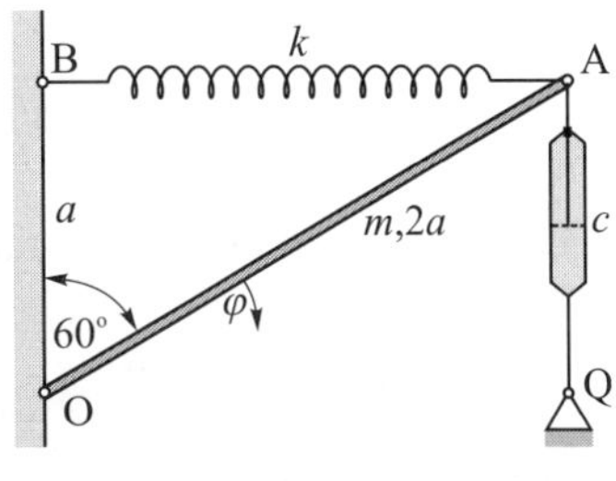

Figure 4.1.1

Solution

This mechanical system is analysed in Example 3.2 – the only difference is the existence of the damper. Thus, the kinetic and potential energy are taken from therein:

$$T = \frac{2}{3} m a^2 \dot{\varphi}^2, \quad V \approx \frac{1}{2}\left(ka^2 - \frac{mga}{2}\right)\varphi^2. \tag{4.1.1}$$

The dissipation function has the form

$$D = \frac{1}{2} c v_{p/c}^2, \tag{4.1.2}$$

where $v_{p/c}$ represents the relative velocity of the piston with respect to the cylinder in the damper. As discussed in Section 2.5, this dissipation function should be determined for the case when the system passes through the equilibrium position. Thus, the velocity $v_{p/c}$ can be obtained by projecting the velocity of the point A (Figure 4.1.2) on the QA direction, that is, the direction of the damper in the equilibrium position:

$$v_{p/c} = v_A \cos 30° = \overline{\text{OA}}\,\dot{\varphi}\frac{\sqrt{3}}{2} = \sqrt{3} a \dot{\varphi}. \tag{4.1.3}$$

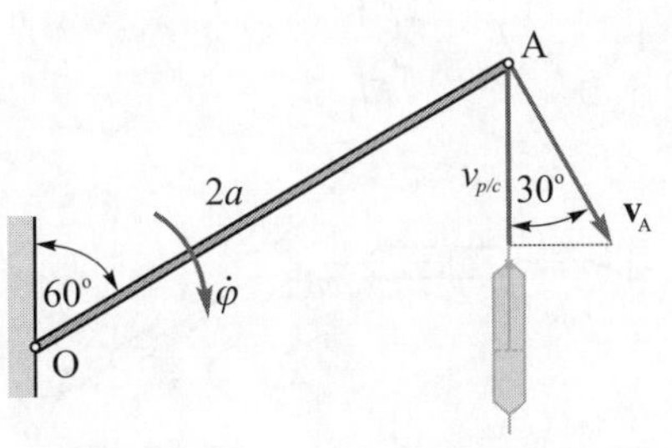

Figure 4.1.2

The dissipation function (4.1.2) now becomes

$$D \approx \frac{1}{2} 3ca^2 \dot{\varphi}^2. \tag{4.1.4}$$

Lagrange's equation (4.2) transforms into

$$\ddot{\varphi} + 2\delta\dot{\varphi} + \omega_n{}^2\varphi = 0, \quad \delta = \frac{9}{8}\frac{c}{m}, \quad \omega_n = \sqrt{\frac{6ak - 3mg}{8am}}. \tag{4.1.5}$$

Using $g = 9.81\,\mathrm{m/s^2}$, one can calculate the critical damping coefficient

$$\delta = \omega_n \quad \Rightarrow \quad c = c_c = \frac{8m}{9}\sqrt{\frac{6ak - 3mg}{8am}} \approx 83.705\,\frac{\mathrm{Ns}}{\mathrm{m}}. \tag{4.1.6}$$

The time–history diagrams corresponding to all three cases (under-damped, critically damped and over-damped) are given in Figure 4.1.3 and clearly illustrate the characteristic explained in the theoretical introduction above.

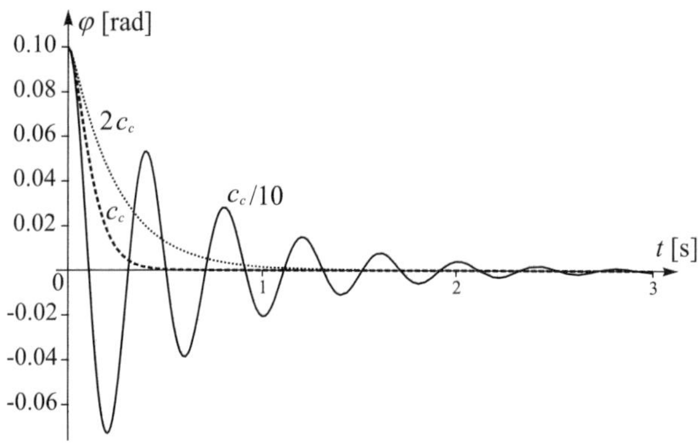

Figure 4.1.3

Example 4.2 A slider of mass m can slide along a frictionless guide and is attached to a spring that has stiffness k. The system oscillates around the position $x = 0$ shown in Figure 4.2.1. The length of the spring is then l, and the angle between the guide and the spring is β. There are two linear viscous dampers: the first one of damping coefficient c_1 is horizontal; the other one is of damping coefficient c_2 and lies in the same direction as the guide. Determine the damped natural frequency of small oscillations as well as the quasiperiod of vibration.

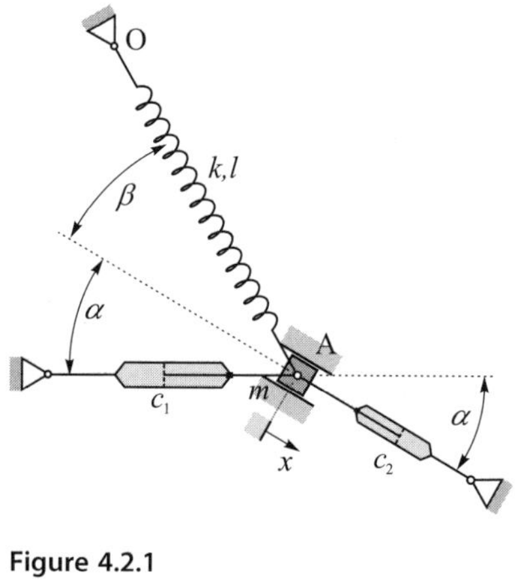

Figure 4.2.1

Solution

This system is an upgrade of Example 3.5, with only two dampers added. Thus, the kinetic and potential energy are the same as obtained in Example 3.5:

$$T = \frac{1}{2}m\dot{x}^2, \quad V \approx \frac{1}{2}\left(k\cos^2\beta + \frac{mg\sin\alpha\sin^2\beta}{l\cos\beta}\right)x^2. \tag{4.2.1}$$

The dissipation function contains two terms, each of which stems from one damper:

$$D = \frac{c_1}{2}\left(v_{p/c}\right)_1^{\,2} + \frac{c_2}{2}\left(v_{p/c}\right)_2^{\,2}, \tag{4.2.2}$$

where $v_{p/c}$ represents the relative velocity of the piston with respect to the cylinder in each damper. These relative velocities should be determined for the case when the system passes through the equilibrium position (Figure 4.2.2) and are given by

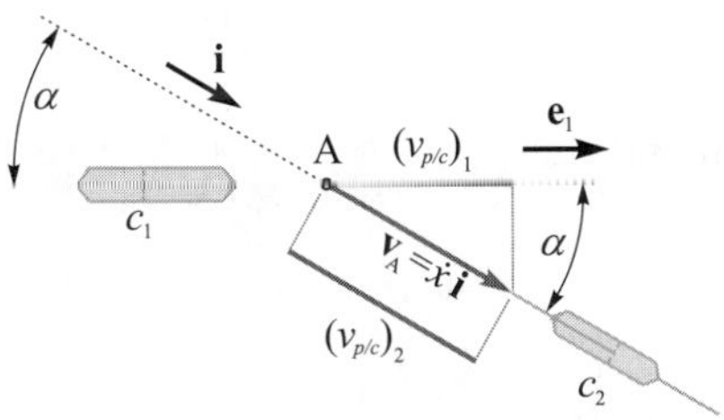

Figure 4.2.2

$$\left(v_{p/c}\right)_1 = \mathbf{v}_A \cdot \mathbf{e}_1 = \dot{x}\cos\alpha, \quad \left(v_{p/c}\right)_2 = \mathbf{v}_A \cdot \mathbf{i} = \dot{x}. \tag{4.2.3}$$

Thus, the overall dissipation function (4.2.2) becomes

$$D \approx \frac{1}{2}\left(c_1\cos^2\alpha + c_2\right)\dot{x}^2. \tag{4.2.4}$$

The differential equation of small oscillations (4.2) is

$$\ddot{x} + 2\delta\dot{x} + \omega_n^{\,2}x = 0, \quad \delta = \frac{1}{2m}\left(c_1\cos^2\alpha + c_2\right), \quad \omega_n = \sqrt{\frac{k}{m}\cos^2\beta + \frac{g\sin\alpha\sin^2\beta}{l\cos\beta}}. \tag{4.2.5}$$

Using Equations (4.10) and (4.8), one can obtain the damped natural frequency and the quasiperiod as follows

$$\widetilde{\omega} = \sqrt{\omega_n^{\,2} - \delta^2} = \sqrt{\frac{k}{m}\cos^2\beta + \frac{g\sin\alpha\sin^2\beta}{l\cos\beta} - \frac{\left(c_1\cos^2\alpha + c_2\right)^2}{4m^2}}, \tag{4.2.6}$$

$$T^* = \frac{2\pi}{\tilde{\omega}} = \frac{2\pi}{\sqrt{\frac{k}{m}\cos^2\beta + \frac{g\sin\alpha\sin^2\beta}{l\cos\beta} - \frac{(c_1\cos^2\alpha + c_2)^2}{4m^2}}}. \quad (4.2.7)$$

Example 4.3 A disc of mass m and radius r rolls without slipping inside a fixed circular surface of radius R. The centre C is attached to a spring AC of stiffness k. The system performs small oscillations around the equilibrium position shown in Figure 4.3.1. When the spring is vertical, it has length l and is predeformed by force F. There is a linear viscous damper BQ of damping coefficient c attached to the disc that is vertical at the equilibrium position. Determine the differential equation of motion for small oscillations as well as the critical value of the damping coefficient.

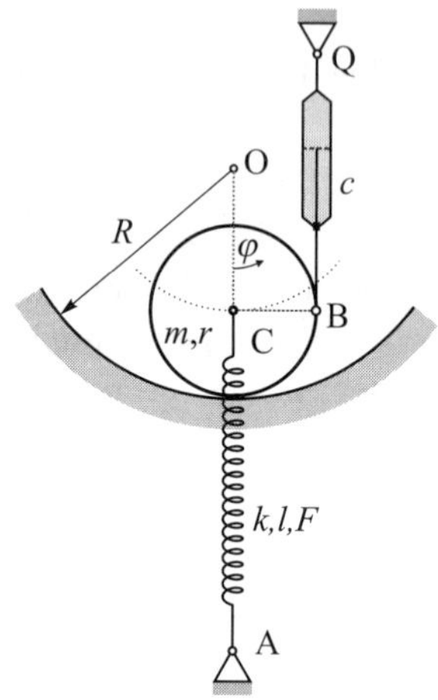

Figure 4.3.1

Solution

This system is considered in Example 3.6, and the kinetic and potential energy are obtained there in the form:

$$T = \frac{3}{4}m(R-r)^2\dot{\varphi}^2, \quad V \approx \frac{R-r}{2}\left[mg + F\left(1 + \frac{R-r}{l}\right)\right]\varphi^2. \quad (4.3.1)$$

Due to the existence of the damper, one just needs to find the dissipation function

$$D = \frac{1}{2}cv_{p/c}^2, \quad (4.3.2)$$

where $v_{p/c}$ represents the relative velocity of the piston with respect to the cylinder. Considering the case when the system passes through the equilibrium position (Figure 4.3.2), one can relate the velocity $v_{p/c}$ to the velocity of the point B as follows:

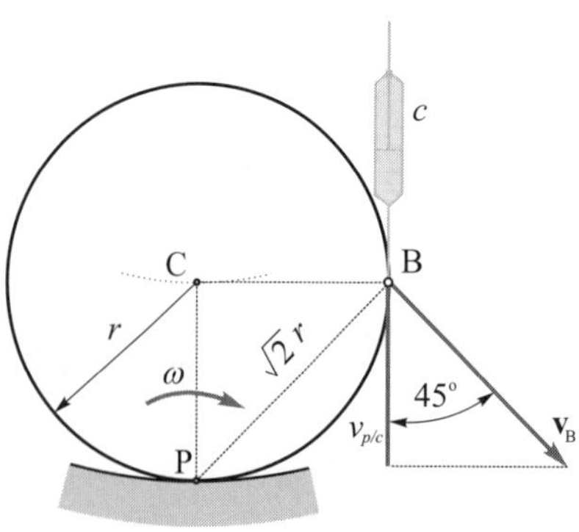

Figure 4.3.2

$$v_{p/c} = v_B \cos 45^\circ = \overline{PB}\,\omega \frac{\sqrt{2}}{2}. \tag{4.3.3}$$

By using the relationship for the angular velocity (3.6.11), the dissipation function (4.3.2) becomes

$$D \approx \frac{c}{2}(R-r)^2 \dot{\varphi}^2. \tag{4.3.4}$$

Lagrange's equation (4.2) gives:

$$\ddot{\varphi} + 2\delta\dot{\varphi} + \omega_n{}^2 \varphi = 0, \quad \delta = \frac{c}{3m}, \quad \omega_n = \sqrt{\frac{2mgl + 2F(l+R-r)}{3m(R-r)l}}. \tag{4.3.5}$$

The system is critically damped when $\delta = \omega_n$, which yields

$$c = c_c = \sqrt{\frac{3m[2mgl + 2F(l+R-r)]}{(R-r)l}}. \tag{4.3.6}$$

Example 4.4 A right-angle link consists of two rigidly connected bars OA, of mass M and length L, and AB, of mass m and length l. The link is pivoted at O and attached to it via a torsional spring of stiffness k. The system can oscillate freely in the vertical plane around the equilibrium position shown in Figure 4.4.1. As presented in this figure, there are two linear viscous dampers of damping coefficients c_1 and c_2. Determine the damped natural frequency of oscillations as well as the quasiperiod.

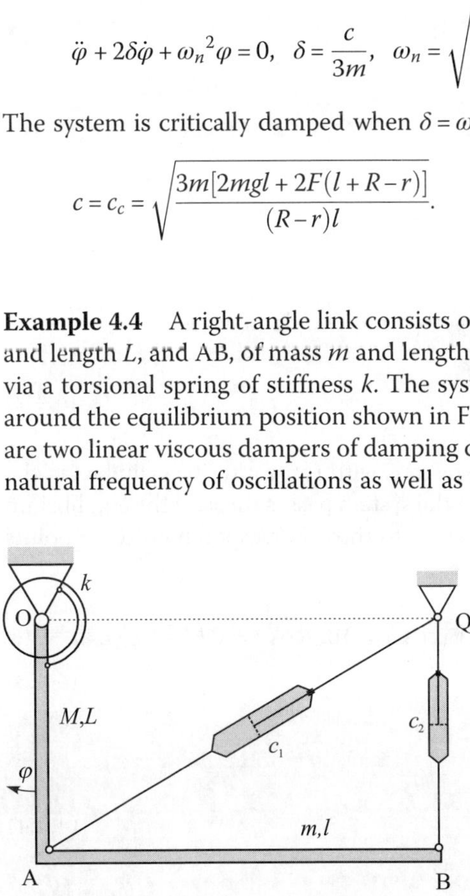

Figure 4.4.1

Solution

This system is considered in Example 3.13, while here only two dampers are added. The potential and kinetic energy are defined by Equations (3.13.4) and (3.13.8):

$$V \approx \frac{1}{2}\left[\left(\frac{M}{2} + m\right)gL + k\right]\varphi^2, \quad T = \frac{1}{2}\left[\left(\frac{M}{3} + m\right)L^2 + \frac{1}{3}ml^2\right]\dot{\varphi}^2. \tag{4.4.1}$$

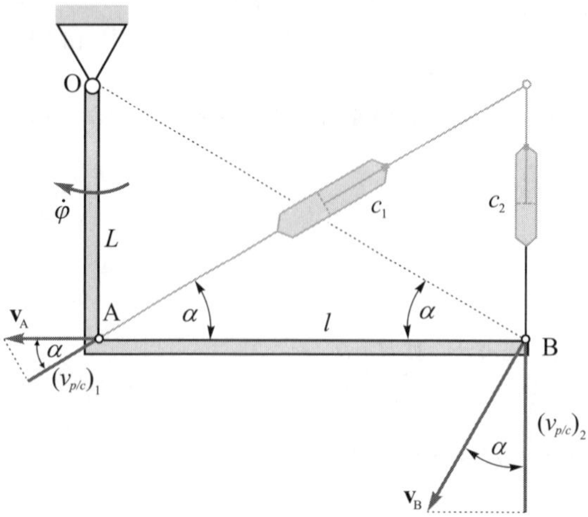

Figure 4.4.2

The overall dissipation function stemming from the two dampers is

$$D = \frac{1}{2}c_1\left(v_{p/c}\right)_1^2 + \frac{1}{2}c_2\left(v_{p/c}\right)_2^2. \tag{4.4.2}$$

To determine the relative velocities of the pistons with respect to the cylinders in the dampers, one needs to consider the case when the system passes through the equilibrium position (Figure 4.4.2), and relate these velocities to those corresponding to the points A and B:

$$\left(v_{p/c}\right)_1 = v_A \cos\alpha = L\dot{\varphi}\cos\alpha, \quad \left(v_{p/c}\right)_2 = v_B \cos\alpha = \overline{OB}\,\dot{\varphi}\cos\alpha = \sqrt{L^2 + l^2}\dot{\varphi}\cos\alpha. \tag{4.4.3}$$

Considering the triangle OAB, one can write:

$$\cos\alpha = \frac{\overline{AB}}{\overline{OB}} = \frac{l}{\sqrt{L^2 + l^2}}, \tag{4.4.4}$$

which defines $\cos\alpha$ in Eqs. (4.4.3).

The dissipation function (4.4.2) now takes the form

$$D \approx \frac{l^2}{2}\left(c_1\frac{L^2}{L^2 + l^2} + c_2\right)\dot{\varphi}^2. \tag{4.4.5}$$

Lagrange's equation (4.2) gives

$$\ddot{\varphi} + 2\delta\dot{\varphi} + \omega_n{}^2\varphi = 0, \; \delta = \frac{3l^2\left[c_1L^2 + c_2(L^2 + l^2)\right]}{2(L^2 + l^2)\left[(M + 3m)L^2 + ml^2\right]}. \tag{4.4.6}$$

Based on Equation (4.10) and (4.4.6), one can obtain the damped natural frequency:

$$\widetilde{\omega} = \sqrt{\frac{3(M+2m)gL+6k}{2(M+3m)L^2+2ml^2} - \frac{9l^4\left[c_1L^2+c_2(L^2+l^2)\right]^2}{4(L^2+l^2)^2\left[(M+3m)L^2+ml^2\right]^2}}, \tag{4.4.7}$$

and determine the quasiperiod from Equation (4.8):

$$\mathrm{T}^* = \frac{2\pi}{\sqrt{\frac{3(M+2m)gL+6k}{2(M+3m)L^2+2ml^2} - \frac{9l^4\left[c_1L^2+c_2(L^2+l^2)\right]^2}{4(L^2+l^2)^2\left[(M+3m)L^2+ml^2\right]^2}}}. \tag{4.4.8}$$

Example 4.5 A disc of mass m and radius r rolls without slipping inside a fixed cylinder of radius R. There is a spring of stiffness k attached to the disc at the point A and a linear viscous damper of damping coefficient c attached at the point B. This system performs small oscillations around the equilibrium position shown in Figure 4.5.1. Determine the differential equation of motion for small oscillations and the quasiperiod of these oscillations.

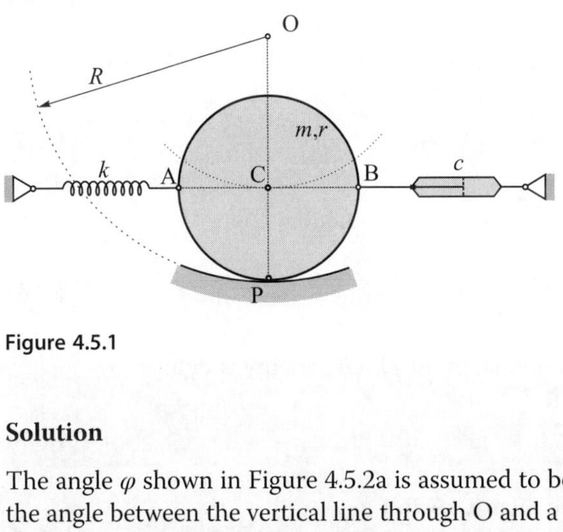

Figure 4.5.1

Solution

The angle φ shown in Figure 4.5.2a is assumed to be the generalised coordinate. This is the angle between the vertical line through O and a 'virtual' rod OC. In accordance with the static equilibrium equation, the spring is not predeformed at the equilibrium and, therefore, its potential energy depends only on the additional deflection λ. The overall potential energy takes the form

$$V \approx mgy_{\mathrm{C}} + \frac{k}{2}\lambda^2, \tag{4.5.1}$$

where

$$y_{\mathrm{C}} = -(R-r)\cos\varphi, \tag{4.5.2}$$

$$\lambda \approx u. \tag{4.5.3}$$

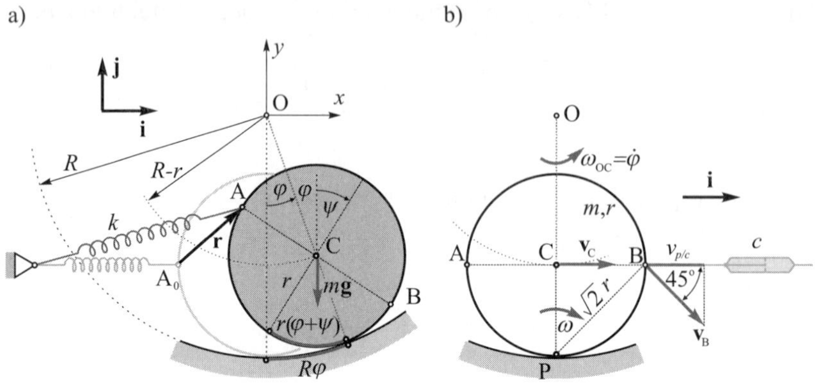

Figure 4.5.2

To determine the deflection u, the angle ψ is introduced (Figure 4.5.2a). The equality of the arcs depicted in this figure leads to

$$R\varphi = r(\varphi + \psi) \Rightarrow \psi = \frac{R-r}{r}\varphi, \tag{4.5.4}$$

where the deflection u can now be obtained as follows:

$$u = \mathbf{r}\cdot\mathbf{i} = r + (R-r)\sin\varphi - r\cos\left(\frac{R-r}{r}\varphi\right) \Rightarrow u \approx (R-r)\varphi. \tag{4.5.5}$$

This gives the following expression for the overall potential energy

$$V \approx \frac{R-r}{2}[mg + k(R-r)]\varphi^2. \tag{4.5.6}$$

The disc executes general plane motion, and its kinetic energy is defined by

$$T = \frac{m}{2}v_C{}^2 + \frac{1}{2}J_C\omega^2, \; J_C = \frac{1}{2}mr^2. \tag{4.5.7}$$

The velocity of the centre C is (see Figure 4.5.2b):

$$\mathbf{v}_C = \overline{OC}\,\omega_{OC}\mathbf{i} = (R-r)\dot{\varphi}\mathbf{i}, \; v_C = (R-r)\dot{\varphi}. \tag{4.5.8}$$

The angular velocity is given by

$$\omega = \frac{v_C}{\overline{CP}} = \frac{R-r}{r}\dot{\varphi}. \tag{4.5.9}$$

Thus, the kinetic energy (4.5.7) becomes

$$T = \frac{3}{4}m(R-r)^2\dot{\varphi}^2. \tag{4.5.10}$$

The dissipation function is defined by

$$D=\frac{1}{2}cv_{p/c}{}^{2}, \tag{4.5.11}$$

where the velocity $v_{p/c}$ while the system passes through the equilibrium can be obtained by relating it to the velocity of the point B, as shown in Figure 4.5.2b:

$$v_{p/c}=v_{\mathrm{B}}\cos 45^{\circ}=\overline{\mathrm{PB}}\,\omega\cos 45^{\circ}=(R-r)\dot{\varphi}. \tag{4.5.12}$$

The dissipation function (4.5.11) is now

$$D=\frac{1}{2}c(R-r)^{2}\dot{\varphi}^{2}. \tag{4.5.13}$$

Lagrange's equation (4.2) gives

$$\ddot{\varphi}+2\delta\dot{\varphi}+\omega_{n}{}^{2}\varphi=0,\quad \delta=\frac{c}{3m},\quad \omega_{n}=\sqrt{\frac{2[mg+k(R-r)]}{3m(R-r)}}. \tag{4.5.14}$$

The quasiperiod is easily recognised as:

$$\mathrm{T}^{*}=\frac{2\pi}{\sqrt{\frac{2[mg+k(R-r)]}{3m(R-r)}-\frac{c^{2}}{9m^{2}}}}. \tag{4.5.15}$$

Example 4.6 A linkage consists of two slender bars: OA, of mass m and length l, and AB, of mass M and length l. The sliding block pivoted at B has negligible mass. There is a linear viscous damper attached to the slider and the mid-point of the bar OA. There is a spring of stiffness k that connects the slider and the point O. The system performs small

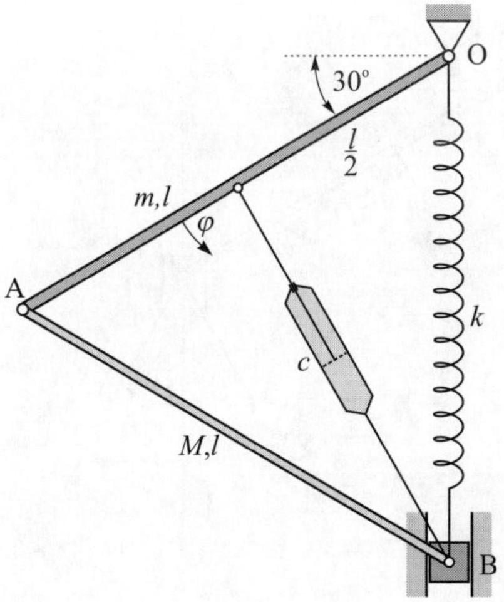

Figure 4.6.1

oscillations around the equilibrium position shown in Figure 4.6.1. Taking the angle φ as the generalised coordinate, determine the differential equation for small oscillations.

Solution

In order to assure the equilibrium position, the spring must be predeformed. Thus, its static deflection f_{st} needs to be obtained first. To determine it, the principle of virtual work will be used. In accordance with Figure 4.6.2a and using $F_{sp} = kf_{st}$ for the spring force, one has

$$\delta A = \delta A^{mg} + \delta A^{Mg} + \delta A^{\mathbf{F}_{sp}} = 0 \;\Rightarrow\; mg\delta y_{\mathrm{S}} + Mg\delta y_{\mathrm{C}} - kf_{st}\delta y_{\mathrm{B}} = 0. \tag{4.6.1}$$

To derive the virtual displacement along the axis, one can write down

$$y_{\mathrm{S}} = \frac{l}{2}\sin(30^\circ + \varphi) \;\Rightarrow\; \delta y_{\mathrm{S}} = \frac{l}{2}\cos(30^\circ + \varphi)\delta\varphi, \tag{4.6.2}$$

$$y_{\mathrm{C}} = \frac{3}{2}l\sin(30^\circ + \varphi) \;\Rightarrow\; \delta y_{\mathrm{C}} = \frac{3}{2}l\cos(30^\circ + \varphi)\delta\varphi, \tag{4.6.3}$$

$$y_{\mathrm{B}} = 2l\sin(30^\circ + \varphi) \;\Rightarrow\; \delta y_{\mathrm{B}} = 2l\cos(30^\circ + \varphi)\delta\varphi. \tag{4.6.4}$$

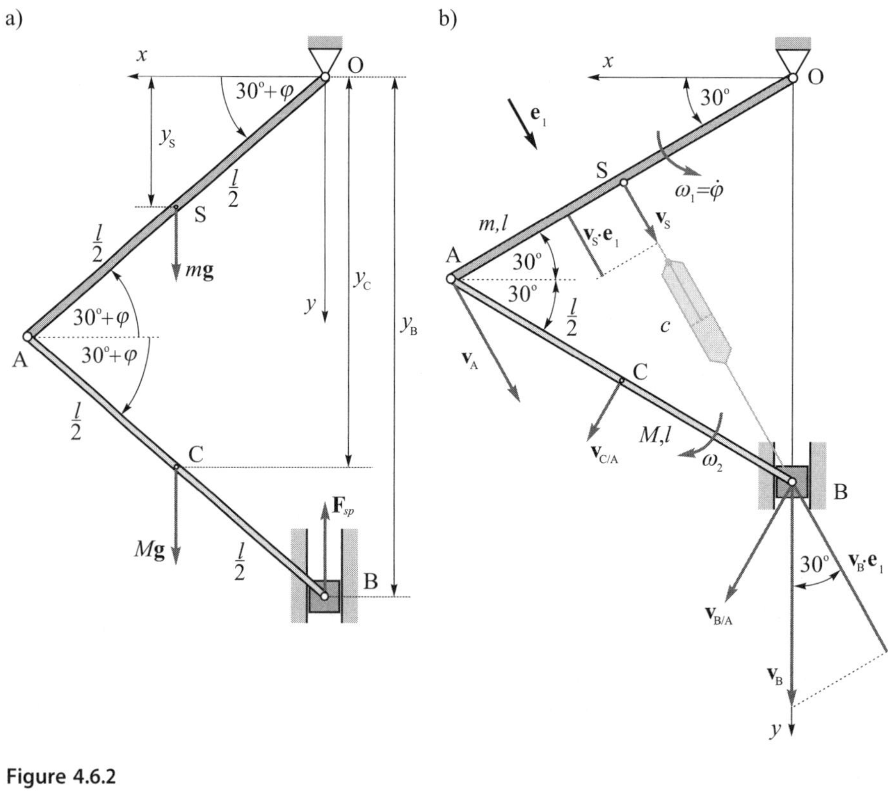

Figure 4.6.2

Introducing Equations (4.6.2)–(4.6.4) into Equation (4.6.1) leads to

$$\left(\frac{1}{2}mg + \frac{3}{2}Mg - 2kf_{st}\right)l\cos(30^\circ + \varphi)\delta\varphi = 0. \tag{4.6.5}$$

Given the fact that $\delta\varphi \neq 0$, Equation (4.6.5) is satisfied for

$$f_{st} = \frac{m + 3M}{4k}g. \tag{4.6.6}$$

Alternatively, one can obtain this static deflection while considering the zero sum of the moments about the point O for the whole system and the zero sum of the moments about the point A for the subsystem formed by the bar AB and the sliding block.

The overall potential energy is now

$$V = -mgy_S - Mgy_C + \frac{k}{2}(f_{st} + \lambda)^2. \tag{4.6.7}$$

The additional deflection of the spring is obtained by defining its exact value and then developing it into series as follows:

$$\lambda = \lambda(\varphi) = y_B(\varphi) - y_B(0) = 2l\sin(30^\circ + \varphi) - 2l\sin 30^\circ \quad \Rightarrow \quad \lambda = \sqrt{3}l\varphi - \frac{l\varphi^2}{2} + \dots \tag{4.6.8}$$

The expressions for the coordinates y_S and y_C (4.6.2) and (4.6.3) can also be developed into series:

$$y_S = const. + \frac{\sqrt{3}}{4}l\varphi - \frac{1}{8}l\varphi^2 + \dots , \tag{4.6.9}$$

$$y_C = const. + \frac{3\sqrt{3}}{4}l\varphi - \frac{3}{8}l\varphi^2 + \dots . \tag{4.6.10}$$

Using Equations (4.6.6)–(4.6.10), the following expression for the potential energy is obtained:

$$V \approx \frac{3}{2}kl^2\varphi^2. \tag{4.6.11}$$

To determine the kinetic energy, one needs to note that the bar OA rotates around the axis through O, while the bar AB performs general plane motion, so that the expression for the overall kinetic energy takes the form

$$T = \frac{1}{2}J_O\dot{\varphi}^2 + \frac{M}{2}v_C{}^2 + \frac{1}{2}J_C\omega_2{}^2, \tag{4.6.12}$$

where the mass moments of inertia are given by

$$J_O = J_S + m\overline{OS}^2 = \frac{1}{12}ml^2 + m\left(\frac{l}{2}\right)^2 = \frac{1}{3}ml^2, \ J_C = \frac{1}{12}Ml^2. \tag{4.6.13}$$

To determine the velocities existing in Equation (4.6.12), one can analyse the system passing through the equilibrium position shown in Figure 4.6.2b. As the bar OA rotates around the axis through O, its points S and A have the following velocities:

$$\mathbf{v}_\mathrm{S} = \frac{l}{2}\dot{\varphi}\mathbf{e}_1, \quad \mathbf{v}_\mathrm{A} = l\dot{\varphi}\mathbf{e}_1. \tag{4.6.14}$$

As the bar AB is in general plane motion, the velocities of their points B and A can be related to each other as follows

$$\left.\begin{aligned} &\mathbf{v}_\mathrm{B} = \mathbf{v}_\mathrm{A} + \mathbf{v}_{\mathrm{B/A}}, \\ x:\ &0 = -l\dot{\varphi}\sin 30^\circ + l\omega_2 \sin 30^\circ \ \Rightarrow\ \omega_2 = \dot{\varphi}, \\ y:\ &v_\mathrm{B} = l\dot{\varphi}\cos 30^\circ + l\dot{\varphi}\cos 30^\circ \ \Rightarrow\ v_\mathrm{B} = \sqrt{3}l\dot{\varphi}. \end{aligned}\right\} \tag{4.6.15}$$

In addition, the velocity of the points C can be obtained by relating it to the velocity of the point A:

$$\begin{aligned} &\mathbf{v}_\mathrm{C} = \mathbf{v}_\mathrm{A} + \mathbf{v}_{\mathrm{C/A}}, \\ x:\ &v_{\mathrm{C}x} = -l\dot{\varphi}\sin 30^\circ + \frac{l}{2}\omega_2 \sin 30^\circ \ \Rightarrow\ v_{\mathrm{C}x} = -\frac{1}{4}l\dot{\varphi}, \\ y:\ &v_{\mathrm{C}y} = l\dot{\varphi}\cos 30^\circ + \frac{l}{2}\omega_2 \cos 30^\circ \ \Rightarrow\ v_{\mathrm{C}y} = \frac{3\sqrt{3}}{4}l\dot{\varphi}, \end{aligned}$$

$${v_\mathrm{C}}^2 = {v_{\mathrm{C}x}}^2 + {v_{\mathrm{C}y}}^2 \ \Rightarrow\ {v_\mathrm{C}}^2 = \frac{7}{4}l^2\dot{\varphi}^2. \tag{4.6.16}$$

Finally, the kinetic energy (4.6.12) takes the form

$$T \approx \frac{2m + 11M}{12}l^2\dot{\varphi}^2. \tag{4.6.17}$$

The dissipation function is defined by

$$D = \frac{1}{2}c{v_{p/c}}^2, \tag{4.6.18}$$

and the velocity $v_{p/c}$ as the system passes through the equilibrium can be calculated as follows (see Figure 4.6.2b):

$$v_{p/c} = \mathbf{v}_\mathrm{B}\cdot\mathbf{e}_1 - \mathbf{v}_\mathrm{S}\cdot\mathbf{e}_1 = v_\mathrm{B}\cos 30^\circ - v_\mathrm{S} \ \Rightarrow\ v_{p/c} = l\dot{\varphi}. \tag{4.6.19}$$

This transforms Equation (4.6.18) into

$$D \approx \frac{1}{2}cl^2\dot{\varphi}^2. \tag{4.6.20}$$

Lagrange's equation (4.2) gives

$$\ddot{\varphi} + 2\delta\dot{\varphi} + {\omega_n}^2\varphi = 0, \quad \delta = \frac{3c}{2m + 11M}, \quad \omega_n = \sqrt{\frac{18k}{2m + 11M}}. \tag{4.6.21}$$

Example 4.7 Two identical slender bars AE and AB of mass m and length $2a$ are freely hinged at A. The bar AE moves inside a frictionless vertical guide and the bar AB slides along a cylinder of radius r (Figure 4.7.1). There are two linear viscous dampers: one has damping coefficient c_1 and links the points E and B; the other is vertical in the equilibrium position shown in Figure 4.7.1 and is attached at B. Based on the static equilibrium equation, determine the radius r as a function of the length a. Then, taking the angle φ as the generalised coordinate, derive the differential equation of motion for small oscillations.

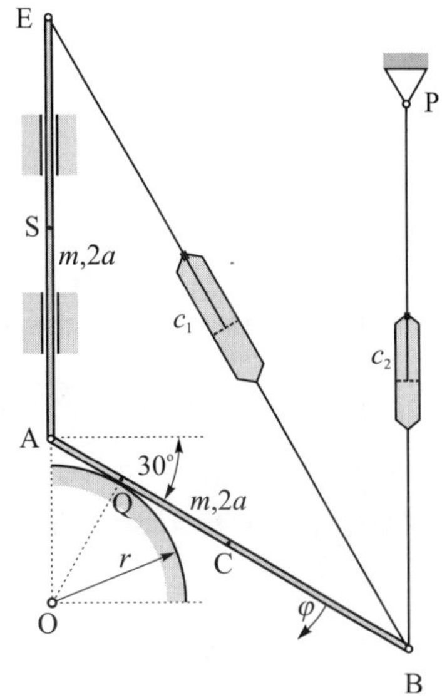

Figure 4.7.1

Solution

To determine the radius r, the principle of virtual work will be used for the system of forces shown in Figure 4.7.2a. Based on this figure, one can write down:

$$\delta A = \delta A^{mg_{AE}} + \delta A^{mg_{AB}} = 0 \quad \Rightarrow \quad -mg\delta y_S - mg\delta y_C = 0. \tag{4.7.1}$$

The coordinates of the points A and C can be used to derive the corresponding virtual displacements:

$$y_S = \frac{r}{\cos(30^\circ + \varphi)} + a \quad \Rightarrow \quad \delta y_S = \frac{r\sin(30^\circ + \varphi)}{\cos^2(30^\circ + \varphi)}\delta\varphi, \tag{4.7.2}$$

$$y_C = \frac{r}{\cos(30^\circ + \varphi)} - a\sin(30^\circ + \varphi) \quad \Rightarrow \quad \delta y_C = \left(\frac{r\sin(30^\circ + \varphi)}{\cos^2(30^\circ + \varphi)} - a\cos(30^\circ + \varphi)\right)\delta\varphi, \tag{4.7.3}$$

so that Equation (4.7.1) becomes

$$-mg\left(2\frac{r\sin(30^\circ + \varphi)}{\cos^2(30^\circ + \varphi)} - a\cos(30^\circ + \varphi)\right)\delta\varphi = 0. \tag{4.7.4}$$

For this expression to be satisfied at the equilibrium position φ=0, the following should be fulfilled

$$2\frac{r\frac{1}{2}}{\left(\frac{\sqrt{3}}{2}\right)^2} - a\frac{\sqrt{3}}{2} = 0 \quad \Rightarrow \quad r = \frac{3\sqrt{3}}{8}a, \tag{4.7.5}$$

which defines the relationship between r and a. Alternatively, the value of the radius can be obtained while considering the equilibrium of forces in the vertical direction for the system as a whole and the zero sum of the moments about the point A for the bar AB.

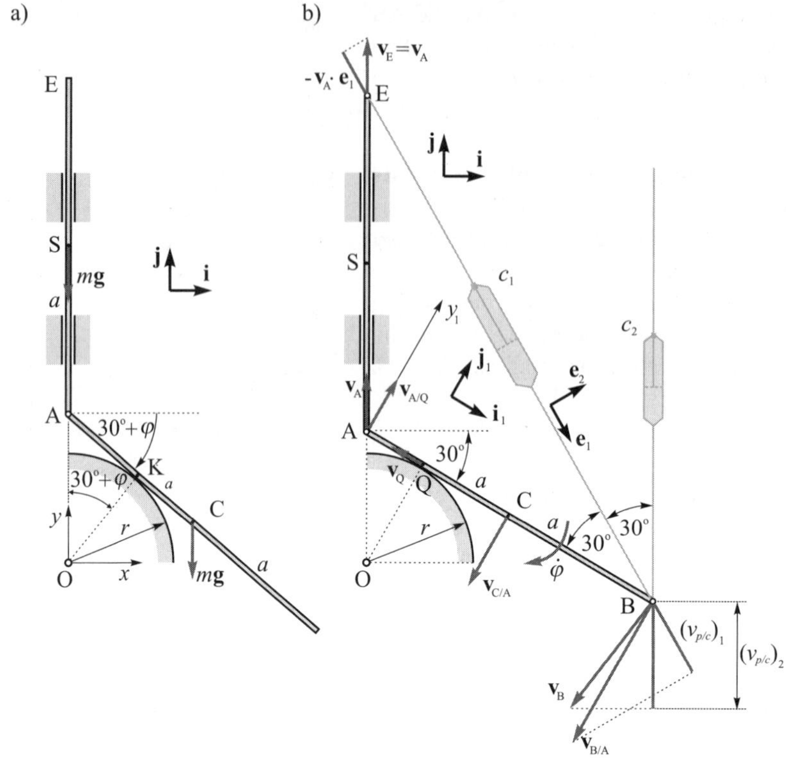

Figure 4.7.2

The gravitational potential energy for the two bars is

$$V = mgy_S + mgy_C, \tag{4.7.6}$$

which, due to Equations (4.7.2), (4.7.3) and (4.7.5), becomes

$$V = const. + mga\left(\frac{3\sqrt{3}}{4\cos(30^\circ + \varphi)} - \sin(30^\circ + \varphi)\right). \tag{4.7.7}$$

Developing this into series, one finds

$$V \approx \frac{3}{2}mga\varphi^2. \tag{4.7.8}$$

To determine the kinetic energy, one should note that the bar AE moves in translation while the bar AB executes general plane motion, so that the overall kinetic energy takes the form

$$T = \frac{m}{2}v_S{}^2 + \frac{m}{2}v_C{}^2 + \frac{1}{2}J_C\dot{\varphi}^2, \tag{4.7.9}$$

where

$$J_C = \frac{1}{12}m(2a)^2 = \frac{1}{3}ma^2. \tag{4.7.10}$$

To determine the velocities of the points S and C, one should consider the case when the system passes through the equilibrium position (Figure 4.7.2b). Thus, by relating the velocities of the points A and B, one has

$$\mathbf{v}_{\mathrm{A}} = \mathbf{v}_{\mathrm{Q}} + \mathbf{v}_{\mathrm{A/Q}}, \quad \mathbf{v}_{\mathrm{Q}} = -v_{\mathrm{Q}}\mathbf{i}_1, \quad \mathbf{v}_{\mathrm{A/Q}} = \overline{\mathrm{AQ}}\dot{\varphi}\mathbf{j}_1, \tag{4.7.11}$$

$$v_{\mathrm{A}}\cos 30^\circ = \overline{\mathrm{AQ}}\,\dot{\varphi}, \quad \overline{\mathrm{AQ}} = r\tan 30^\circ = \frac{3}{8}a \;\Rightarrow\; \mathbf{v}_{\mathrm{A}} = \frac{\sqrt{3}}{4}a\dot{\varphi}\mathbf{j}. \tag{4.7.12}$$

Translatory motion of the bar AE imposes

$$\mathbf{v}_{\mathrm{S}} = \mathbf{v}_{\mathrm{E}} = \frac{\sqrt{3}}{4}a\dot{\varphi}\mathbf{j}. \tag{4.7.13}$$

The velocity of the point C can be obtained from the following scalar product:

$$\begin{aligned} &v_{\mathrm{C}}^2 = \mathbf{v}_{\mathrm{C}}\cdot\mathbf{v}_{\mathrm{C}} = \left(\mathbf{v}_{\mathrm{A}} + \mathbf{v}_{\mathrm{C/A}}\right)\cdot\left(\mathbf{v}_{\mathrm{A}} + \mathbf{v}_{\mathrm{C/A}}\right) = v_{\mathrm{A}}^2 + v_{\mathrm{C/A}}^2 + 2\mathbf{v}_{\mathrm{A}}\cdot\mathbf{v}_{\mathrm{C/A}}, \\ &\mathbf{v}_{\mathrm{C/A}} = a\dot{\varphi}\mathbf{j}_1 \;\Rightarrow v_{\mathrm{C}}^2 = \frac{7}{16}a^2\dot{\varphi}^2. \end{aligned} \tag{4.7.14}$$

The velocity of the point B can be obtained by relating it to the velocity of the point A:

$$\mathbf{v}_{\mathrm{B}} = \mathbf{v}_{\mathrm{A}} + \mathbf{v}_{\mathrm{B/A}}, \quad \mathbf{v}_{\mathrm{B/A}} = -2a\dot{\varphi}\mathbf{j}_1 \;\Rightarrow\; \mathbf{v}_{\mathrm{B}} = \frac{\sqrt{3}}{4}a\dot{\varphi}\mathbf{j} - 2a\dot{\varphi}\mathbf{j}_1. \tag{4.7.15}$$

Besides this, the velocity of the piston with respect to the cylinder in the first damper stems from the projections of the velocities of the points E and B on the direction of the damper in the equilibrium position:

$$\left(v_{p/c}\right)_1 = \mathbf{v}_{\mathrm{B}}\cdot\mathbf{e}_1 - \mathbf{v}_{\mathrm{E}}\cdot\mathbf{e}_1 = \left(\mathbf{v}_{\mathrm{A}} + \mathbf{v}_{\mathrm{B/A}}\right)\cdot\mathbf{e}_1 - \mathbf{v}_{\mathrm{A}}\cdot\mathbf{e}_1 = \mathbf{v}_{\mathrm{B/A}}\cdot\mathbf{e}_1 \;\Rightarrow\; \left(v_{p/c}\right)_1 = a\dot{\varphi}. \tag{4.7.16}$$

The velocity of the piston with respect to the cylinder in the second damper can be obtained as the projection of the velocity of the point B on the direction of the damper in the equilibrium position:

$$\begin{aligned} &\left(v_{p/c}\right)_2 = -\mathbf{v}_B\cdot\mathbf{j} = -\left(\frac{\sqrt{3}}{4}a\dot{\varphi}\mathbf{j} - 2a\dot{\varphi}\mathbf{j}_1\right)\cdot\mathbf{j} = -\frac{\sqrt{3}}{4}a\dot{\varphi} + 2a\dot{\varphi}\cos 30^\circ \\ &\Rightarrow\; \left(v_{p/c}\right)_2 = \frac{3\sqrt{3}}{4}a\dot{\varphi}. \end{aligned} \tag{4.7.17}$$

By using these expressions for the velocities, the kinetic energy is obtained in the form

$$T \approx \frac{23}{48}ma^2\dot{\varphi}^2. \tag{4.7.18}$$

The dissipative function

$$D = \frac{c_1}{2}\left(v_{p/c}\right)_1^2 + \frac{c_2}{2}\left(v_{p/c}\right)_2^2, \tag{4.7.19}$$

is

$$D \approx \frac{16c_1 + 27c_2}{32} a^2 \dot{\varphi}^2. \qquad (4.7.20)$$

Lagrange's equation (4.2) leads to

$$\ddot{\varphi} + 2\delta\dot{\varphi} + \omega_n{}^2 \varphi = 0, \quad \delta = \frac{48c_1 + 81c_2}{92m}, \quad \omega_n = \sqrt{\frac{72g}{23a}}. \qquad (4.7.21)$$

Example 4.8 The identical sliders A and B of mass m can move without friction along the guides OE and OQ. They are linked with a weightless rod. The system can oscillate around the equilibrium position shown in Figure 4.8.1, and contains also a spring of stiffness k attached to the points B and C, as well as a linear viscous damper of damping coefficient c, which is horizontal in this position. Assuming that the angle φ is the generalised coordinate, determine the differential equation of motion for small oscillations.

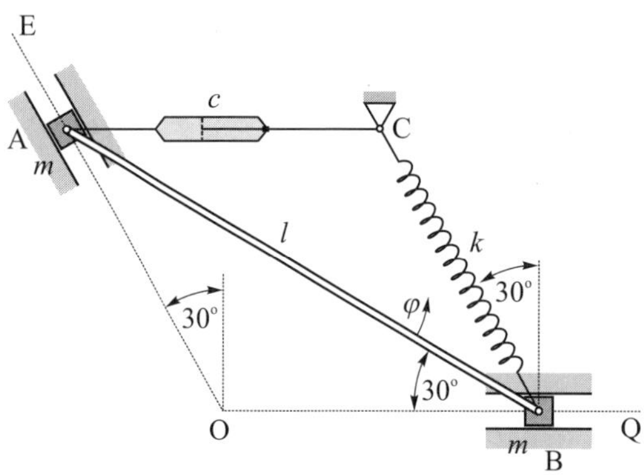

Figure 4.8.1

Solution

In order to assure the equilibrium position shown, the spring needs to be predeformed. The first task is to obtain its static deflection f_{st}. To that end, the free-body diagram is formed for the system as a whole (Figure 4.8.2). The spring force $F_{sp} = kf_{st}$ can be obtained from the following equilibrium equation:

$$\sum M_P = 0 \;\Rightarrow\; mgl\cos 30^\circ - kf_{st}\overline{PB}\sin 30^\circ = 0. \qquad (4.8.1)$$

Recognising that $\overline{PB} = l$, one has

$$f_{st} = \frac{\sqrt{3}mg}{k}. \qquad (4.8.2)$$

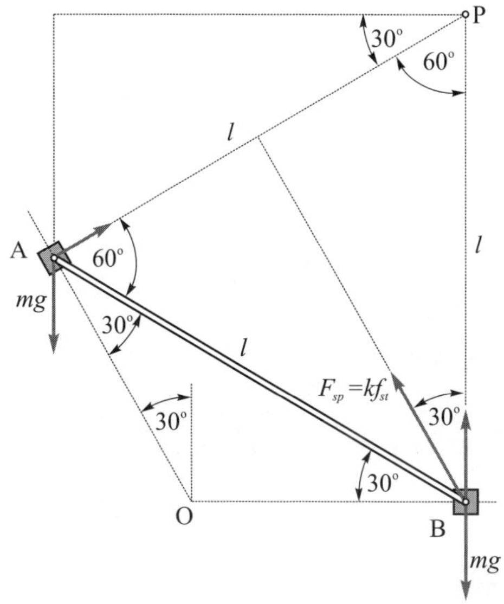

Figure 4.8.2

The expression for the overall potential energy (Figure 4.8.3) reads as

$$V = -mgw\cos 30^\circ + \frac{k}{2}(f_{st} + \lambda)^2. \tag{4.8.3}$$

Given the fact that the additional deflection can be defined in terms of the deflections in two orthogonal directions, one can write

$$\lambda \approx u + \frac{1}{2l_{st}}v^2, \quad l_{st} = \frac{l}{\sqrt{3}}. \tag{4.8.4}$$

To determine these deflections, one can first consider the displacements of the points A and B, which are depicted by the coordinates w and z in Figure 4.8.3a. Thus, the Sine Law for the triangle OAB gives:

$$\frac{\frac{l}{\sqrt{3}} + z}{\sin(30^\circ + \varphi)} = \frac{l}{\sin 120^\circ} \quad \Rightarrow \quad z = \frac{2l}{\sqrt{3}}\sin(30^\circ + \varphi) - \frac{l}{\sqrt{3}} \quad \Rightarrow \quad z \approx l\varphi - \frac{\sqrt{3}}{6}l\varphi^2, \tag{4.8.5}$$

$$\frac{\frac{l}{\sqrt{3}} - w}{\sin(30^\circ - \varphi)} = \frac{l}{\sin 120^\circ} \quad \Rightarrow \quad w = \frac{l}{\sqrt{3}} - \frac{2l}{\sqrt{3}}\sin(30^\circ - \varphi) \quad \Rightarrow \quad w \approx l\varphi + \frac{\sqrt{3}}{6}l\varphi^2. \tag{4.8.6}$$

The deflection u is now given by

$$u = z\sin 30^\circ \quad \Rightarrow \quad u \approx \frac{l}{2}\varphi - \frac{\sqrt{3}}{12}l\varphi^2, \tag{4.8.7}$$

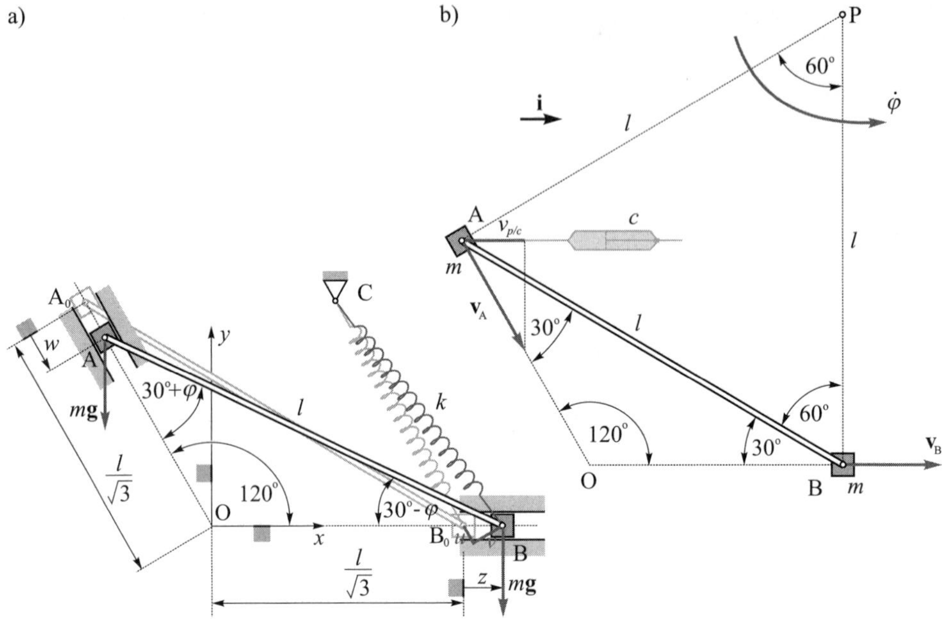

Figure 4.8.3

while the deflection v is

$$v = z\cos 30° \Rightarrow \quad v \approx \frac{\sqrt{3}}{2} l\varphi. \tag{4.8.8}$$

The additional deflection (4.8.4) takes the form

$$\lambda \approx \frac{l}{2}\varphi + \frac{7\sqrt{3}}{24} l\varphi^2. \tag{4.8.9}$$

The overall potential energy (4.8.3) now reads as:

$$V \approx \frac{l}{8}(5mg + kl)\varphi^2. \tag{4.8.10}$$

The kinetic energy consists of two terms,

$$T = \frac{1}{2} m v_A{}^2 + \frac{1}{2} m v_B{}^2. \tag{4.8.11}$$

Given the equal distance of the points A and B from the instantaneous centre of zero velocity P (Figure 3.8.3b), their velocities are equal

$$v_B = v_A = l\dot{\varphi}, \tag{4.8.12}$$

and the kinetic energy becomes

$$T \approx ml^2\dot{\varphi}^2. \tag{4.8.13}$$

The dissipation function

$$D = \frac{1}{2}cv_{p/c}{}^2, \tag{4.8.14}$$

can be obtained knowing that the velocity $v_{p/c}$ (Figure 4.8.3b) corresponds to the projection of the velocity of the point A on the horizontal direction

$$v_{p/c} = \mathbf{v}_A \cdot \mathbf{i} = l\dot{\varphi}\cos 60^\circ \;\; \Rightarrow \;\; v_{p/c} = l\dot{\varphi}/2. \tag{4.8.15}$$

The dissipative function is, thus, given by

$$D \approx \frac{c}{8}l^2\dot{\varphi}^2. \tag{4.8.16}$$

Finally, Lagrange's equation (4.2) yields

$$\ddot{\varphi} + 2\delta\dot{\varphi} + \omega_n{}^2\varphi = 0, \;\; \delta = \frac{c}{16m}, \;\; \omega_n = \sqrt{\frac{5mg + kl}{8ml}} \tag{4.8.17}$$

Example 4.9 A triple pulley consists of three wheels which are rigidly attached to each other and have a radii R_1, R_2 and R_3, as shown in Figure 4.9.1. Its mass moment of inertia for the axis through O is J_O. There is a resistance torque M_w acting on the pulley that is proportional to its angular velocity with damping coefficient c ($M_w = c\omega$). There are two ropes wrapped around the inner wheels with two blocks of mass m at their ends. A third rope is wrapped around the outer wheel and then attached to a horizontal spring of

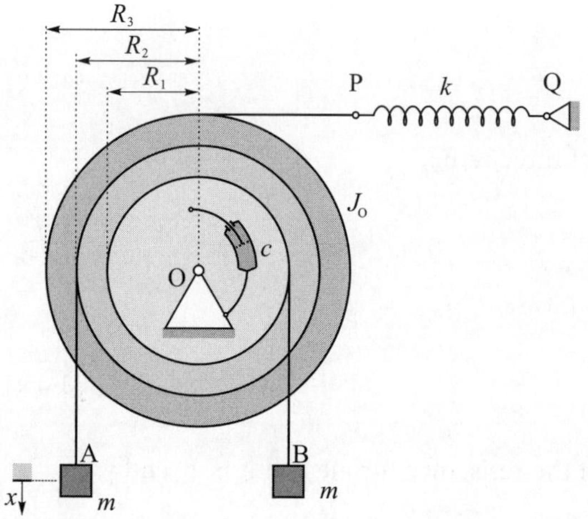

Figure 4.9.1

stiffness k. The generalised coordinate x defines the displacement of the block A. There is no slipping between the ropes and the wheels. Assuming: $R_1 = 0.06$ m, $R_2 = 0.14$ m, $R_3 = 0.20$ m, $m = 10$ kg, $J_O = 0.05$ kgm^2, $g = 9.81$ m/s^2, $c = 0.5$ Nms and $k = 1000$ N/m, determine the motion $x(t)$ for the initial conditions $x(0) = a = 0.03$ m and $\dot{x}(0) = 0$, and show it graphically. In addition, plot how the forces in these three ropes change with time.

Solution

It is important to note first that the spring is predeformed in the static equilibrium position shown in Figure 4.9.1. Its static deflection can be obtained from the free-body diagram in Figure 4.9.2a and the zero sum of the moments about the point O:

$$\sum M_O = 0 \;\Rightarrow\; mgR_2 - mgR_1 - kf_{st}R_3 = 0 \;\Rightarrow\; f_{st} = \frac{R_2 - R_1}{R_3}\frac{mg}{k}. \tag{4.9.1}$$

The overall potential energy (Figure 4.9.2b) can be represented as

$$V = -mgx + mgz + \frac{k}{2}(f_{st} + y)^2, \tag{4.9.2}$$

where z and y define the position of the block with respect to their equilibrium positions. To determine them in terms of the generalised coordinate, one can first note that:

$$\omega = \dot{\varphi}, \quad v_A = \dot{x}, \quad v_B = \dot{z}, \quad v_P = \dot{y}, \tag{4.9.3}$$

while the kinematic relationships (Figure 4.9.2b) give:

$$\omega = \frac{v_A}{R_2} \;\Rightarrow\; \dot{\varphi} = \frac{\dot{x}}{R_2}, \quad v_B = R_1\omega \;\Rightarrow\; \dot{z} = \frac{R_1}{R_2}\dot{x}, \quad v_P = R_3\omega \;\Rightarrow\; \dot{y} = \frac{R_3}{R_2}\dot{x}. \tag{4.9.4}$$

Integrating these and using the fact that their values are zero when $x = 0$, one has

$$\varphi = \frac{x}{R_2}, \quad z = \frac{R_1}{R_2}x, \quad y = \frac{R_3}{R_2}x. \tag{4.9.5a–c}$$

The potential energy (4.9.2) becomes

$$V \approx \frac{1}{2}k\frac{R_3{}^2}{R_2{}^2}x^2. \tag{4.9.6}$$

The overall kinetic energy consists of three terms

$$T = \frac{1}{2}mv_A{}^2 + \frac{1}{2}mv_B{}^2 + \frac{1}{2}J_O\omega^2, \tag{4.9.7}$$

and given (4.9.3) and (4.9.4), takes the form

$$T = \frac{1}{2}\left(m + m\frac{R_1{}^2}{R_2{}^2} + \frac{J_O}{R_2{}^2}\right)\dot{x}^2. \tag{4.9.8}$$

The dissipation function stems from the resistance torque and it is given by

$$D = \frac{1}{2}c\omega^2, \tag{4.9.9}$$

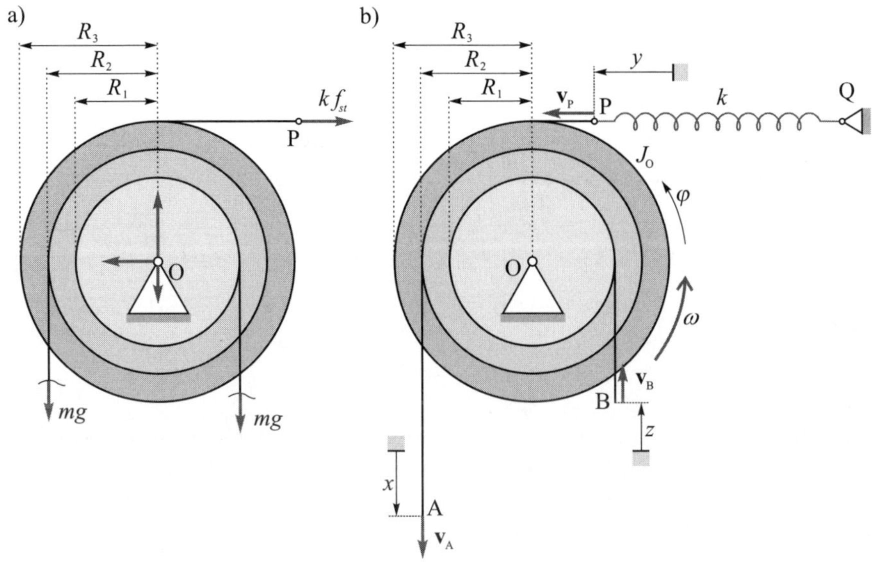

Figure 4.9.2

which, due to (4.9.3) and (4.9.4), transforms into

$$D = \frac{1}{2} c \frac{\dot{x}^2}{R_2{}^2}. \tag{4.9.10}$$

Lagrange's equation (4.2), gives

$$\ddot{x} + 2\delta\dot{x} + \omega_n{}^2 x = 0, \tag{4.9.11}$$

with

$$\delta = \frac{c}{2\left(J_O + mR_1{}^2 + mR_2{}^2\right)} = 0.88652\ \text{s}^{-1}, \tag{4.9.12}$$

$$\omega_n = R_3 \sqrt{\frac{k}{J_O + mR_1{}^2 + mR_2{}^2}} = 11.90983\ \text{s}^{-1}. \tag{4.9.13}$$

The solution of Equation (4.9.11) for the initial conditions given can be represented as

$$x(t) = ae^{-\delta t}\left(\frac{\delta}{\sqrt{\omega_n{}^2 - \delta^2}}\ \sin\sqrt{\omega_n{}^2 - \delta^2}\,t + \cos\sqrt{\omega_n{}^2 - \delta^2}\,t\right), \tag{4.9.14}$$

that is, as

$$x(t) = 0.03e^{-0.8865t}[0.0746\ \ \sin(11.8768t) + \cos(11.8768t)]. \tag{4.9.15}$$

This solution is plotted in Figure 4.9.3.

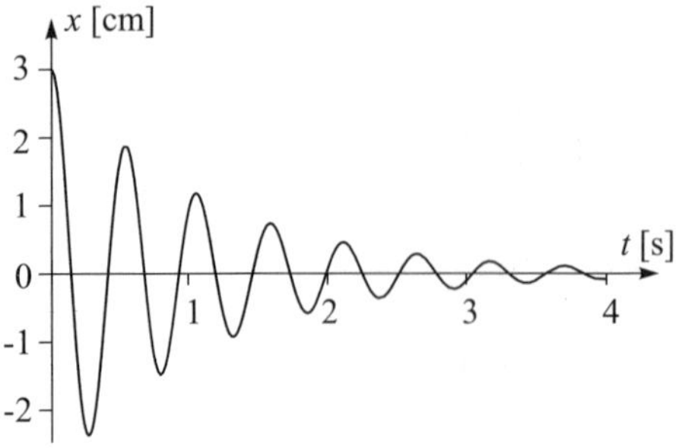

Figure 4.9.3

To determine the forces in the ropes, one cannot use Lagrange's equation (4.2), but needs to turn to the differential equations of motion, separately for the blocks and the pulley. The free-body diagrams are shown in Figure 4.9.4a–c. Newton's second law of motion for the block A has the form

$$m\ddot{x} = mg - F_1, \tag{4.9.16}$$

and, together with Equation (4.9.14) and its second derivative, gives the force

$$F_1(t) = mg - mae^{-\delta t}\left(\frac{\delta\omega_n^{\ 2}}{\sqrt{\omega_n^{\ 2}-\delta^2}}\ \sin\sqrt{\omega_n^{\ 2}-\delta^2}t - \omega_n^{\ 2}\cos\sqrt{\omega_n^{\ 2}-\delta^2}t\right). \tag{4.9.17}$$

Newton's second law of motion for the block B reads as

$$m\ddot{z} = -mg + F_2. \tag{4.9.18}$$

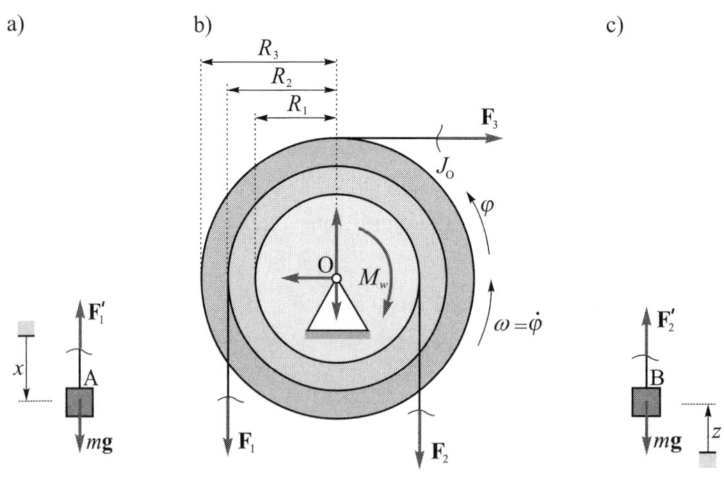

Figure 4.9.4

After using Equation (4.9.5b), differentiating it twice and substituting Equation (4.9.14) into it, one can derive

$$F_2(t) = mg + m\frac{R_1}{R_2}ae^{-\delta t}\left(\frac{\delta\omega_n{}^2}{\sqrt{\omega_n{}^2-\delta^2}}\sin\sqrt{\omega_n{}^2-\delta^2}t - \omega_n{}^2\cos\sqrt{\omega_n{}^2-\delta^2}t\right). \tag{4.9.19}$$

The force in the third rope can be obtained in two ways. In the first approach, it can be calculated easily as it balances the force in the spring:

$$F_3(t) = F_{sp}(t) = k(f_{st} + y(t)), \tag{4.9.20}$$

which gives

$$F_3(t) = \frac{R_2 - R_1}{R_3}mg + k\frac{R_3}{R_2}ae^{-\delta t}\left(\frac{\delta}{\sqrt{\omega_n{}^2-\delta^2}}\sin\sqrt{\omega_n{}^2-\delta^2}t + \cos\sqrt{\omega_n{}^2-\delta^2}t\right). \tag{4.9.21}$$

The second approach involves the differential equation of motion for the pulley

$$J_O\ddot{\varphi} = F_1(t)R_2 - F_2(t)R_1 - F_3(t)R_3 - c\dot{\varphi}(t), \tag{4.9.22}$$

which also requires the substitution of all the expressions for the forces found previously and the angular velocity and angular acceleration found from Equations (4.9.5a) and (4.9.14).

The way that the forces in the three ropes change with time is plotted in Figure 4.9.5.

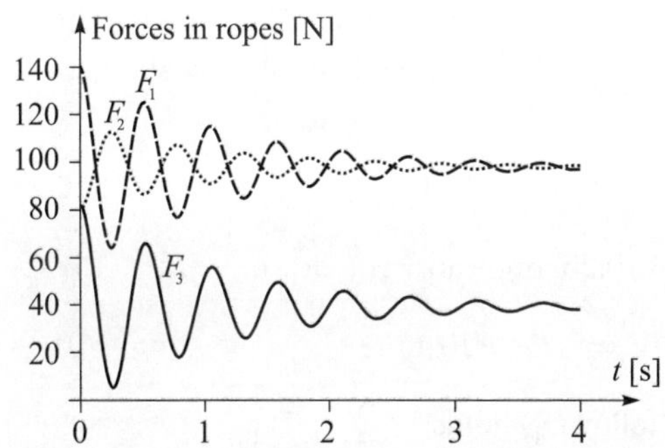

Figure 4.9.5

Example 4.10 Two identical slender bars AO_1 and BO_2 of mass m and length $2l$ are pinned at their fixed ends, where $\overline{O_1O_2} = 2l\sqrt{3}/3$. They are also linked by a slender rod of mass M and length $2L = 2l\sqrt{3}/3$. The system oscillates around the equilibrium position shown in Figure 4.10.1. The spring RC of stiffness k is extended by the force F. There is another spring EQ, which has the same stiffness. Two linear viscous dampers PR and CE have the same damping coefficient c. Taking the angle φ as the generalised coordinate, determine the differential equation of small oscillations.

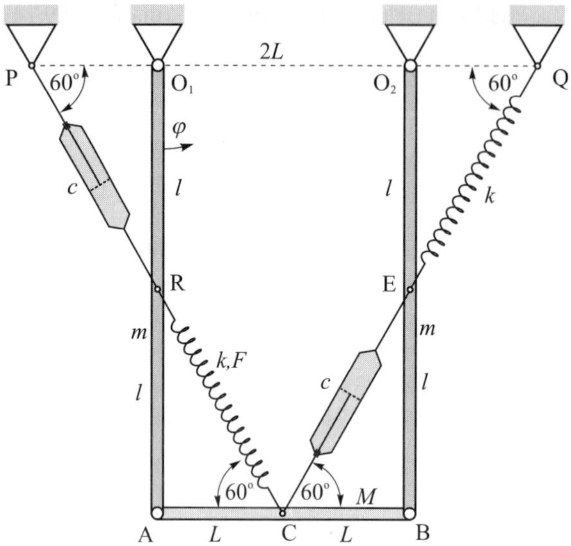

Figure 4.10.1

Solution

Let us first determine the static deflection of the spring RC:

$$f_{st1} = \frac{F}{k}. \tag{4.10.1}$$

To obtain the static deflection of the spring EQ, one needs to consider the free-body diagrams in Figure 4.10.2a-c and write down the equilibrium equations. For the system shown in Figure 4.10.2a, one has

$$\sum M_{O_1} = 0 \;\Rightarrow\; F\cos 60° l - F_{Ax} 2l = 0 \;\Rightarrow\; F_{Ax} = F/4. \tag{4.10.2}$$

While considering the system in Figure 4.10.2b, one can write down:

$$\sum F_x = 0 \;\Rightarrow\; F_{Ax} - F\cos 60° + F_{Bx} = 0 \;\Rightarrow\; F_{Bx} = F/4, \tag{4.10.3}$$

while for the system in Figure 4.10.2c, the following holds:

$$\sum M_{O_2} = 0 \;\Rightarrow\; kf_{st2}\cos 60° l - F_{Bx} 2l = 0 \;\Rightarrow\; f_{st2} = F/k. \tag{4.10.4}$$

The lengths of the springs in the equilibrium position are (Figure 4.10.1)

$$l_{st1} = l_{st2} = \frac{l}{\sin 60°} = \frac{2l\sqrt{3}}{3}. \tag{4.10.5}$$

The overall potential energy is (see Figure 4.10.3a)

$$V = -mgy_R - mgy_E - Mgy_C + \frac{k}{2}(f_{st1} + \lambda_1)^2 + \frac{k}{2}(f_{st2} + \lambda_2)^2, \tag{4.10.6}$$

where the coordinates of the points R, E and C are

$$y_{\mathrm{R}} = y_{\mathrm{E}} = l\cos\varphi, \quad y_{\mathrm{C}} = 2l\cos\varphi, \quad \cos\varphi \approx 1 - \frac{\varphi^2}{2}. \tag{4.10.7}$$

The additional deflection of the first spring can be derived as follows

$$\begin{aligned}&\lambda_1 = \overline{\mathrm{RC}}(\varphi) - \overline{\mathrm{RC}}(0), \quad \overline{\mathrm{RC}}(\varphi) = \sqrt{l^2 + L^2 - 2lL\cos(90^\circ + \varphi)} \;\Rightarrow \\ &\overline{\mathrm{RC}}(\varphi) \approx \frac{2\sqrt{3}}{3}l + \frac{l}{2}\varphi - \frac{\sqrt{3}}{16}l\varphi^2 \;\Rightarrow\; \lambda_1 \approx \frac{l}{2}\varphi - \frac{\sqrt{3}}{16}l\varphi^2.\end{aligned} \tag{4.10.8}$$

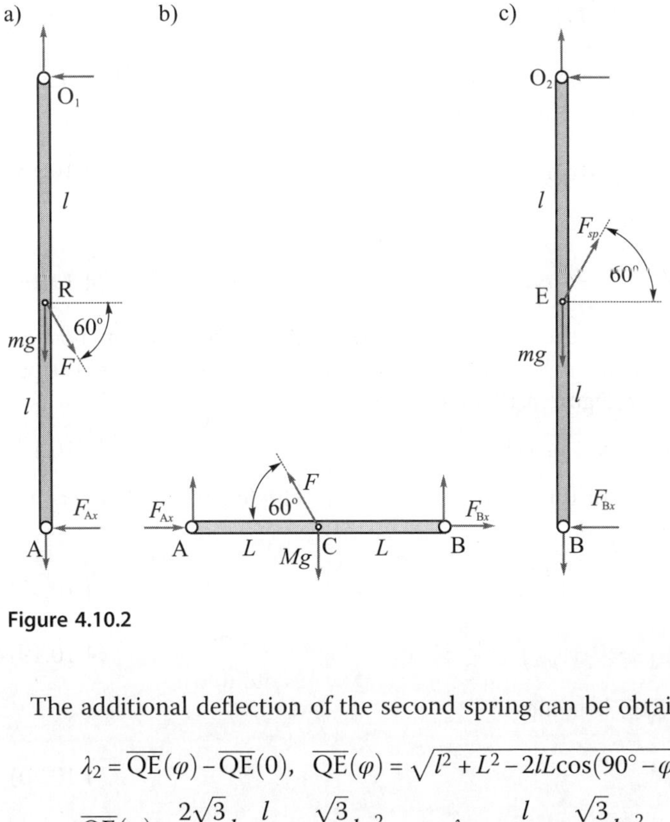

Figure 4.10.2

The additional deflection of the second spring can be obtained in the following way

$$\begin{aligned}&\lambda_2 = \overline{\mathrm{QE}}(\varphi) - \overline{\mathrm{QE}}(0), \quad \overline{\mathrm{QE}}(\varphi) = \sqrt{l^2 + L^2 - 2lL\cos(90^\circ - \varphi)} \;\Rightarrow \\ &\overline{\mathrm{QE}}(\varphi) \approx \frac{2\sqrt{3}}{3}l - \frac{l}{2}\varphi - \frac{\sqrt{3}}{16}l\varphi^2 \;\Rightarrow\; \lambda_2 \approx -\frac{l}{2}\varphi - \frac{\sqrt{3}}{16}l\varphi^2.\end{aligned} \tag{4.10.9}$$

This approach involved the use of the exact expressions for the additional deflections and their approximations in series. Alternatively, one can use

$$\lambda_1 \approx u_1 + \frac{1}{2l_{st1}}{v_1}^2, \quad \lambda_2 \approx u_2 + \frac{1}{2l_{st2}}{v_2}^2, \tag{4.10.10}$$

recognising that

$$u_1 = (\mathbf{r}_{\mathrm{C}} - \mathbf{r}_{\mathrm{R}}) \cdot \mathbf{e}_1, \quad v_1 = (\mathbf{r}_{\mathrm{C}} - \mathbf{r}_{\mathrm{R}}) \cdot \mathbf{e}_2, \quad u_2 = \mathbf{r}_{\mathrm{E}} \cdot \boldsymbol{\mu}_1, \quad v_2 = \mathbf{r}_{\mathrm{E}} \cdot \boldsymbol{\mu}_2, \tag{4.10.11}$$

where

$$\begin{aligned}\mathbf{r}_{\mathrm{R}} = \mathbf{r}_{\mathrm{E}} &= l\sin\varphi\ \mathbf{i} - l(1-\cos\varphi)\ \mathbf{j} \approx l\varphi\ \mathbf{i} - l\frac{\varphi^2}{2}\ \mathbf{j},\\ \mathbf{r}_{\mathrm{C}} &= 2l\sin\varphi\ \mathbf{i} - 2l(1-\cos\varphi)\ \mathbf{j} \approx 2l\varphi\ \mathbf{i} - l\varphi^2\mathbf{j},\end{aligned} \tag{4.10.12}$$

and

$$\begin{aligned}&\mathbf{i}\cdot\mathbf{e}_1 = \cos 60^\circ,\ \ \mathbf{j}\cdot\mathbf{e}_1 = \cos 30^\circ,\ \ \mathbf{i}\cdot\mathbf{e}_2 = \cos 30^\circ,\ \ \mathbf{j}\cdot\mathbf{e}_2 = \cos 120^\circ,\\ &\mathbf{i}\cdot\boldsymbol{\mu}_1 = \cos 120^\circ,\ \ \mathbf{j}\cdot\boldsymbol{\mu}_1 = \cos 30^\circ,\ \ \mathbf{i}\cdot\boldsymbol{\mu}_2 = \cos 30^\circ,\ \ \mathbf{j}\cdot\boldsymbol{\mu}_2 = \cos 60^\circ.\end{aligned} \tag{4.10.13}$$

In both cases, the potential energy takes the form

$$V \approx \left[(m+M)gl + \frac{kl^2}{4} - \frac{\sqrt{3}}{8}Fl\right]\varphi^2. \tag{4.10.14}$$

As the bars AO_1 and BO_2 perform rotational motion and the bar AB translatory motion, the overall kinetic energy has the form:

$$T = \frac{1}{2}J_{O_1}\dot{\varphi}^2 + \frac{1}{2}J_{O_2}\dot{\varphi}^2 + \frac{1}{2}Mv_{\mathrm{C}}^2, \tag{4.10.15}$$

where the mass moments of inertia are given by

$$J_{O_1} = J_{O_2} = \frac{1}{12}m(2l)^2 + ml^2 = \frac{4}{3}ml^2. \tag{4.10.16}$$

Analysing the velocities while the system passes through the equilibrium positon enables one to define the kinetic energy completely as well as to determine the dissipation function. Thus, based on Figure 4.10.3b, one has

$$\mathbf{v}_{\mathrm{R}} = \mathbf{v}_{\mathrm{E}} = l\dot{\varphi}\ \mathbf{i},\quad \mathbf{v}_{\mathrm{A}} = \mathbf{v}_{\mathrm{B}} = \mathbf{v}_{\mathrm{C}} = 2l\dot{\varphi}\ \mathbf{i}. \tag{4.10.17}$$

The relative velocities of the pistons with respect to the cylinders in the dampers are found from the projections

$$\left(v_{p/c}\right)_1 = \mathbf{v}_{\mathrm{R}}\cdot\mathbf{e}_1 \ \Rightarrow\ \left(v_{p/c}\right)_1 = \frac{l\dot{\varphi}}{2}. \tag{4.10.18}$$

$$\left(v_{p/c}\right)_2 = (\mathbf{v}_{\mathrm{E}} - \mathbf{v}_{\mathrm{C}})\cdot\boldsymbol{\mu}_1 \ \Rightarrow\ \left(v_{p/c}\right)_2 = \frac{l\dot{\varphi}}{2}. \tag{4.10.19}$$

The kinetic energy (4.10.15) becomes

$$T = \left(\frac{4}{3}m + 2M\right)l^2\dot{\varphi}^2. \tag{4.10.20}$$

The dissipative function

$$D = \frac{1}{2}c\left(v_{p/c}\right)_1^2 + \frac{1}{2}c\left(v_{p/c}\right)_2^2, \tag{4.10.21}$$

is

$$D \approx \frac{1}{4}cl^2\dot{\varphi}^2. \tag{4.10.22}$$

Lagrange's equation (4.2) gives

$$\ddot{\varphi} + 2\delta\dot{\varphi} + \omega_n^2\varphi = 0, \tag{4.10.23}$$

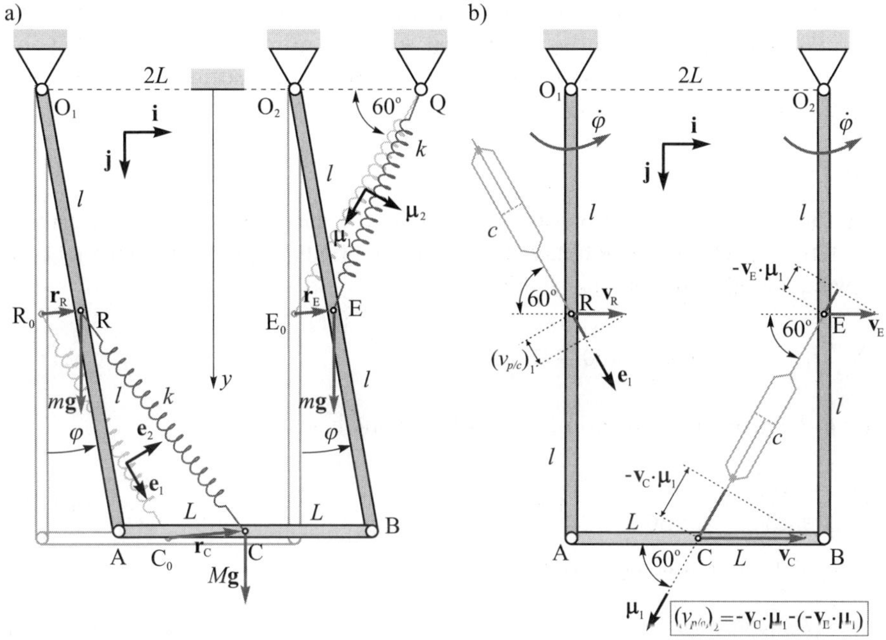

Figure 4.10.3

with

$$\delta = \frac{3c}{16(2m+3M)}, \quad \omega_n = \sqrt{\frac{3(m+M)g}{2(2m+3M)l} + \frac{3k}{8(2m+3M)} - \frac{3\sqrt{3}F}{16(2m+3M)l}}. \tag{4.10.24}$$

Example 4.11 The mechanism under consideration is shown in its equilibrium position in Figure 4.11.1. It consists of a slender bar AB of mass m and length l, a horizontal spring RA of stiffness k and a linear viscous damper BQ with damping coefficient c. The bar is pinned at A to a weightless slider, which moves without friction along a horizontal

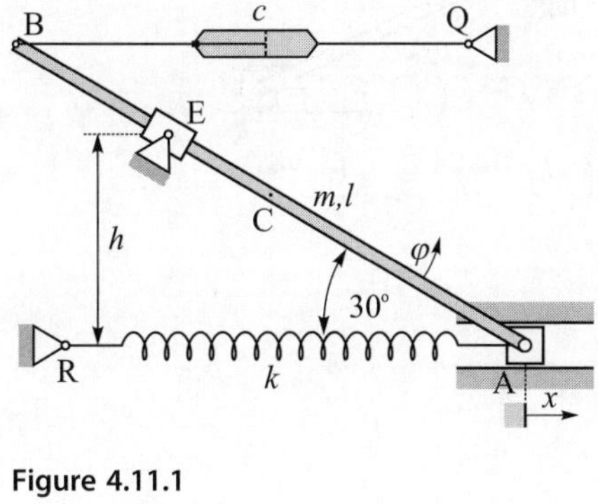

Figure 4.11.1

guide. The bar passes through the weightless slider E, which is hinged at the height h with respect to the guide. Determine the differential equation of motion choosing first the angle φ as the generalised coordinate, and then choosing the displacement x of the slider at A as the generalised coordinate.

Solution

To make the equilibrium position shown possible, the spring must be predeformed. First, its static deflection will be obtained. For the free-body diagram in Figure 4.11.2a, the following equilibrium equation holds:

$$\sum M_P = 0 \;\Rightarrow\; mg\frac{l}{2}\cos 30^\circ - kf_{st}\overline{PA} = 0. \tag{4.11.1}$$

Based on the geometry shown in Figure 4.11.2a, one has

$$\overline{PA} = \frac{\overline{AE_1}}{\sin 30^0} = \frac{h}{\sin^2 30^\circ} \;\Rightarrow\; \overline{PA} = 4h, \tag{4.11.2}$$

and the static deflection is found to be

$$f_{st} = \frac{\sqrt{3}mgl}{16kh}. \tag{4.11.3}$$

Based on Figure 4.11.2b, one can express the potential energy as follows

$$V = mg\frac{l}{2}\sin(30^\circ - \varphi) + \frac{k}{2}(f_{st} + \lambda)^2. \tag{4.11.4}$$

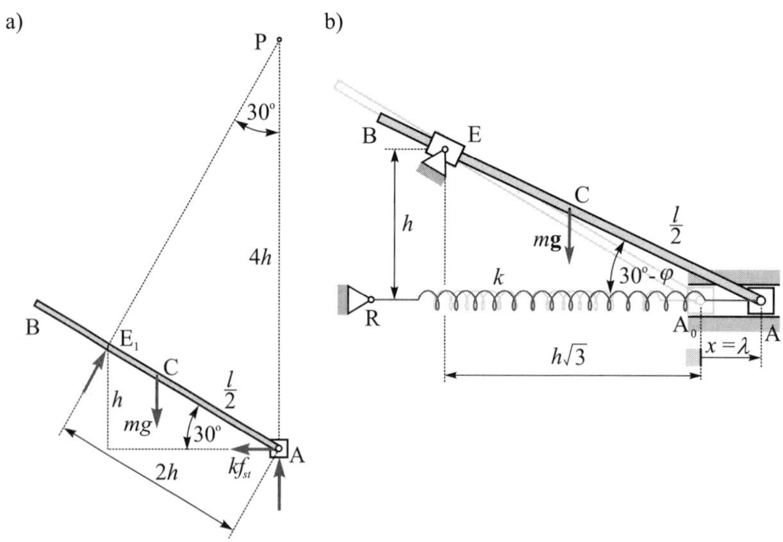

Figure 4.11.2

The additional deflection λ corresponds to the displacement of the point A, and can be obtained as follows:

$$\cot(30^\circ - \varphi) = \frac{h\sqrt{3}+\lambda}{h} \quad \Rightarrow \quad \lambda = h\cot(30^\circ - \varphi) - h\sqrt{3} \quad \Rightarrow \quad \lambda \approx 4h\varphi + 4\sqrt{3}h\varphi^2. \tag{4.11.5}$$

Thus, the potential energy takes the form

$$V \approx \frac{1}{8}\left(5mgl + 64kh^2\right)\varphi^2. \tag{4.11.6}$$

The bar AB executes general plane motion and its kinetic energy is defined by

$$T = \frac{m}{2}v_C{}^2 + \frac{1}{2}J_C\omega^2, \tag{4.11.7}$$

where

$$J_C = \frac{1}{12}ml^2. \tag{4.11.8}$$

The expression for the velocity v_C is (see Figure 4.11.3a):

$$v_C = \overline{PC}\dot{\varphi}, \quad \overline{PC}^2 = (4h)^2 + \left(\frac{l}{2}\right)^2 - 2\cdot 4h\cdot\frac{l}{2}\cdot\cos 60^\circ = (4h)^2 + \left(\frac{l}{2}\right)^2 - 2hl. \tag{4.11.9}$$

The kinetic energy now becomes

$$T \approx \frac{m}{6}\left(l^2 + 48h^2 - 6hl\right)\dot{\varphi}^2. \tag{4.11.10}$$

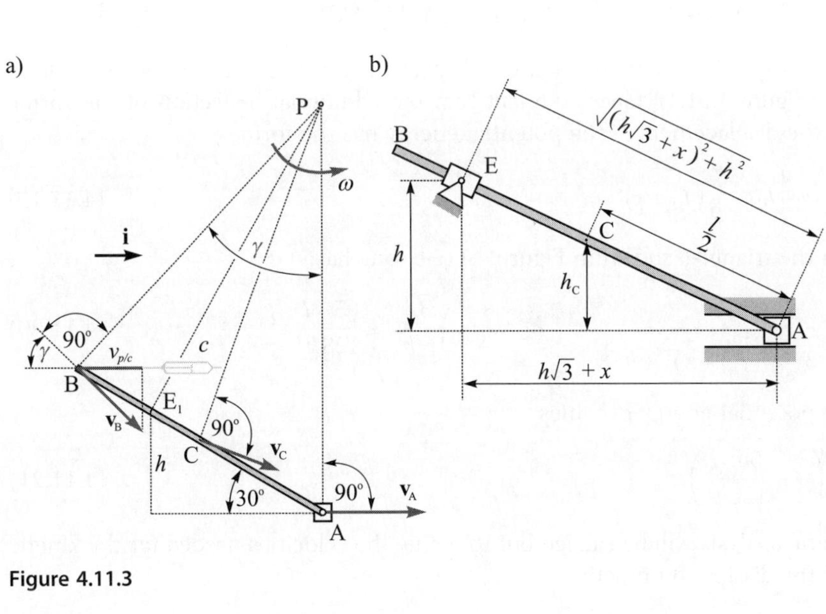

Figure 4.11.3

The dissipation function is defined by

$$D=\frac{1}{2}cv_{p/c}{}^{2}, \tag{4.11.11}$$

where

$$v_{p/c}=v_{\mathrm{B}}\cos\gamma=\dot{\varphi}\overline{\mathrm{PB}}\cos\gamma. \tag{4.11.12}$$

The velocity v_{B} is:

$$v_{\mathrm{B}}=\overline{\mathrm{PB}}\dot{\varphi},\quad \overline{\mathrm{PB}}^{2}=(4h)^{2}+l^{2}-2\cdot 4h\cdot l\cdot\cos 60^{\circ}=(4h)^{2}+l^{2}-4hl. \tag{4.11.13}$$

The term $\overline{\mathrm{PB}}\cos\gamma$ follows from

$$l^{2}=(4h)^{2}+\overline{\mathrm{PB}}^{2}-8h\overline{\mathrm{PB}}\cos\gamma\ \Rightarrow\ \overline{\mathrm{PB}}\cos\gamma=\frac{(4h)^{2}+\overline{\mathrm{PB}}^{2}-l^{2}}{8h}=\frac{8h-l}{2}. \tag{4.11.14}$$

Finally, the velocity $v_{p/c}$ is found to be

$$v_{p/c}=v_{\mathrm{B}}\cos\gamma=\dot{\varphi}\overline{\mathrm{PB}}\cos\gamma=\dot{\varphi}\frac{8h-l}{2}\ \Rightarrow\ v_{p/c}=\frac{8h-l}{2}\dot{\varphi}. \tag{4.11.15}$$

The dissipative function (4.11.11) can now be presented as

$$D\approx\frac{c}{8}(8h-l)^{2}\dot{\varphi}^{2}. \tag{4.11.16}$$

Lagrange's equation (4.2) gives

$$\ddot{\varphi}+2\delta\dot{\varphi}+\omega_{n}{}^{2}\varphi=0, \tag{4.11.17}$$

with

$$\delta=\frac{3c(8h-l)^{2}}{8m(l^{2}+48h^{2}-6hl)},\quad \omega_{n}=\frac{1}{2}\sqrt{\frac{3(5mgl+64kh^{2})}{m(l^{2}+48h^{2}-6hl)}}. \tag{4.11.18}$$

Let us now solve the problem when the displacement x is chosen as the generalised coordinate (Figure 4.11.1). Given the fact that the additional deflection of the spring is equal to the displacement x, the potential energy has the form

$$V=mg\frac{l}{2}h_{\mathrm{C}}+\frac{k}{2}(f_{st}+x)^{2}. \tag{4.11.19}$$

Based on the triangles shown in Figure 4.11.3b, one has:

$$h_{\mathrm{C}}=\frac{hl}{2\sqrt{\left(h\sqrt{3}+x\right)^{2}+h^{2}}}\ \Rightarrow\ h_{\mathrm{C}}\approx\frac{l}{4}-\frac{\sqrt{3}\,l}{16\,h}x+\frac{5}{128}\frac{l}{h^{2}}x^{2} \tag{4.11.20}$$

Thus the potential energy becomes

$$V\approx\left(\frac{k}{2}+\frac{5mgl}{128h^{2}}\right)x^{2}. \tag{4.11.21}$$

A kinematic analysis will be carried out to define the velocities needed for the kinetic energy and the dissipation function.

First, one can consider the relationship between the velocities of the points E_1 and A to derive (see Figure 4.11.4):

$$\mathbf{v}_{E_1} = \mathbf{v}_A + \mathbf{v}_{E_1/A}, \quad v_A = \dot{x}, \quad v_{E_1/A} = \overline{AE_1}\omega, \tag{4.11.22}$$

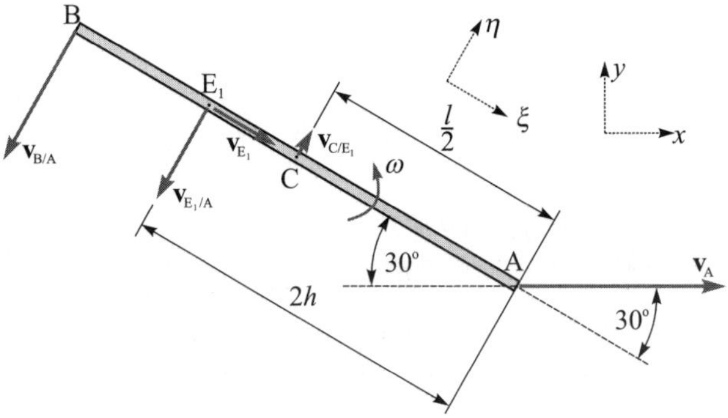

Figure 4.11.4

$$\eta: \quad 0 = \dot{x}\sin 30^\circ - 2h\omega \quad \Rightarrow \quad \omega = \dot{x}/4h, \tag{4.11.23}$$

$$\xi: \quad v_{E_1} = \dot{x}\cos 30^\circ + 0 \quad \Rightarrow \quad v_{E_1} = \dot{x}\sqrt{3}/2. \tag{4.11.24}$$

Then, the relationship between the velocities of the points E_1 and C yields:

$$\mathbf{v}_C = \mathbf{v}_{E_1} + \mathbf{v}_{C/E_1}, \quad v_{C/E_1} = \overline{CE_1}\,\omega = \left(2h - \frac{l}{2}\right)\frac{\dot{x}}{4h}, \tag{4.11.25}$$

$$\eta: \quad v_{C\eta} = 0 + \left(2h - \frac{l}{2}\right)\frac{\dot{x}}{4h} \quad \Rightarrow \quad v_{C\eta} = \left(2h - \frac{l}{2}\right)\frac{\dot{x}}{4h}, \tag{4.11.26}$$

$$\xi: \quad v_{C\xi} = \dot{x}\frac{\sqrt{3}}{2} + 0 \quad \Rightarrow \quad v_{C\xi} = \dot{x}\frac{\sqrt{3}}{2}, \tag{4.11.27}$$

$$v_C{}^2 = v_{C\xi}{}^2 + v_{C\eta}{}^2 = \left(\frac{3}{4} + \frac{4h^2 + l^2/4 - 2hl}{16h^2}\right)\dot{x}^2, \tag{4.11.28}$$

The kinetic energy can be transformed to the following form:

$$T \approx \frac{1}{2}m\left(16h^2 + \frac{l^2}{4} - 2hl\right)\frac{\dot{x}^2}{16h^2} + \frac{ml^2}{24}\frac{\dot{x}^2}{16h^2}. \tag{4.11.29}$$

The relative velocity existing in the dissipation function can be obtained from the following analysis:

$$\mathbf{v}_B = \mathbf{v}_A + \mathbf{v}_{B/A}, \quad v_{B/A} = \overline{AB}\,\omega = \frac{l}{4h}\dot{x}, \tag{4.11.30}$$

$$x: \quad v_{Bx} = \dot{x} - \frac{l}{4h}\dot{x}\sin 30^\circ \quad \Rightarrow \quad v_{Bx} = v_{p/c} = \left(1 - \frac{l}{8h}\right)\dot{x}, \tag{4.11.31}$$

so that the dissipation function becomes

$$D \approx \frac{1}{2}c\frac{(8h-l)^2}{64h^2}\dot{x}^2. \tag{4.11.32}$$

Lagrange's equation for the generalised coordinate x has the form

$$\ddot{x} + 2\delta\dot{x} + \omega_n{}^2\varphi = 0, \tag{4.11.33}$$

where

$$\delta = \frac{3c(8h-l)^2}{8m(l^2+48h^2-6hl)}, \quad \omega_n = \frac{1}{2}\sqrt{\frac{3(5mgl+64kh^2)}{m(l^2+48h^2-6hl)}}. \tag{4.11.34}$$

By comparing the parameters in Equation (4.11.34) with the ones in Equation (4.11.18), one can see that they are identical, as they should be regardless of the choice of the generalised coordinate.

5

Forced Vibration of Single-Degree-of-Freedom Systems

Chapter Outline

This chapter starts with a brief overview of Lagrange's equation for forced undamped and damped vibration, the quantities existing in it, and the resulting equation of motion in general form. Seven examples are solved in detail. Special attention is now given to the external excitation, which can have the form of either base excitation or force excitation. The resulting response is also found and illustrated graphically.

Chapter Objectives

- To present Lagrange's equation of the second kind for forced undamped and damped vibration
- To demonstrate how external excitation, which can have the form of either base excitation or force excitation, should be included into Lagrange's equation
- To illustrate the forced response and frequency–amplitude diagrams of certain oscillatory systems

Theoretical Introduction

Lagrange's equation for one-degree-of-freedom forced damped systems has the form

$$\frac{d}{dt}\frac{\partial T}{\partial \dot{q}} - \frac{\partial T}{\partial q} + \frac{\partial D}{\partial \dot{q}} + \frac{\partial V}{\partial q} = Q_q. \tag{5.1}$$

When the system performs small oscillations around the equilibrium position $q = 0$, this Lagrange's equation can be simplified to

$$\frac{d}{dt}\frac{\partial T}{\partial \dot{q}} + \frac{\partial D}{\partial \dot{q}} + \frac{\partial V}{\partial q} = Q_q, \tag{5.2}$$

where the kinetic energy takes the form

$$T = \frac{1}{2}\widetilde{T}\dot{q}^2, \tag{5.3}$$

Mechanical Vibration: Fundamentals with Solved Examples, First Edition.
Ivana Kovačić and Dragi Radomirović.

with the coefficient $\widetilde{T}$ being constant and to be calculated while the system passes through the equilibrium position: $\widetilde{T}(q) = \widetilde{T}(0)$.

The potential energy in Equation (5.2) has the general form

$$V = \frac{1}{2}\widetilde{V}q^2, \tag{5.4}$$

where its coefficient $\widetilde{V}$ corresponds to

$$\widetilde{V} = \left(\frac{\mathrm{d}^2 V}{\mathrm{d}q^2}\right)_{q=0}. \tag{5.5}$$

The dissipative function D in Equation (5.2) has the general form

$$D = \frac{1}{2}\widetilde{D}\dot{q}^2, \tag{5.6}$$

where $\widetilde{D}$ is a constant and it should also be calculated while the system passes through the equilibrium position.

The generalised force Q_q corresponds to the coefficient in the expression for the virtual work δA next to the variation of the generalised coordinate (see Section 2.4 for details):

$$\delta A = Q_q\ \delta q. \tag{5.7}$$

Note that, in the general case, this Q_q will depend both on the generalised coordinate and time. However, for the sake of deriving the equation of motion for small oscillations, one needs to use $q = 0$, that is, to consider the case when the system passes through the equilibrium position.

Substituting (5.3), (5.4), (5.6) and Q_q into Equation (5.2), one can derive

$$\ddot{q} + 2\delta\dot{q} + {\omega_n}^2 q = p. \tag{5.8}$$

where ω_n represents the natural frequency and is defined by ${\omega_n}^2 = \widetilde{V}/\widetilde{T}$, the damping ratio is $2\delta = \widetilde{D}/\widetilde{T}$, while $p = Q_q/\widetilde{T}$. Depending on the form of $p = p(t)$, Equation (5.8) has a different particular integral and this will be illustrated thoroughly in the following examples.

Example 5.1 A slider of mass m can slide along a frictionless guide and is attached to a spring that has stiffness k as in Example 3.5. The system oscillates around the position $x = 0$ shown in Figure 5.1.1. The length of the spring is then l, and the angle between the guide and the spring is β. The slider is subjected to an external harmonic excitation $\mathbf{F}(t) = F_0 \cos\Omega t\mathbf{i}$, in which F_0 is the constant force amplitude and Ω is the constant driving frequency. Determine the differential equation of motion for small oscillations and show its solution graphically as well as the particular solution and frequency-amplitude diagram assuming: $l = 2\,\text{m}$, $m = 6\,\text{kg}$, $k = 900\,\text{N/m}$, $\alpha = 30°$, $\beta = 30°$, $F_0 = 120\,\text{N}$, $\Omega = 7\,\text{s}^{-1}$, $x(0) = 0.1$ m, $\dot{x}(0) = 0$.

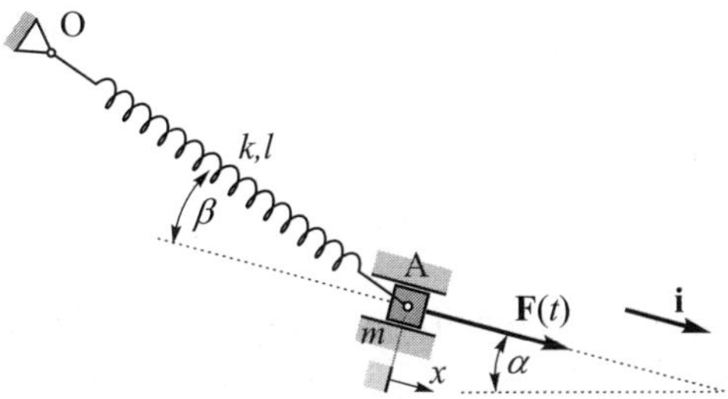

Figure 5.1.1

Solution

The kinetic and potential energy are the same as in Example 3.5 and have the form

$$T = \frac{1}{2}m\dot{x}^2, \quad V \approx \frac{1}{2}\left(k\cos^2\beta + \frac{mg\sin\alpha\,\sin^2\beta}{l\cos\beta}\right)x^2. \tag{5.1.1}$$

In order to find the generalised force, one needs to consider the virtual work done by the external force $\mathbf{F}(t)$:

$$\delta A^{\mathrm{F}} = \mathbf{F}\cdot\delta\mathbf{r}_{\mathrm{A}} = F_0\cos\Omega t\;\mathbf{i}\cdot\delta x\mathbf{i} = F_0\cos\Omega t\delta x, \tag{5.1.2}$$

where the coefficient in front of δx yields the following generalised force:

$$Q_x = F_0\cos\Omega t. \tag{5.1.3}$$

Lagrange's equation (5.2) gives the following differential equation of motion

$$\ddot{x} + {\omega_n}^2 x = h\cos\Omega t, \tag{5.1.4}$$

where:

$$\omega_n = \sqrt{\frac{k}{m}\cos^2\beta + \frac{g\sin\alpha\,\sin^2\beta}{l\cos\beta}} \cong 10.64\ \mathrm{s}^{-1}, \quad h = \frac{F_0}{m} = 20\ \mathrm{ms}^{-2}. \tag{5.1.5}$$

The complete solution of Equation (5.1.4) is the sum of the complementary solution x_c, which is the general solution of Equation (5.1.4) with the right-hide side set to zero, and the particular solution x_p which is any solution to the complete equation that depends on the expression on the right-hand side. Let us first find the latter. The particular solution is assumed to resemble that of the force term:

$$x_p(t) = A\cos\Omega t, \tag{5.1.6}$$

where A is the amplitude to be calculated. Differentiating this expression twice and substituting it together with Equation (5.1.6) into Equation (5.1.4), one obtains

$$A = \frac{h}{{\omega_n}^2 - \Omega^2}. \tag{5.1.7}$$

The complementary solution has the form discussed in Section 3. The complete solution must satisfy the initial conditions and is found to be

$$x(t) = x_c + x_p = (x(0) - A)\cos\omega_n t + A\cos\Omega t, \tag{5.1.8}$$

where A has been calculated to be $A = 0.3115$ m. Both the complete solution and the particular solution are plotted in Figure 5.1.2.

The so-called frequency–amplitude diagram is presented in Figure 5.1.3: it shows how the absolute value of the amplitude of the particular solution changes with the forcing frequency as given by Equation (5.1.7). When $\Omega = \omega_n$, the resonance phenomenon occurs and the amplitude tends to infinity. For $\Omega >> \omega_n$, this amplitude has a decreasing trend.

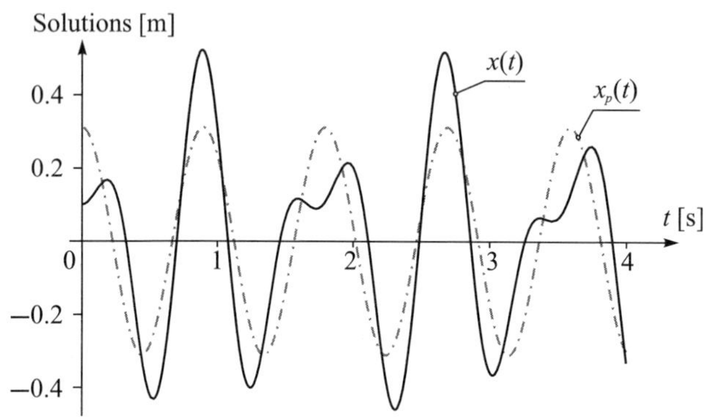

Figure 5.1.2

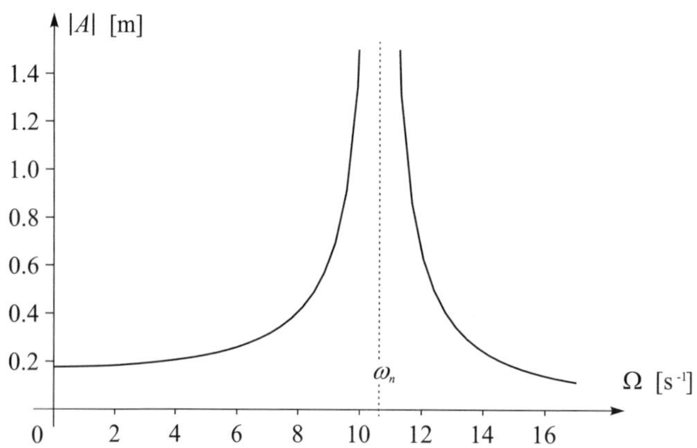

Figure 5.1.3

Example 5.2 A linkage consists of two slender bars: OA of mass m and length l and AB of mass M and length $L = l$ as in Example 4.6, but with the damper removed. The sliding block pivoted at B has negligible mass. The system performs small oscillations around the equilibrium position shown in Figure 5.2.1. There is an external torque acting on the bar OA: $\mathfrak{M}(t) = \mathfrak{M}_0 \sin \Omega t$. Taking the angle φ as the generalised coordinate, determine the differential equation for small forced oscillations and find its particular solution.

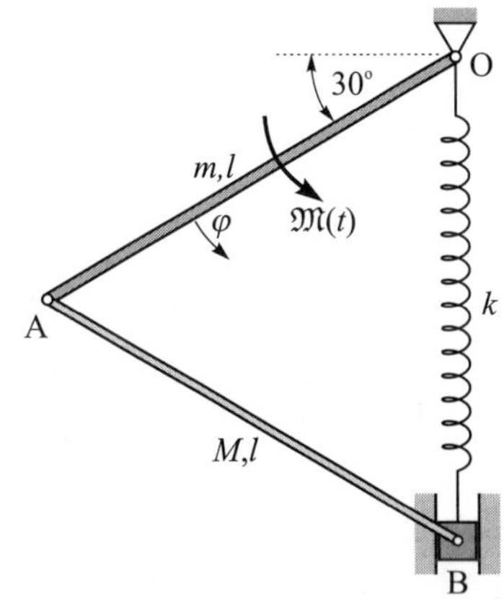

Figure 5.2.1

Solution

The kinetic energy and the potential energy are obtained in Example 4.6, and their forms are used here:

$$T \approx \frac{2m + 11M}{12} l^2 \dot{\varphi}^2, \quad V \approx \frac{3}{2} k l^2 \varphi^2. \tag{5.2.1}$$

The virtual work stemming from the external excitation is given by

$$\delta A = \delta A^{\mathfrak{M}} = \mathfrak{M} \delta \varphi = \mathfrak{M}_0 \sin \Omega t \delta \varphi. \tag{5.2.2}$$

The coefficient in front of $\delta\varphi$ represents the generalised force

$$Q_\varphi = \mathfrak{M}_0 \sin \Omega t. \tag{5.2.3}$$

Lagrange's equation (5.2) leads to the following differential equation of motion

$$\ddot{\varphi} + {\omega_n}^2 \varphi = h \sin \Omega t, \tag{5.2.4}$$

where

$$\omega_n = \sqrt{\frac{18k}{2m + 11M}}, \quad h = \frac{6\mathfrak{M}_0}{(2m + 11M) l^2}. \tag{5.2.5}$$

The particular integral has the form

$$x_p(t) = A \sin \Omega t, \tag{5.2.6}$$

where

$$A = \frac{h}{{\omega_n}^2 - \Omega^2} = \frac{6\mathfrak{M}_0}{(2m + 11M) l^2} \frac{1}{\dfrac{18k}{2m + 11M} - \Omega^2}. \tag{5.2.7}$$

Example 5.3 The ends of a light bar of length L move without friction in vertical and horizontal slots while being attached to a horizontal spring of stiffness k, as in Example 3.3. The masses of the sliders at A and B are, respectively, M and m. The system is shown in its equilibrium position in Figure 5.3.1. The slider A is subjected to an external force $\mathbf{F}_1(t) = F_{10} \sin \Omega_1 t \mathbf{i}$ and the slider B to an external force $\mathbf{F}_2(t) = F_{20} \sin(\Omega_2 t + \alpha) \mathbf{j}$.

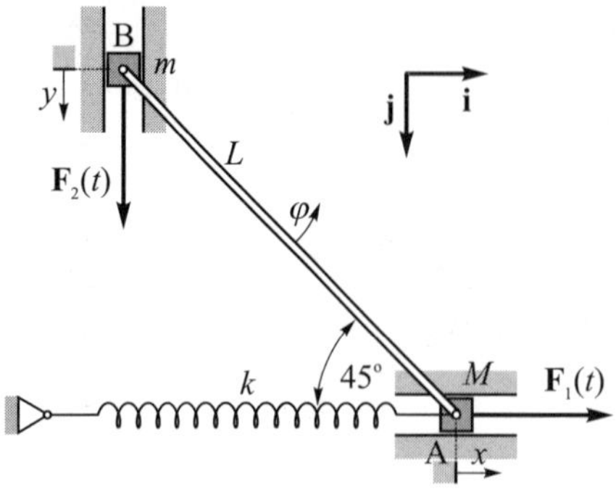

Figure 5.3.1

Determine the generalised force for three choices of the generalised coordinate x, y and φ and its approximations for small vibrations. Form the equation of motion for small oscillations in terms of the generalised coordinate φ. Plot its complete and particular solution, assuming: $L = 1\,\text{m}$, $M = 3\,\text{kg}$, $m = 2\,\text{kg}$, $k = 900\,\text{N/m}$, $F_{10} = 100\,\text{N}$, $F_{20} = 10\,\text{N}$, $\Omega_1 = 2\omega_n$, $\Omega_2 = 1.1\omega_n$, $\alpha = \pi/6$, $x(0) = 0$, $\dot{x}(0) = 0$.

Solution

The virtual work stemming from the external forces $\mathbf{F}_1(t)$ and $\mathbf{F}_2(t)$ is given by

$$\delta A = \delta A^{\mathbf{F}_1} + \delta A^{\mathbf{F}_2} = F_{10}\sin(\Omega_1 t)\delta x + F_{20}\sin(\Omega_2 t + \alpha)\delta y. \tag{5.3.1}$$

Let us first analyse the case when the coordinate x is chosen to be the generalised coordinate. One needs to express $\delta y = \delta y(\delta x)$, that is, to find $y = y(x)$. Given Equation (3.3.3)

$$y(x) = L\frac{\sqrt{2}}{2} - \sqrt{L^2 - \left(L\frac{\sqrt{2}}{2} + x\right)^2}, \tag{5.3.2}$$

one can obtain

$$\delta y = \frac{L\sqrt{2} + 2x}{\sqrt{2L^2 - 4L\sqrt{2}x - 4x^2}}\delta x. \tag{5.3.3}$$

Substituting Equation (5.3.3) into Equation (5.3.1), one derives

$$\delta A = \left(F_{10}\sin\Omega_1 t + \frac{L\sqrt{2} + 2x}{\sqrt{2L^2 - 4L\sqrt{2}x - 4x^2}}F_{20}\sin(\Omega_2 t + \alpha)\right)\delta x, \tag{5.3.4}$$

which gives

$$Q_x = F_{10}\sin\Omega_1 t + \frac{L\sqrt{2} + 2x}{\sqrt{2L^2 - 4L\sqrt{2}x - 4x^2}}F_{20}\sin(\Omega_2 t + \alpha). \tag{5.3.5}$$

Approximating this for the case when the system passes through the equilibrium position, one has

$$Q_x = F_{10}\sin\Omega_1 t + F_{20}\sin(\Omega_2 t + \alpha). \tag{5.3.6}$$

If the generalised coordinate is given by the coordinate y, one can express Equation (3.3.8) as

$$x(y) = \sqrt{L^2 - \left(L\frac{\sqrt{2}}{2} - y\right)^2} - L\frac{\sqrt{2}}{2}, \tag{5.3.7}$$

which gives

$$\delta x = \frac{L\sqrt{2} - 2y}{\sqrt{2L^2 + 4L\sqrt{2}y - 4y^2}}\delta y. \tag{5.3.8}$$

The expression for the virtual work (5.3.1) now becomes

$$\delta A = \left(\frac{L\sqrt{2} - 2y}{\sqrt{2L^2 + 4L\sqrt{2}y - 4y^2}}F_{10}\sin\Omega_1 t + F_{20}\sin(\Omega_2 t + \alpha)\right)\delta y. \tag{5.3.9}$$

The generalised force Q_y is

$$Q_y = \frac{L\sqrt{2} - 2y}{\sqrt{2L^2 + 4L\sqrt{2}y - 4y^2}}F_{10}\sin\Omega_1 t + F_{20}\sin(\Omega_2 t + \alpha). \tag{5.3.10}$$

For $y = 0$, this simplifies to

$$Q_y = F_{10}\sin\Omega_1 t + F_{20}\sin(\Omega_2 t + \alpha). \tag{5.3.11}$$

Note that the expressions for the generalised force Q_x and Q_y are identical, which is expected as the equations of motion for $q = x$ and $q = y$ are of the same form (see Example 3.4).

Finally, let us analyse the case when the generalised coordinate is chosen to be the angle φ. Given Equations (3.3.12) and (3.3.13), one has

$$x(\varphi) = L\sin(45^\circ + \varphi) - L\frac{\sqrt{2}}{2}, \quad y(\varphi) = L\frac{\sqrt{2}}{2} - L\cos(45^\circ + \varphi), \tag{5.3.12}$$

which yield

$$\delta x = L\cos(45^\circ + \varphi)\delta\varphi, \quad \delta y = L\sin(45^\circ + \varphi)\delta\varphi. \tag{5.3.13}$$

Introducing these into Equation (5.3.1) leads to

$$\delta A = [L\cos(45^\circ + \varphi)F_{10}\sin\Omega_1 t + L\sin(45^\circ + \varphi)F_{20}\sin(\Omega_2 t + \alpha)]\delta\varphi. \tag{5.3.14}$$

The generalised force Q_φ is recognised as

$$Q_\varphi = L\cos(45^\circ + \varphi)F_{10}\sin\Omega_1 t + L\sin(45^\circ + \varphi)F_{20}\sin(\Omega_2 t + \alpha). \tag{5.3.15}$$

Its approximate expression that corresponds to the case when the system passes through the equilibrium position is

$$Q_\varphi = L\frac{\sqrt{2}}{2}F_{10}\sin\Omega_1 t + L\frac{\sqrt{2}}{2}F_{20}\sin(\Omega_2 t + \alpha). \tag{5.3.16}$$

The kinetic and potential energy are given by Equations (3.3.16) and (3.3.14):

$$T \approx \frac{1}{2}\frac{(M+m)L^2}{2}\dot{\varphi}^2, \quad V \approx \frac{1}{2}\left(\frac{kL^2}{2} - \sqrt{2}mgL\right)\varphi^2. \tag{5.3.17}$$

Lagrange's equation (5.2) gives

$$\ddot{\varphi} + {\omega_n}^2\varphi = h_1\sin\Omega_1 t + h_2\sin(\Omega_2 t + \alpha), \tag{5.3.18}$$

where:

$$\omega_n = \sqrt{\frac{kL - 2\sqrt{2}mg}{(M+m)L}}, \quad h_1 = \frac{\sqrt{2}F_{10}}{(M+m)L}, \quad h_2 = \frac{\sqrt{2}F_{20}}{(M+m)L}. \tag{5.3.19}$$

The particular solution is assumed as

$$\varphi_p = \varphi_{p1} + \varphi_{p2}, \quad \varphi_{p1} = A_1\sin\Omega_1 t, \quad \varphi_{p2} = A_2\sin(\Omega_2 t + \alpha), \tag{5.3.20}$$

where:

$$A_1 = \frac{h_1}{{\omega_n}^2 - {\Omega_1}^2}, \quad A_2 = \frac{h_2}{{\omega_n}^2 - {\Omega_2}^2}. \tag{5.3.21}$$

This particular solution, as well as φ_{p1} and φ_{p2}, are plotted in Figure 5.3.2 for the given values of the system parameters.

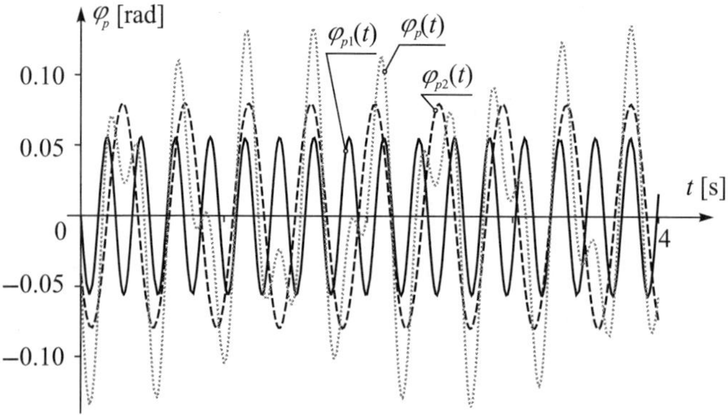

Figure 5.3.2

The complete solution of Equation (5.3.18) that satisfies the initial conditions is given by

$$\varphi(t) = A_1 \sin\Omega_1 t + A_2 \sin(\Omega_2 t + \alpha) - \frac{h_2\left(\omega_n^2 - \Omega_1^2\right)\sin\alpha\cos\omega_n t}{\omega_n^4 - \omega_n^2\Omega_2^2 - \omega_n^2\Omega_1^2 + \Omega_1^2\Omega_2^2} - \frac{h_2\Omega_2\left(\omega_n^2 - \Omega_1^2\right)\cos\alpha\sin\omega_n t}{\left(\omega_n^4 - \omega_n^2\Omega_2^2 - \omega_n^2\Omega_1^2 + \Omega_1^2\Omega_2^2\right)\omega_n} - \frac{h_1\Omega_1\left(\omega_n^2 - \Omega_2^2\right)\sin\omega_n t}{\left(\omega_n^4 - \omega_n^2\Omega_2^2 - \omega_n^2\Omega_1^2 + \Omega_1^2\Omega_2^2\right)\omega_n}. \quad (5.3.22)$$

This solution is shown in Figure 5.3.3.

Figure 5.3.4 contains the frequency–amplitude graphs for both the excitations. The values of the amplitudes corresponding to $\Omega_1 = 2\omega_n$ and $\Omega_2 = 1.1\omega_n$ are also depicted.

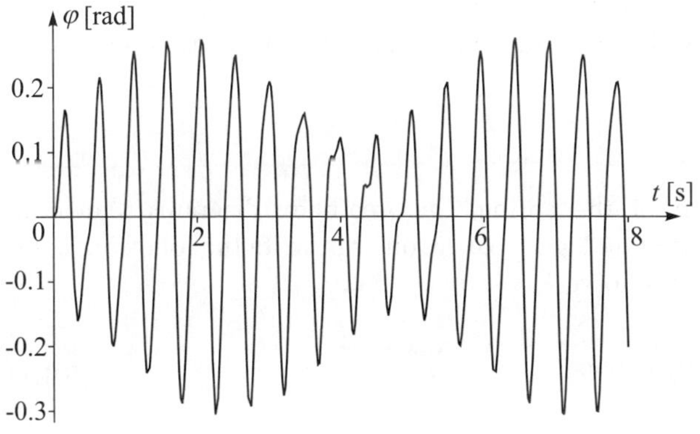

Figure 5.3.3

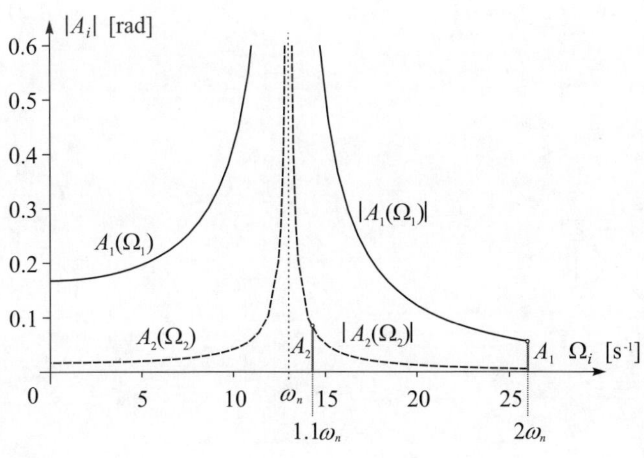

Figure 5.3.4

Example 5.4 The mechanism shown in Figure 5.4.1 is placed in the vertical plane and consists of a semi-circular slender bar OA of mass m and radius R, a straight slender bar AB of mass M and length $L = 2R\sqrt{2}$ and a weightless slider that moves along a frictionless horizontal guide. Its equilibrium position is shown in Figure 5.4.1. The slider is subjected to an external force $\mathbf{F}(t) = F_0 \sin\Omega t\mathbf{i}$. The generalised coordinate x defines the displacement of the slider with respect to its equilibrium position. Determine the differential equation of motion for small oscillations. Plot the solution $x(t)$ assuming: $x(0) = 0$, $\dot{x}(0) = 0$, $m = 1$ kg, $M = 1.2$ kg, $R = 0.2$ m, $k = 500$ N/m and $F_0 = 1$ N. Assume that the excitation frequency can have different values: $\Omega = 0.4\omega_n$, $\Omega = 0.6\omega_n$, $\Omega = 0.8\omega_n$, $\Omega = 0.94\omega_n$, $\Omega = 0.96\omega_n$, $\Omega = \omega_n$, $\Omega = 1.2\omega_n$, $\Omega = 1.4\omega_n$ and $\Omega = 1.6\omega_n$. Determine also the equation of motion for small oscillations in terms of the generalised coordinate being the angle of rotation of the curved bar and compare it to the one corresponding to x.

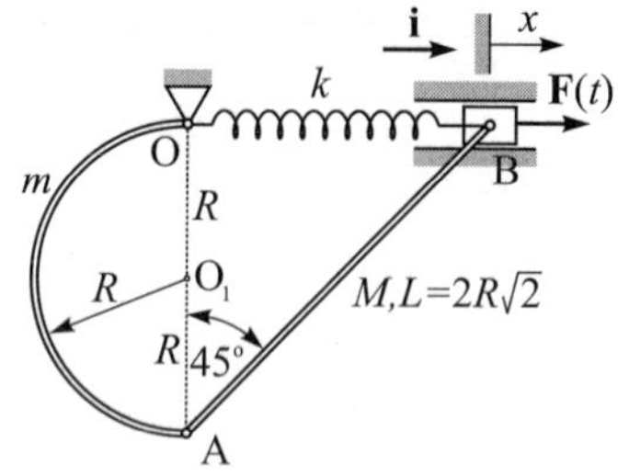

Figure 5.4.1

Solution

To obtain the static deflection of the spring, the free-body diagrams shown in Figure 5.4.2a,b are used. First, based on Figure 5.4.2a, one can write the following equilibrium equation and determine the force X_A:

$$\sum M_O = 0 \;\Rightarrow\; mg\frac{2R}{\pi} - X_A 2R = 0 \;\Rightarrow\; X_A = \frac{mg}{\pi}. \tag{5.4.1}$$

Further, all the forces along the x-direction must be in balance for the bar AB in the equilibrium position (Figure 5.4.2b), so that the following holds

$$\sum F_x = 0 \;\Rightarrow\; X_A - kf_{st} = 0 \;\Rightarrow\; f_{st} = \frac{mg}{\pi k}. \tag{5.4.2}$$

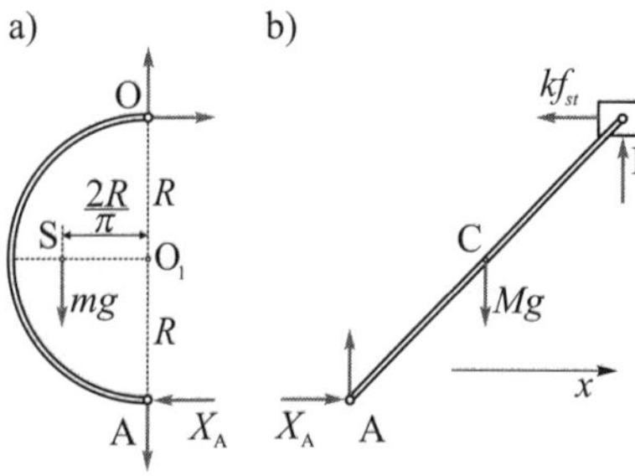

Figure 5.4.2

The potential energy of the system (see Figure 5.4.3) is

$$V = -mgy_S - Mgy_C + \frac{k}{2}(f_{st} + x)^2. \tag{5.4.3}$$

Noting that the position of the centre of mass of the curved bar is $\overline{O_1S} = 2R/\pi$, one can recognise

$$y_S = R\cos\varphi + \frac{2R}{\pi}\sin\varphi, \quad y_C = y_{O_1} = R\cos\varphi, \tag{5.4.4}$$

where the angle φ is introduced (see Figure 5.4.3) and it should be expressed in terms of the generalised coordinate. To that end, one can focus on the triangle OAB and write down the Law of Cosines:

$$\begin{aligned} &\overline{AB}^2 = \overline{OA}^2 + \overline{OB}^2 - 2\overline{OA}\cdot\overline{OB}\cos(90^\circ - \varphi) \Rightarrow \\ &8R^2 = 4R^2 + (2R + x)^2 - 4R(2R + x)\sin\varphi, \end{aligned} \tag{5.4.5}$$

which yields

$$\sin\varphi = \frac{x^2 + 4Rx}{4R(2R + x)} \Rightarrow \cos\varphi = \sqrt{1 - \left(\frac{x^2 + 4Rx}{4R(2R + x)}\right)^2}. \tag{5.4.6}$$

Developing these two expressions into series, one derives

$$\sin\varphi(x) = \frac{1}{2R}x - \frac{1}{8R^2}x^2 + \ldots, \quad \cos\varphi(x) = 1 - \frac{1}{8R^2}x^2 + \ldots. \tag{5.4.7}$$

Thus, the potential energy takes the form

$$V \approx \frac{1}{2}\left(k + \frac{g}{4R}\left(m\left(1 + \frac{2}{\pi}\right) + M\right)\right)x^2. \tag{5.4.8}$$

To obtain the kinetic energy, one should first note that the curved bar executes rotation around the pivot O, while the straight bar executes general plane motion. Therefore, the expression for the kinetic energy is

$$T = \frac{1}{2}J_O(\omega_{OA})^2 + \frac{1}{2}Mv_C{}^2 + \frac{1}{2}J_C(\omega_{AB})^2. \tag{5.4.9}$$

Knowing that $J_{O_1} = mR^2$, the parallel-axis theorem gives

$$\begin{aligned} &J_{O_1} = J_S + m\overline{O_1S}^2, \quad J_O = J_S + m\overline{OS}^2 \Rightarrow \\ &J_O - J_{O_1} = m\left(\overline{OS}^2 - \overline{O_1S}^2\right) \Rightarrow \\ &J_O = J_{O_1} + mR^2 \Rightarrow J_O = 2mR^2. \end{aligned} \tag{5.4.10}$$

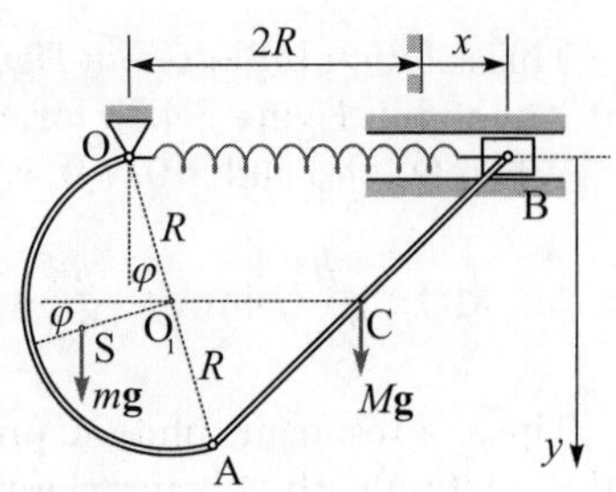

Figure 5.4.3

Considering the system as it passes through the equilibrium position, one can write the following equalities (Figure 5.4.4)

$$v_B = v_C = v_A = v = \dot{x}, \quad \omega_{AB} = 0, \quad \omega_{OA} = \frac{v_A}{\overline{OA}} \Rightarrow \omega_{OA} = \frac{\dot{x}}{2R}, \tag{5.4.11}$$

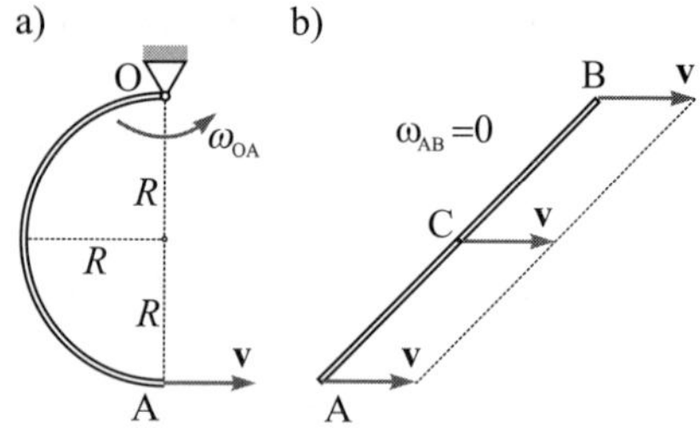

Figure 5.4.4

leading to the following expression for the kinetic energy:

$$T = \frac{1}{2}\left(\frac{m}{2} + M\right)\dot{x}^2. \tag{5.4.12}$$

To obtain the generalised force, the virtual work is considered

$$\delta A = \delta A^{F} = F_0 \sin \Omega t \delta x. \tag{5.4.13}$$

The generalised force is recognised as

$$Q_x = F_0 \sin \Omega t. \tag{5.4.14}$$

Lagrange's equation (5.2) gives

$$\ddot{x} + \omega_n{}^2 x = h \sin \Omega t, \tag{5.4.15}$$

where

$$\omega_n = \sqrt{\frac{4Rk + g(m(1 + 2/\pi) + M)}{2R(m + 2M)}}, \quad h = \frac{2F_0}{m + 2M}. \tag{5.4.16}$$

In the case when $\Omega \neq \omega_n$, the general solution satisfying the initial conditions $x(0) = 0$ and $\dot{x}(0) = 0$, is

$$x(t) = \frac{h\Omega}{\omega_n\left(\Omega^2 - \omega_n{}^2\right)} \sin \omega t - \frac{h}{\Omega^2 - \omega_n{}^2} \sin \Omega t. \tag{5.4.17}$$

This solution is plotted in Figure 5.4.5a for excitation frequencies of $0.4\omega_n$, $0.6\omega_n$ and $0.8\omega_n$, and in Figure 5.4.5b for excitation frequencies of $1.2\omega_n$, $1.4\omega_n$ and $1.6\omega_n$.

When $\Omega = \omega_n$ and $x(0) = 0$, $\dot{x}(0) = 0$, the solution of Equation (5.4.15) has the form

$$x(t) = \frac{h}{2\omega_n{}^2} \sin \omega_n t - \frac{ht}{2\omega_n} \cos \omega_n t. \tag{5.4.18}$$

This is a resonant solution, presented in Figure 5.4.6. In addition, this figure contains the solution with Ω being close to the natural frequency ($\Omega = 0.94\omega_n$ and $\Omega = 0.96\omega_n$), when beating oscillations take place.

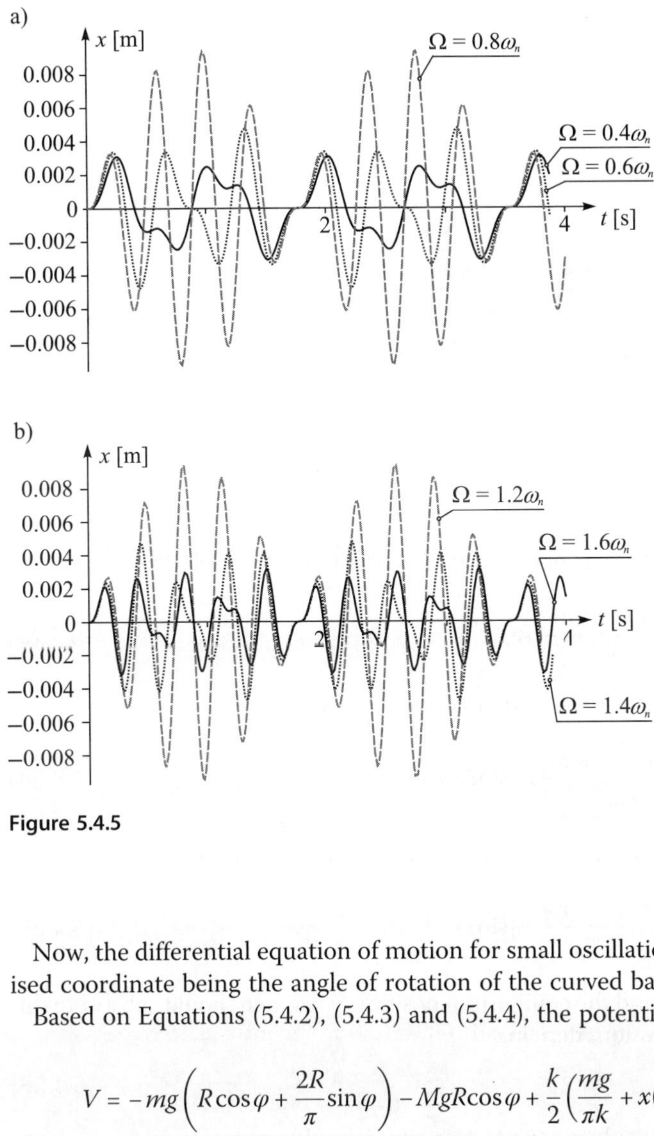

Figure 5.4.5

Now, the differential equation of motion for small oscillations in terms of the generalised coordinate being the angle of rotation of the curved bar φ is to be considered.

Based on Equations (5.4.2), (5.4.3) and (5.4.4), the potential energy takes the form

$$V = -mg\left(R\cos\varphi + \frac{2R}{\pi}\sin\varphi\right) - MgR\cos\varphi + \frac{k}{2}\left(\frac{mg}{\pi k} + x(\varphi)\right)^2. \tag{5.4.19}$$

Solving Equation (5.4.5), one can derive

$$x = 2R\left(\sin\varphi + \sqrt{1+\sin^2\varphi} - 1\right) \quad \Rightarrow \quad x(\varphi) = 2R\varphi + R\varphi^2 + \ldots \tag{5.4.20}$$

Finally, approximating $\sin\varphi = \varphi + \ldots$, $\cos\varphi = 1 - \varphi^2/2 + \ldots$, the potential energy (5.4.19) becomes

$$V \approx \frac{1}{2}\left(4kR^2 + gR\left(m\left(1 + \frac{2}{\pi}\right) + M\right)\right)\varphi^2. \tag{5.4.21}$$

Given Equation (5.4.11), the kinetic energy (5.4.12) transforms into

$$T = (m + 2M)R^2\dot{\varphi}^2. \tag{5.4.22}$$

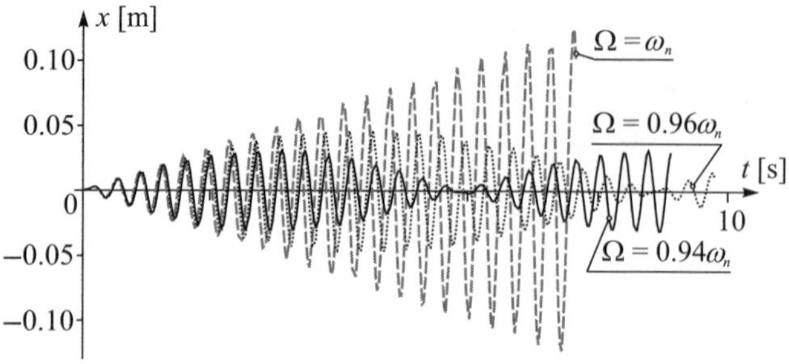

Figure 5.4.6

To obtain the generalised force, one can use Equation (5.4.20) to derive

$$\delta x = 2R\cos\varphi\frac{\sqrt{2-\cos^2\varphi}+\sin\varphi}{\sqrt{2-\cos^2\varphi}}\delta\varphi, \tag{5.4.23}$$

and compose the expression for the virtual work as follows:

$$\delta A = 2R\cos\varphi\frac{\sqrt{2-\cos^2\varphi}+\sin\varphi}{\sqrt{2-\cos^2\varphi}}F_0\sin\Omega t\delta\varphi. \tag{5.4.24}$$

It is easy to recognise the coefficient in front of $\delta\varphi$ as the generalised force

$$Q_\varphi = 2R\cos\varphi\frac{\sqrt{2-\cos^2\varphi}+\sin\varphi}{\sqrt{2-\cos^2\varphi}}F_0\sin\Omega t. \tag{5.4.25}$$

For small oscillations around the equilibrium position $\varphi = 0$, one should substitute this value in the previous expression, deriving

$$Q_\varphi = 2RF_0\sin\Omega t. \tag{5.4.26}$$

Lagrange's equation (5.2) with respect to this generalised coordinate φ has the form

$$\ddot{\varphi} + \omega_n{}^2\varphi = h_1\sin\Omega t, \tag{5.4.27}$$

where

$$\omega_n = \sqrt{\frac{4Rk + g(m(1+2/\pi)+M)}{2R(m+2M)}}, \quad h_1 = \frac{F_0}{R(m+2M)}. \tag{5.4.28}$$

Comparing Equations (5.4.15) and (5.4.16) with Equations (5.4.27) and (5.4.28), one can determine the relationship between x and φ for small oscillations:

$$x(t) \approx 2R\varphi(t). \tag{5.4.29}$$

Example 5.5 The curved bar AB of mass m shown in Figure 5.5.1 is pinned to the weightless slider which can move along a horizontal frictionless guide. The slider's position s_A is defined by $s_A(t) = s_0 \sin \Omega t$, where s_0 and Ω are constant. The bar and the slider are also linked to each another by a torsional spring of stiffness k. The spring is predeformed so that when the slider does not move, the system is at rest in the position shown, with the AB direction being vertical. Assuming that the angle φ is the generalised coordinate, determine the differential equation of motion for small oscillations. Plot the solution $\varphi(t)$ for the initial conditions $\varphi(0) = 0$ and $\dot{\varphi}(0) = 0$, assuming: $m = 4$ kg, $R = 0.6$ m, $k = 200$ Nm and $s_0 = 0.01$ m for two cases $\Omega = 2\omega_n$ and $\Omega = 0.5\omega_n$. In addition, plot the particular solution and the frequency-amplitude curve.

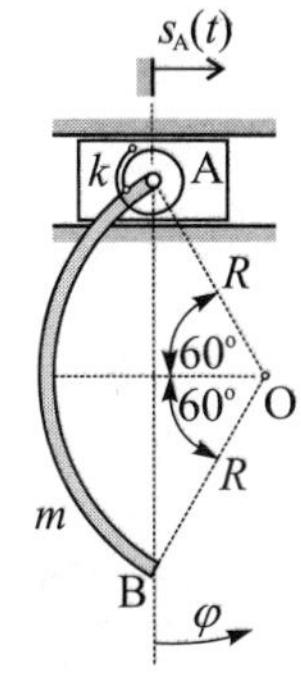

Figure 5.5.1

Solution

The potential energy (Figure 5.5.2a) is defined by

$$V = \frac{1}{2}k(\alpha_{st} + \varphi)^2 - mgy_C. \tag{5.5.1}$$

The zero sum of the moments about the point A for the bar is written based on the free-body diagram presented in Figure 5.5.2b:

$$\sum M_A = 0 \;\Rightarrow\; mg \cdot \overline{CD} - k\alpha_{st} = 0, \tag{5.5.2}$$

where

$$\overline{CD} = \overline{OC} - \overline{OD} = \frac{R \sin 60^\circ}{\pi/3} - R\cos 60^\circ = \frac{R}{2}\left(\frac{3\sqrt{3}}{\pi} - 1\right), \tag{5.5.3}$$

yielding

$$\alpha_{st} = \frac{mgR}{2k}\left(\frac{3\sqrt{3}}{\pi} - 1\right). \tag{5.5.4}$$

To determine the coordinate y_C in Equation (5.5.1), one can use Figure 5.5.2a and express it as follows:

$$y_C = \overline{AD}\cos\varphi + \overline{CD}\sin\varphi = \frac{R\sqrt{3}}{2}\cos\varphi + \frac{R}{2}\left(\frac{3\sqrt{3}}{\pi} - 1\right)\sin\varphi. \tag{5.5.5}$$

Developing the cosine and sine functions into series, one can obtain the following expression for the potential energy:

$$V \approx \frac{1}{2}\left(k + \frac{mgR\sqrt{3}}{2}\right)\varphi^2. \tag{5.5.6}$$

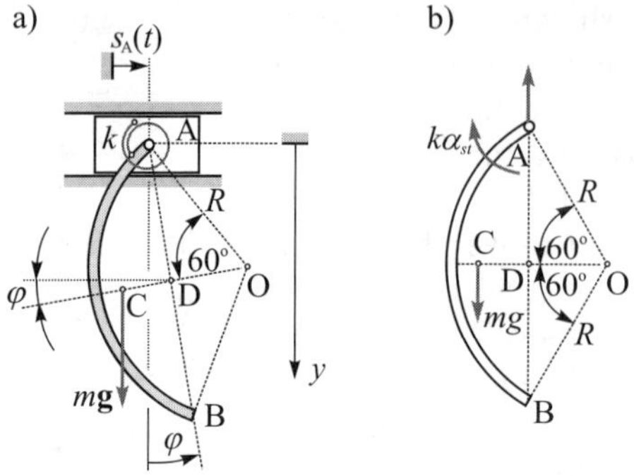

Figure 5.5.2

Since the bar executes general planar motion, its kinetic energy is defined by

$$T = \frac{1}{2} m v_C{}^2 + \frac{1}{2} J_C \omega^2, \tag{5.5.7}$$

where $\omega = \dot{\varphi}$.

Knowing $J_O = mR^2$, one can use the parallel-axis theorem to obtain J_C:

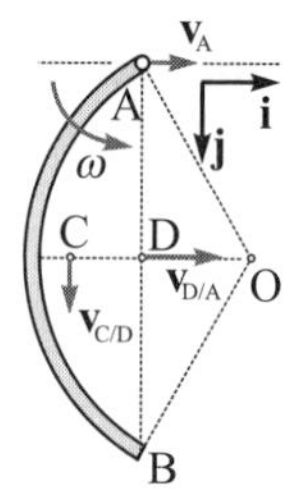

Figure 5.5.3

$$J_C = J_O - m\overline{OC}^2 = mR^2 - m\left(\frac{3\sqrt{3}}{2\pi}R\right)^2 \quad \Rightarrow \quad J_C = mR^2\left(1 - \frac{27}{4\pi^2}\right). \tag{5.5.8}$$

To determine the velocity v_C, one can analyse Figure 5.5.3, which shows the system passing through the equilibrium position:

$$\mathbf{v}_C = \mathbf{v}_A + \mathbf{v}_{D/A} + \mathbf{v}_{C/D}, \quad \mathbf{v}_A = \dot{s}_A \mathbf{i}, \quad \mathbf{v}_{D/A} = \overline{AD}\,\dot{\varphi}\mathbf{i}, \quad \mathbf{v}_{C/D} = \overline{CD}\,\dot{\varphi}\mathbf{j}. \tag{5.5.9}$$

Thus, one can write

$$v_C{}^2 = \mathbf{v}_C \cdot \mathbf{v}_C = \left[\left(\dot{s}_A + \overline{AD}\,\dot{\varphi}\right)\mathbf{i} + \overline{CD}\,\dot{\varphi}\mathbf{j}\right] \cdot \left[\left(\dot{s}_A + \overline{AD}\,\dot{\varphi}\right)\mathbf{i} + \overline{CD}\,\dot{\varphi}\mathbf{j}\right] \quad \Rightarrow$$

$$v_C{}^2 = \left(\dot{s}_A + \frac{R\sqrt{3}}{2}\dot{\varphi}\right)^2 + \left(\frac{R}{2}\left(\frac{3\sqrt{3}}{\pi} - 1\right)\right)^2 \dot{\varphi}^2. \tag{5.5.10}$$

This gives the kinetic energy

$$T = \frac{m}{2}\left[\left(\dot{s}_A + \frac{R\sqrt{3}}{2}\dot{\varphi}\right)^2 + \frac{R^2}{4}\left(\frac{3\sqrt{3}}{\pi} - 1\right)^2 \dot{\varphi}^2\right] + \frac{1}{2} mR^2\left(1 - \frac{27}{4\pi^2}\right)\dot{\varphi}^2. \tag{5.5.11}$$

For Lagrange's equation (5.2), one needs to find the partial derivative with respect to $\dot{\varphi}$. This gives

$$\frac{\partial T}{\partial \dot{\varphi}} = m\left[\frac{R\sqrt{3}}{2}\left(\dot{s}_A + \frac{R\sqrt{3}}{2}\dot{\varphi}\right) + \frac{R^2}{4}\left(\frac{3\sqrt{3}}{\pi} - 1\right)^2 \dot{\varphi}\right] + mR^2\left(1 - \frac{27}{4\pi^2}\right)\dot{\varphi}. \tag{5.5.12}$$

However, when finding the total time derivative in Lagrange's equation (5.2), one needs to take into account that both of them change in time, which yields

$$\frac{d}{dt}\frac{\partial T}{\partial \dot{\varphi}} = -m\frac{R\sqrt{3}}{2}s_0\Omega^2\sin\Omega t + mR^2\left(2-\frac{3\sqrt{3}}{2\pi}\right)\ddot{\varphi}. \tag{5.5.13}$$

The following differential equation of motion is derived:

$$\ddot{\varphi} + \omega_n{}^2\varphi = h\Omega^2\sin\Omega t, \tag{5.5.14}$$

where

$$\omega_n = \frac{1}{R}\sqrt{\frac{2\pi k + \sqrt{3}\pi mgR}{m\left(4\pi - 3\sqrt{3}\right)}},\quad h = \frac{\sqrt{3}\pi s_0}{R\left(4\pi - 3\sqrt{3}\right)}. \tag{5.5.15}$$

When $\Omega \neq \omega_n$, the solution of Equation (5.5.14) satisfying $\varphi(0) = 0$ and $\dot{\varphi}(0) = 0$ is

$$\varphi(t) = \frac{h\Omega^3}{\omega_n\left(\Omega^2 - \omega_n{}^2\right)}\sin\omega_n t - \frac{h\Omega^2}{\Omega^2 - \omega_n{}^2}\sin\Omega t. \tag{5.5.16}$$

The time–history diagrams corresponding to $\Omega = 0.5\omega_n$ and $\Omega = 2\omega_n$ are presented in Figure 5.5.4.

The particular integral is

$$\varphi_p(t) = -\frac{h\Omega^2}{\Omega^2 - \omega_n{}^2}\sin\Omega t, \tag{5.5.17}$$

while its coefficient defines the frequency–amplitude equation

$$|A(\Omega)| = \left|\frac{h\Omega^2}{\Omega^2 - \omega_n{}^2}\right|. \tag{5.5.18}$$

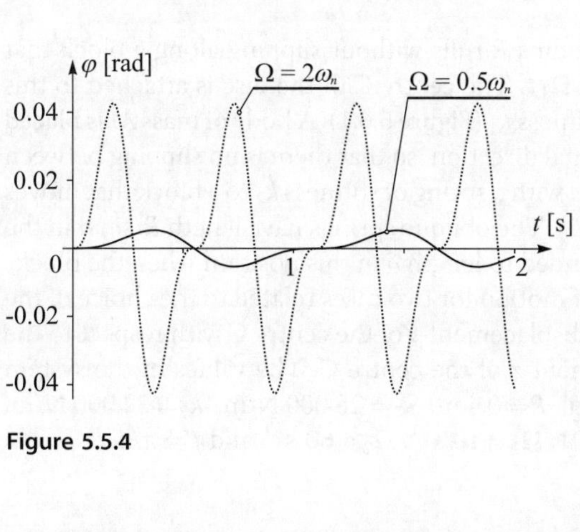

Figure 5.5.4

The time–history diagrams of the particular solution corresponding to $\Omega = 0.5\omega_n$ and $\Omega = 2\omega_n$ are shown in Figure 5.5.5.

The frequency–amplitude diagram is presented in Figure 5.5.6. The amplitudes corresponding to $\Omega = 0.5\omega_n$ and $\Omega = 2\omega_n$ are also depicted.

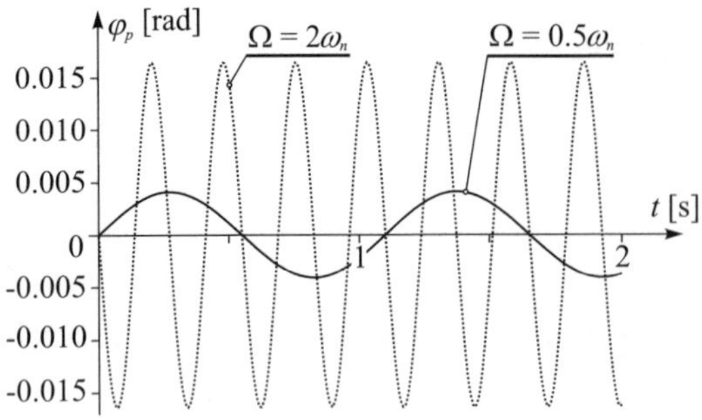

Figure 5.5.5

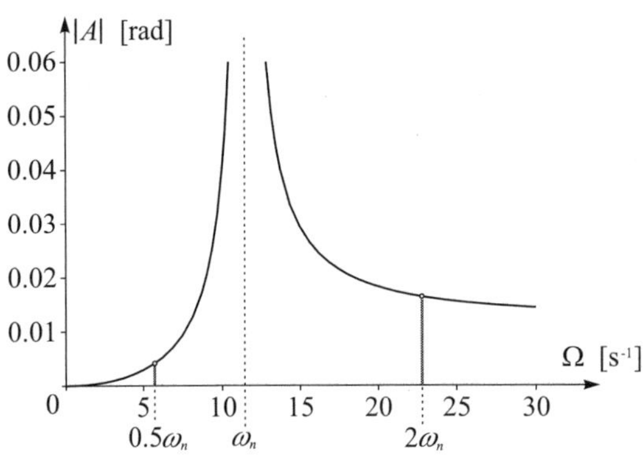

Figure 5.5.6

Example 5.6 A disc of mass m and radius R rolls without slipping along a block that moves in accordance with $s_1(t) = s_{10} \sin \Omega_1 t$. The centre C of the disc is attached to this block with two springs AC and CB of stiffness k_1 (Figure 5.6.1). A body of mass M is placed on the disc and can move in the horizontal direction, so that there is no slipping between these two elements. The body is attached with a spring of stiffness k_2 to a block that moves in accordance with $s_2(t) = s_{20} \sin(\Omega_2 t + \beta)$. The oblique springs have length $R/\sin \alpha$ in the static equilibrium position and are extended to length δ in this position when the blocks do not move. Determine the equation of motion for two cases related to the choice of the generalised coordinate: (1) the relative displacement y of the centre C with respect to the lower block; (2) the absolute displacement z of the centre C. The values of the system parameters are: $m = 40$ kg, $M = 15$ kg, $R = 0.4$ m, $k_1 = 28\,000$ N/m, $k_2 = 22\,000$ N/m, $\alpha = 30°$, $s_{10} = 3$ mm, $s_{20} = 4$ mm, $\alpha = 0.01$, $\Omega_1 = 10$ s^{-1}, $\Omega_2 = 60$ s^{-1} and $\beta = \pi/4$.

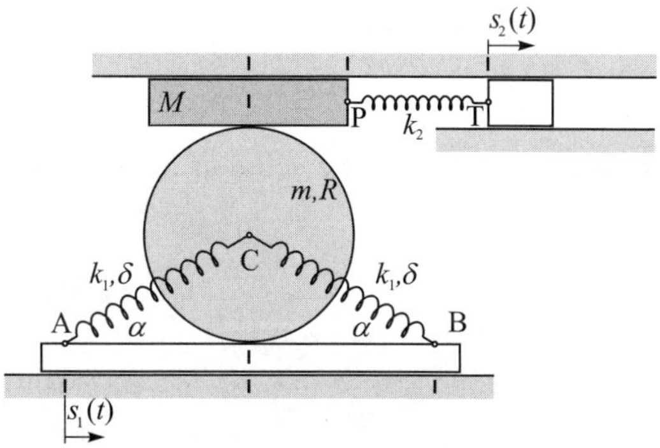

Figure 5.6.1

Solution

The kinetic energy will be determined first for both choices of the generalised coordinate.

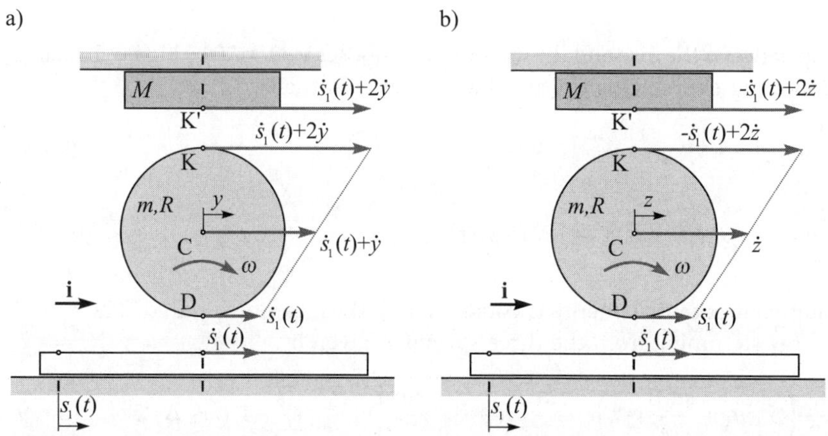

Figure 5.6.2

The overall kinetic energy stemming from the disc and the massive body takes the form

$$T = \frac{1}{2} m v_{\mathrm{C}}^2 + \frac{1}{2} J_{\mathrm{C}} \omega^2 + \frac{1}{2} M v_{\mathrm{K'}}^2, \tag{5.6.1}$$

where $J_{\mathrm{C}} = mR^2/2$, while the velocities of the centre C in both cases are

$$\mathbf{v}_{\mathrm{C}} = (\dot{s}_1(t) + \dot{y})\,\mathbf{i}, \tag{5.6.2i}$$

$$\mathbf{v}_{\mathrm{C}} = \dot{z}\,\mathbf{i}. \tag{5.6.2ii}$$

There is no slipping between the lower block and the disc, which imposes

$$\mathbf{v}_{\mathrm{D}} = \dot{s}_1(t)\,\mathbf{i}. \tag{5.6.3}$$

The relationship between the velocities of the points C and D reads as

$$\mathbf{v}_C = \mathbf{v}_D + \mathbf{v}_{C/D}, \quad \mathbf{v}_{C/D} = R\omega\mathbf{i}. \tag{5.6.4}$$

Projecting this relationship on the horizontal axes gives the solution in two cases considered:

$$\omega = \frac{\dot{y}}{R}, \tag{5.6.5i}$$

$$\omega = \frac{\dot{z} - \dot{s}_1(t)}{R}. \tag{5.6.5ii}$$

In addition, the velocities of the points K and C satisfy

$$\mathbf{v}_K = \mathbf{v}_C + \mathbf{v}_{K/C}, \quad \mathbf{v}_{K/C} = R\omega\mathbf{i}, \tag{5.6.6}$$

which again can be projected on the horizontal axes to derive two solutions

$$\mathbf{v}_K = (\dot{s}_1(t) + 2\dot{y})\mathbf{i}, \tag{5.6.7i}$$

$$\mathbf{v}_K = (-\dot{s}_1(t) + 2\dot{z})\mathbf{i}. \tag{5.6.7ii}$$

No slipping between the disc and the upper body imposes $\mathbf{v}_{K'} = \mathbf{v}_K$ (see Figure 5.6.2a, b), giving the following expressions for the kinetic energy in the two cases:

$$T = \frac{1}{2}m(\dot{s}_1(t) + \dot{y})^2 + \frac{1}{4}m\dot{y}^2 + \frac{1}{2}M(\dot{s}_1(t) + 2\dot{y})^2, \tag{5.6.8i}$$

$$T = \frac{1}{2}m\dot{z}^2 + \frac{1}{4}m(\dot{z} - \dot{s}_1(t))^2 + \frac{1}{2}M(-\dot{s}_1(t) + 2\dot{z})^2. \tag{5.6.8ii}$$

The gravitational potential energy is constant and, thus, can be omitted. The overall potential energy stemming from the three springs is given by

$$V = \frac{k_1}{2}\left(\delta + u_1 + \frac{\sin\alpha}{2R}v_1^2\right)^2 + \frac{k_1}{2}\left(\delta + u_2 + \frac{\sin\alpha}{2R}v_2^2\right)^2 + \frac{k_2}{2}(s_2(t) - \Delta)^2, \tag{5.6.9}$$

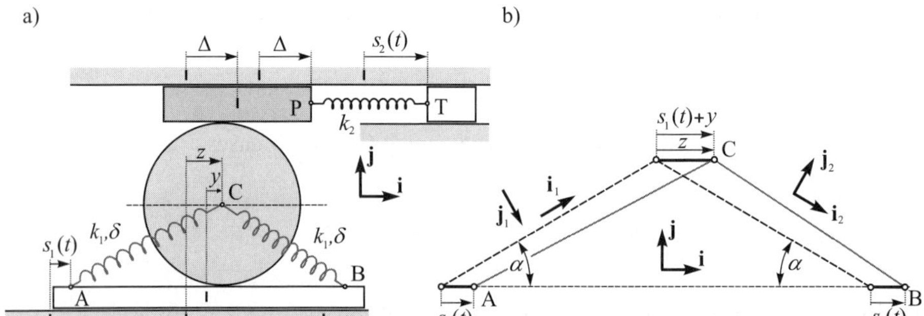

Figure 5.6.3

where Δ stands for the displacement on the left end of the horizontal spring (see Figure 5.6.3a).

To define the deflections u_1, u_2, v_1 and v_2 in Equation (5.6.9), one can use Figure 5.6.3b and write down for the first case:

$$u_1 = ((s_1(t) + y)\mathbf{i} - s_1(t)\mathbf{i})\cdot\mathbf{i}_1 = y\cos\alpha, \quad v_1 = ((s_1(t) + y)\mathbf{i} - s_1(t)\mathbf{i})\cdot\mathbf{j}_1 = y\sin\alpha,$$
$$u_2 = (s_1(t)\mathbf{i} - (s_1(t) + y)\mathbf{i})\cdot\mathbf{i}_2 = -y\cos\alpha, \quad v_2 = ((s_1(t) + y)\mathbf{i} - s_1(t)\mathbf{i})\cdot\mathbf{j}_2 = y\sin\alpha, \tag{5.6.10i}$$

and for the second case:

$$u_1 = (z\mathbf{i} - s_1(t)\mathbf{i})\cdot\mathbf{i}_1 = (z - s_1(t))\cos\alpha, \quad v_1 = (z\mathbf{i} - s_1(t)\mathbf{i})\cdot\mathbf{j}_1 = (z - s_1(t))\sin\alpha,$$
$$u_2 = (s_1(t)\mathbf{i} - z\mathbf{i})\cdot\mathbf{i}_2 = -(z - s_1(t))\cos\alpha, \quad v_2 = (z\mathbf{i} - s_1(t)\mathbf{i})\cdot\mathbf{j}_2 = (z - s_1(t))\sin\alpha. \tag{5.6.10ii}$$

Using $\Delta = s_1(t) + 2y$, the potential energy for the first case becomes

$$V = \frac{k_1}{2}\left(2\cos^2\alpha + \frac{2\delta\sin^3\alpha}{R}\right)y^2 + \frac{k_2}{2}(s_2(t) - s_1(t) - 2y)^2. \tag{5.6.11i}$$

Using $\Delta = -s_1(t) + 2z$, the potential energy for the second case is

$$V = \frac{k_1}{2}\left(2\cos^2\alpha + \frac{2\delta\sin^3\alpha}{R}\right)(z - s_1(t))^2 + \frac{k_2}{2}(s_2(t) + s_1(t) - 2z)^2. \tag{5.6.11ii}$$

The differential equations for small oscillations in terms of two different generalised coordinates are derived to be

$$\ddot{y} + \omega_n^2 y = h_1 \sin\Omega_1 t + h_2 \sin(\Omega_2 t + \beta), \tag{5.6.12i}$$

$$\ddot{z} + \omega_n^2 z = H_1 \sin\Omega_1 t + h_2 \sin(\Omega_2 t + \beta), \tag{5.6.12ii}$$

where

$$\omega_n = \sqrt{\frac{ak_1 + 4k_2}{(3/2)m + 4M}}, \quad a = 2\cos^2\alpha + \frac{2\delta\sin^3\alpha}{R},$$
$$h_1 = \frac{-2k_2 + (m + 2M)\Omega_1{}^2}{(3/2)m + 4M}s_{10}, \quad h_2 = \frac{2k_2 s_{20}}{(3/2)m + 4M}, \tag{5.6.13}$$
$$H_1 = \frac{ak_1 + 2k_2 - ((m/2) + 2M)\Omega_1{}^2}{(3/2)m + 4M}s_{10}.$$

Solving the equation of motion (5.6.12i) for the initial conditions

$$y(0) = 0, \quad \dot{y}(0) = 0, \tag{5.6.14i}$$

one can obtain

$$y(t) = -\frac{h_2}{\omega_n^2 - \Omega_2^2}\sin\beta\cos\omega_n t - \left(\frac{h_1\Omega_1}{(\omega_n^2 - \Omega_1^2)\omega_n} + \frac{h_2\Omega_2}{(\omega_n^2 - \Omega_2^2)\omega_n}\cos\beta\right)\sin\omega_n t$$
$$+ \frac{h_1}{\omega_n^2 - \Omega_1^2}\sin\Omega_1 t + \frac{h_2}{\omega_n^2 - \Omega_2^2}\sin(\Omega_2 t + \beta). \tag{5.6.15i}$$

Solving the differential equation of motion (5.6.12ii) for the initial conditions

$$z(0) = 0, \quad \dot{z}(0) = \Omega_1 s_{10}, \tag{5.6.14ii}$$

the solution for $z(t)$ is found to be

$$z(t) = -\frac{h_2}{\omega_n^2 - \Omega_2^2}\sin\beta\cos\omega_n t - \left(\frac{h_1\Omega_1}{(\omega_n^2 - \Omega_1^2)\omega_n} + \frac{h_2\Omega_2}{(\omega_n^2 - \Omega_2^2)\omega_n}\cos\beta\right)\sin\omega_n t$$
$$+ \frac{H_1}{\omega_n^2 - \Omega_1^2}\sin\Omega_1 t + \frac{h_2}{\omega_n^2 - \Omega_2^2}\sin(\Omega_2 t + \beta). \tag{5.6.15ii}$$

Given the relationship between the coordinates, $z = y + s_1(t)$, one can subtract Equation (5.6.15i) from Equation (5.6.15ii) to derive:

$$z(t) - y(t) = \frac{s_{10}\left(\omega_n^2 - \Omega_1^2\right)}{\omega_n^2 - \Omega_1^2}\sin\Omega_1 t = s_{10}\sin\Omega_1 t, \tag{5.6.16}$$

which correctly gives $s_1(t)$.

The solutions for motion (5.6.15i) and (5.6.15ii) and their difference are presented in Figure 5.6.4.

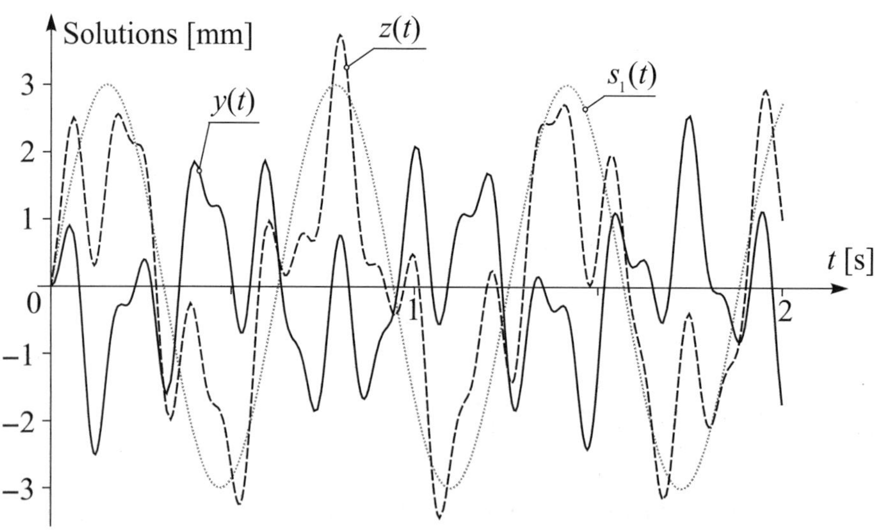

Figure 5.6.4

Example 5.7 A right-angle link consists of two rigidly connected bars AB and BE of mass m and length l. Their end points A and E move, respectively, along vertical and horizontal frictionless guides. A spring of stiffness k and the length l_{st} in the static equilibrium position shown is attached to the point A (Figure 5.7.1). A linear viscous damper with damping coefficient c is attached to the point E. The external force $\mathbf{F}(t) = F_0 \cos \Omega t \, \mathbf{i}$ acts at the point E. Assuming $\alpha = 30^{\circ}$ and $\beta = 30^{\circ}$, determine the differential equation of motion for small oscillations.

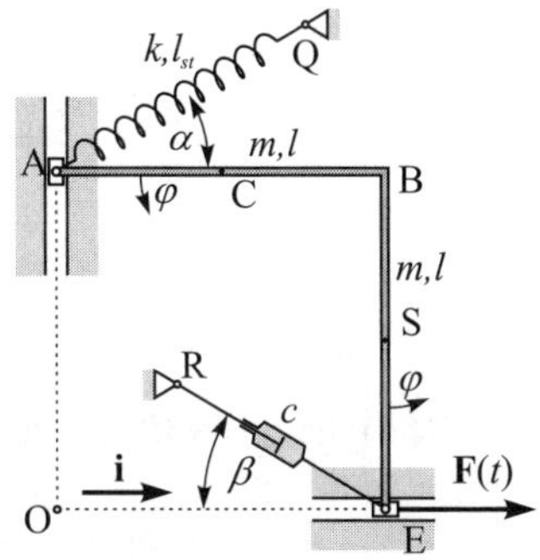

Figure 5.7.1

Solution

The static deflection of the spring is obtained from the equilibrium equation written based on the free-body diagram presented in Figure 5.7.2:

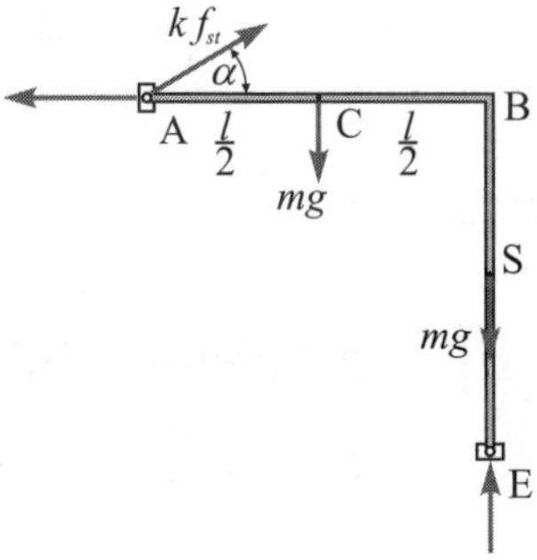

Figure 5.7.2

$$\sum M_B = 0 \;\Rightarrow\; mg\frac{l}{2} - kf_{st}\sin\alpha \cdot l = 0, \tag{5.7.1}$$

which, duc to $\alpha - 30^{\circ}$, gives

$$f_{st} = \frac{mg}{k}. \tag{5.7.2}$$

The overall potential energy is defined by (Figure 5.7.3a)

$$V = \frac{k}{2}(f_{st} + \lambda)^2 + mgy_C + mgy_S. \tag{5.7.3}$$

Let us find now the exact and approximate expressions for the displacement of the points A and E, which are labelled by z and w, respectively (Figure 5.7.3a):

$$z = l - (l\cos\varphi - l\sin\varphi) \;\Rightarrow\; z \approx l\varphi + \frac{1}{2}l\varphi^2, \tag{5.7.4}$$

$$w = (l\cos\varphi + l\sin\varphi) - l \;\Rightarrow\; w \approx l\varphi - \frac{1}{2}l\varphi^2. \tag{5.7.5}$$

The coordinates existing in the potential energy (5.7.3) can be expressed as follows:

$$y_C = l\cos\varphi - \frac{l}{2}\sin\varphi \approx l - \frac{l}{2}\varphi - \frac{l}{2}\varphi^2, \quad y_S = \frac{l}{2}\cos\varphi \approx \frac{l}{2} - \frac{l}{4}\varphi^2. \tag{5.7.6}$$

The additional deflection of the spring is (Figure 5.7.3b):

$$\lambda \approx u + \frac{1}{2l_{st}}v^2, \quad u = z\sin\alpha, \quad v = z\cos\alpha \;\Rightarrow\; \lambda \approx \frac{1}{2}\left(l\varphi + \frac{1}{2}l\varphi^2\right) + \frac{1}{2l_{st}}\left(l\varphi\frac{\sqrt{3}}{2}\right)^2, \tag{5.7.7}$$

which yields

$$\lambda \approx \frac{l}{2}\varphi + \frac{1}{4}\left(\frac{3}{2}\frac{l}{l_{st}} + 1\right) l\varphi^2. \tag{5.7.8}$$

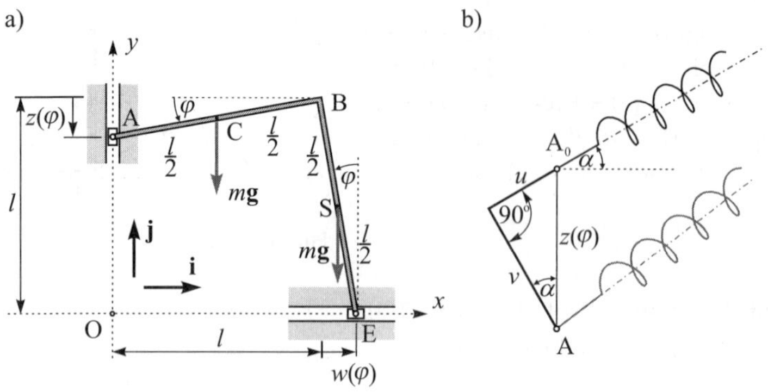

Figure 5.7.3

Thus, the potential energy becomes

$$V \approx \frac{1}{2}\left[mg\left(\frac{3}{4}\frac{l}{l_{st}} - 1\right) + \frac{kl}{4}\right] l\varphi^2. \tag{5.7.9}$$

The kinetic energy is defined by

$$T = \frac{1}{2}mv_C{}^2 + \frac{1}{2}J_C\dot{\varphi}^2 + \frac{1}{2}mv_S{}^2 + \frac{1}{2}J_S\dot{\varphi}^2, \tag{5.7.10}$$

where $J_C = J_S = ml^2/12$. To find the velocities existing in the kinetic energy, one can use Figure 5.7.4, deducing:

$$v_C = v_S = \frac{l}{2}\dot{\varphi}, \tag{5.7.11}$$

where $\dot{\varphi}$ stands for the angular velocity of the link.

Now, the kinetic energy is found to be

$$T = \frac{1}{3}ml^2\dot{\varphi}^2. \tag{5.7.12}$$

The dissipation function is given by

$$D = \frac{c}{2}v_{p/c}{}^2. \tag{5.7.13}$$

To find the relative velocity of the piston with respect to the cylinder $v_{p/c}$, one can first determine the velocity of the point E (Figure 5.7.4):

$$\mathbf{v}_E = l\dot{\varphi}\mathbf{i}, \tag{5.7.14}$$

and then recognise that

$$v_{p/c} = \mathbf{v}_E \cdot \mathbf{e} = l\dot{\varphi}\cos 30^o \quad \Rightarrow \quad v_{p/c} = \left(\sqrt{3}/2\right) l\dot{\varphi}. \tag{5.7.15}$$

The dissipation function takes the form

$$D \approx \frac{3}{8} c l^2 \dot{\varphi}^2. \tag{5.7.16}$$

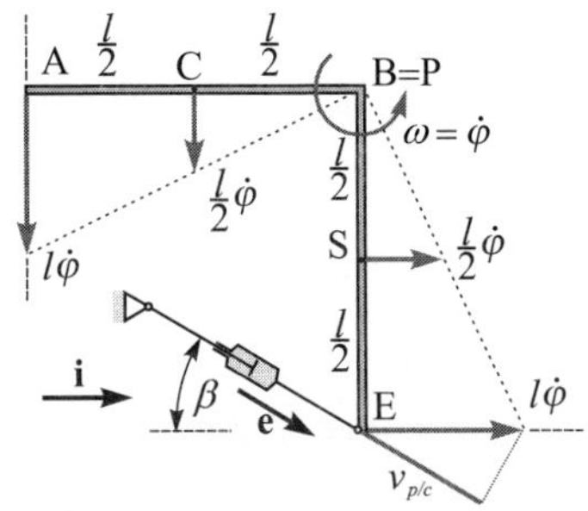

Figure 5.7.4

The virtual work is defined by

$$\delta A = \delta A^{\mathbf{F}} = F_0 \cos\Omega t \delta x_E = F_0 \cos\Omega t \ \delta w. \tag{5.7.17}$$

To express this in terms of the generalised coordinate, one can use Equation (5.7.5) and derive

$$\delta w = l(-\sin\varphi + \cos\varphi)\delta\varphi \approx l(-\sin 0 + \cos 0)\delta\varphi = l\delta\varphi. \tag{5.7.18}$$

Thus, the generalised force is

$$Q_\varphi = F_0 l \cos\Omega t. \tag{5.7.19}$$

Lagrange's equation gives

$$\ddot{\varphi} + 2\delta\dot{\varphi} + \omega_n{}^2\varphi = h\cos\Omega t, \tag{5.7.20}$$

where

$$\delta = \frac{9c}{16m}, \quad \omega_n = \sqrt{\frac{3kl - (12 - 9l/l_{st})mg}{8ml}}, \quad h = \frac{3F_0}{2ml}. \tag{5.7.21}$$

6

Free Undamped Vibration of Two-Degree-of-Freedom Systems

Chapter Outline

This chapter is concerned with the use of Lagrange's equations to obtain the equation of motion of free undamped two-degree-of-freedom oscillatory systems that perform small vibrations. Seven examples are presented, illustrating how the differential equations of motion can be derived. In addition, it is shown how the natural frequencies can be found from the characteristic equation. The mode shape ratios are also defined and calculated, enabling one to present graphically the corresponding mode shapes. Solutions for motion are plotted. One example is focused on principal coordinates and their importance and meaning for motion presentation and characteristics.

Chapter Objectives

- To present Lagrange's equations of the second kind for free undamped two-degree-of-freedom systems
- To illustrate a variety of examples in which they yield the differential equations of motion
- To demonstrate how the characteristic determinant and the characteristic equation can be obtained
- To illustrate how the natural frequencies and mode shape ratios can be determined
- To present graphically the corresponding mode shapes
- To demonstrate the importance and meaning of principal coordinates

Theoretical Introduction

Lagrange's equations for two-degree-of-freedom free conservative systems have the form

$$\frac{d}{dt}\frac{\partial T}{\partial \dot{q}_i} - \frac{\partial T}{\partial q_i} + \frac{\partial V}{\partial q_i} = 0, \quad i = 1, 2. \tag{6.1}$$

If the system performs small oscillations around the equilibrium position $q_1 = 0$ and $q_2 = 0$, these Lagrange's equations can be simplified to

$$\frac{d}{dt}\frac{\partial T}{\partial \dot{q}_i} + \frac{\partial V}{\partial q_i} = 0, \quad i = 1, 2. \tag{6.2}$$

Mechanical Vibration: Fundamentals with Solved Examples, First Edition.
Ivana Kovačić and Dragi Radomirović.

In this case, the kinetic energy takes the form

$$T = \frac{1}{2}\left(\widetilde{T}_{11}\dot{q}_1^2 + 2\widetilde{T}_{12}\dot{q}_1\dot{q}_2 + \widetilde{T}_{22}\dot{q}_2^2\right), \tag{6.3}$$

where:

$$\widetilde{T}_{ij}(q_1, q_2) \approx \widetilde{T}_{ij}(0,0), \tag{6.4}$$

while the potential energy has the form:

$$V(q_1, q_2) \approx \frac{1}{2}\left[\left(\frac{\partial^2 V}{\partial q_1^2}\right)_{(0,0)} q_1^2 + 2\left(\frac{\partial^2 V}{\partial q_1 \partial q_2}\right)_{(0,0)} q_1 q_2 + \left(\frac{\partial^2 V}{\partial q_2^2}\right)_{(0,0)} q_2^2\right], \tag{6.5}$$

or, it can be represented as:

$$V(q_1, q_2) \approx \frac{1}{2}\left[\widetilde{V}_{11} q_1^2 + 2\widetilde{V}_{12} q_1 q_2 + \widetilde{V}_{22} q_2^2\right]. \tag{6.6}$$

Substituting (6.3) and (6.6) into (6.2), one can derive the following system of coupled differential equations:

$$\begin{aligned} \widetilde{T}_{11}\ddot{q}_1 + \widetilde{T}_{12}\ddot{q}_2 + \widetilde{V}_{11}q_1 + \widetilde{V}_{12}q_2 &= 0, \\ \widetilde{T}_{12}\ddot{q}_1 + \widetilde{T}_{22}\ddot{q}_2 + \widetilde{V}_{12}q_1 + \widetilde{V}_{22}q_2 &= 0. \end{aligned} \tag{6.7}$$

Its particular solutions can be sought in the form

$$\begin{aligned} q_1 &= A_1 \cos(\omega t - \alpha), \\ q_2 &= A_2 \cos(\omega t - \alpha). \end{aligned} \tag{6.8}$$

By substituting (6.8) into (6.7), one derives

$$\begin{aligned} A_1\left(\widetilde{V}_{11} - \omega^2\widetilde{T}_{11}\right) + A_2\left(\widetilde{V}_{12} - \omega^2\widetilde{T}_{12}\right) &= 0, \\ A_1\left(\widetilde{V}_{12} - \omega^2\widetilde{T}_{12}\right) + A_2\left(\widetilde{V}_{22} - \omega^2\widetilde{T}_{22}\right) &= 0. \end{aligned} \tag{6.9}$$

This system has a non-trivial solution provided that its determinant (the so-called *characteristic determinant*) is equal to zero:

$$\Delta\left(\omega^2\right) = \begin{vmatrix} \widetilde{V}_{11} - \omega^2\widetilde{T}_{11} & \widetilde{V}_{12} - \omega^2\widetilde{T}_{12} \\ \widetilde{V}_{12} - \omega^2\widetilde{T}_{12} & \widetilde{V}_{22} - \omega^2\widetilde{T}_{22} \end{vmatrix} = 0, \tag{6.10}$$

which yields the so-called *characteristic equation* or *frequency equation*:

$$\left(\widetilde{T}_{11}\widetilde{T}_{22} - \widetilde{T}_{12}^2\right)\omega^4 - \left(\widetilde{T}_{11}\widetilde{V}_{22} + \widetilde{T}_{22}\widetilde{V}_{11} - 2\widetilde{T}_{12}\widetilde{V}_{12}\right)\omega^2 + \left(\widetilde{V}_{11}\widetilde{V}_{22} - \widetilde{V}_{12}^2\right) = 0. \tag{6.11}$$

In general, two positive real roots of this equation ω_1^2 and ω_2^2 give the squares of two natural frequencies. Note that the natural frequencies are ordered so that the first one is lower than the second one.

In addition, what characterises this type of motion is the fact that the system (6.9) gives the ratio of amplitudes

$$b = \frac{A_2}{A_1} = -\frac{\widetilde{V}_{11} - \omega^2 \widetilde{T}_{11}}{\widetilde{V}_{12} - \omega^2 \widetilde{T}_{12}} = -\frac{\widetilde{V}_{12} - \omega^2 \widetilde{T}_{12}}{\widetilde{V}_{22} - \omega^2 \widetilde{T}_{22}}. \tag{6.12}$$

Two associated *mode shape ratios* can be obtained by calculating the values of b for each frequency, that is, as:

$$b_1 = b\left(\omega_1^2\right) = \frac{A_{2,1}}{A_{1,1}}, \quad b_2 = b\left(\omega_2^2\right) = \frac{A_{2,2}}{A_{1,2}}. \tag{6.13}$$

The general solution for motion is then

$$\begin{aligned} q_1 &= A_{1,1}\cos(\omega_1 t - \alpha_1) + A_{1,2}\cos(\omega_2 t - \alpha_2), \\ q_2 &= b_1 A_{1,1}\cos(\omega_1 t - \alpha_1) + b_2 A_{1,2}\cos(\omega_2 t - \alpha_2), \end{aligned} \tag{6.14}$$

where the constants $A_{1,1}$, $A_{1,2}$, α_1, α_2 can be calculated from the initial conditions $q_1(0)$, $q_2(0)$, $\dot{q}_1(0)$, $\dot{q}_2(0)$. The character of the resulting response is related to the values of the two natural frequencies. Two similar natural frequencies give beat oscillations whose frequency is the average of the two natural frequencies. The condition for true periodicity in the combined motion is that the periods of the component motions are commensurable. This means that there exist two integers n_1 and n_2 such that

$$\mathrm{T} = n_1\ \mathrm{T}_1 = n_2\ \mathrm{T}_2. \tag{6.15}$$

The period of the combined motion is then the value of T as obtained above, using the smallest integral values of n_1 and n_2.

The following examples illustrate how Lagrange's equation of motion can be used in the described way for two-degree-of-freedom free conservative systems.

Example 6.1 A double pendulum consists of two slender bars OA and AB shown in Figure 6.1.1. The bar OA of mass m and length l is hinged at O. The bar AB has mass $M = \xi m$ and length $L = \xi l$, where ξ is an unknown constant. Determine the constant ξ so that the system has one mode shape ratio being equal to -1 while it oscillates at the higher natural frequency. Determine both natural frequencies and show the mode shapes graphically, assuming $l = 1$ m, $g = 9.81$ m/s^2. Show graphically the solutions for motion satisfying $\varphi(0) = 0.1$, $\psi(0) = 0$, $\dot{\varphi}(0) = 0$, $\dot{\psi}(0) = 0$.

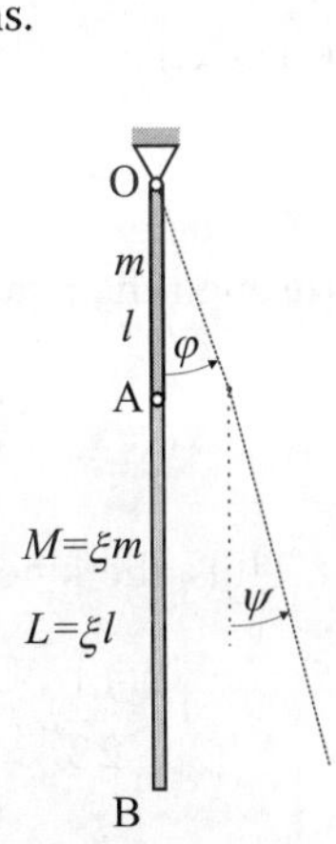

Figure 6.1.1

Solution

The exact expression for the system potential energy has the form (Figure 6.1.2a)

$$V = -mgy_S - \xi mgy_C = -mg\frac{l}{2}\cos\varphi - \xi mg\left(l\cos\varphi + \frac{\xi l}{2}\cos\psi\right). \tag{6.1.1}$$

Developing the cosine into a series:

$$\cos q \approx 1 - q^2/2, \tag{6.1.2}$$

the approximate expression for the potential energy is obtained as

$$V \approx mgl\frac{\varphi^2}{4} + \xi mgl\frac{\varphi^2}{2} + \xi^2 mgl\frac{\psi^2}{4}. \tag{6.1.3}$$

The bar OA executes rotational motion and the bar AB is in general plane motion, so that the kinetic energy is defined by

$$T = \frac{1}{2} J_O \dot{\varphi}^2 + \frac{1}{2} \xi m v_C{}^2 + \frac{1}{2} J_C \dot{\psi}^2, \tag{6.1.4}$$

where the mass moments of inertia are given by

$$J_O = \frac{1}{12} ml^2 + m\left(\frac{l}{2}\right)^2 = \frac{1}{3} ml^2, \; J_C = \frac{1}{12} \xi m (\xi l)^2. \tag{6.1.5}$$

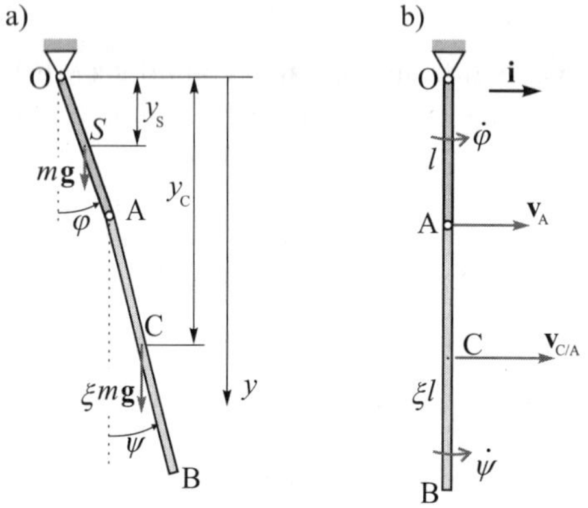

Figure 6.1.2

To determine the velocity of the point C, one should consider the system passing through the equilibrium position (Figure 6.1.2b) to determine

$$\mathbf{v}_C = \mathbf{v}_A + \mathbf{v}_{C/A} = l\dot{\varphi}\mathbf{i} + \frac{\xi l}{2}\dot{\psi}\mathbf{i} \;\Rightarrow\; v_C{}^2 = \mathbf{v}_C \cdot \mathbf{v}_C = \left(l\dot{\varphi} + \frac{\xi l}{2}\dot{\psi}\right)^2. \tag{6.1.6}$$

Thus, the kinetic energy becomes

$$T \approx \frac{1}{2}\frac{1}{3} ml^2 \dot{\varphi}^2 + \frac{1}{2} \xi m l^2 \left(\dot{\varphi} + \frac{\xi}{2}\dot{\psi}\right)^2 + \frac{1}{2}\frac{1}{12} \xi^3 m l^2 \dot{\psi}^2. \tag{6.1.7}$$

Lagrange's equations (6.2) give

$$\frac{1}{3} l\ddot{\varphi} + \xi l \left(\ddot{\varphi} + \frac{\xi}{2}\ddot{\psi}\right) + \frac{1}{2} g\varphi + \xi g \varphi = 0, \tag{6.1.8}$$

$$\frac{1}{2}\xi^2 l \left(\ddot{\varphi} + \frac{\xi}{2}\ddot{\psi}\right) + \frac{1}{12}\xi^3 l \ddot{\psi} + \frac{1}{2}\xi^2 g \psi = 0. \tag{6.1.9}$$

The solution of this system of equations can be represented as

$$\varphi(t) = \varphi_0 \sin(\omega t + \beta), \;\; \psi(t) = \psi_0 \sin(\omega t + \beta), \tag{6.1.10}$$

where ω and β are constants, while φ_0 and ψ_0 are the amplitudes of φ and ψ. Finding their second time derivative

$$\ddot{\varphi}(t) = -\varphi_0\omega^2 \sin(\omega t + \beta), \quad \ddot{\psi}(t) = -\psi_0\omega^2 \sin(\omega t + \beta), \tag{6.1.11}$$

and substituting it together with (6.1.10) into (6.1.8) and (6.1.9), one can derive:

$$\left(\frac{1}{2}g + \xi g - \frac{1}{3}l\omega^2 - \xi l\omega^2\right)\varphi_0 + \left(-\frac{1}{2}\xi^2 l\omega^2\right)\psi_0 = 0, \tag{6.1.12}$$

$$\left(-\frac{1}{2}\xi^2 l\omega^2\right)\varphi_0 + \left(\frac{1}{2}\xi^2 g - \frac{1}{3}\xi^3 l\omega^2\right)\psi_0 = 0. \tag{6.1.13}$$

By forming the corresponding characteristic determinant and equating it with zero, the following characteristic equations can be derived

$$\left(\frac{1}{9}\xi^3 + \frac{1}{12}\xi^4\right)l^2\omega^4 + \left(-\frac{1}{6}\xi^2 - \frac{2}{3}\xi^3 - \frac{1}{3}\xi^4\right)gl\omega^2 + \frac{1}{4}\xi^2 g^2 + \frac{1}{2}\xi^3 g^2 = 0. \tag{6.1.14}$$

Its roots are found to be

$$\omega_1^2 = \frac{1}{4\xi + 3\xi^2}\left(3 + 12\xi + 6\xi^2 - 3\sqrt{1 + 4\xi + 9\xi^2 + 10\xi^3 + 4\xi^4}\right)\frac{g}{l}, \tag{6.1.15}$$

$$\omega_2^2 = \frac{1}{4\xi + 3\xi^2}\left(3 + 12\xi + 6\xi^2 + 3\sqrt{1 + 4\xi + 9\xi^2 + 10\xi^3 + 4\xi^4}\right)\frac{g}{l}. \tag{6.1.16}$$

Based on Equation (6.1.13), one can derive the following mode shape ratio:

$$b_1 = \frac{\psi_0}{\varphi_0} = \frac{3\xi^2 l\,\omega_1^2}{3\xi^2 g - 2\xi^3 l\,\omega_1^2}, \tag{6.1.17}$$

while the second one is obtained for the second natural frequency

$$b_2 = \frac{\psi_0}{\varphi_0} = \frac{3\xi^2 l\,\omega_2^2}{3\xi^2 g - 2\xi^3 l\,\omega_2^2}. \tag{6.1.18}$$

To satisfied the requirement $b_2 = -1$, Equations (6.1.16) and (6.1.18) are used, yielding

$$-3\frac{1 + 4\xi + 2\xi^2 + \xi\sqrt{1 + 2\xi + 4\xi^2} + \sqrt{1 + 2\xi + 4\xi^2}}{\xi\left(-2 + 5\xi + 4\xi^2 + 2\xi\sqrt{1 + 2\xi + 4\xi^2} + 2\sqrt{1 + 2\xi + 4\xi^2}\right)} = -1. \tag{6.1.19}$$

Solving this numerically, one can calculate $\xi = 1.820852385$. For this value of the parameter ξ, Equations (6.1.15) and (6.1.17) give $b_1 = 1.4$. Using again Equations (6.1.15) and (6.1.16) with $\xi = 1.820852385$, the two natural frequencies are found to be:

$$\omega_1 = 0.7201540219\sqrt{\frac{g}{l}}, \quad \omega_2 = 2.162185715\sqrt{\frac{g}{l}}. \tag{6.1.20}$$

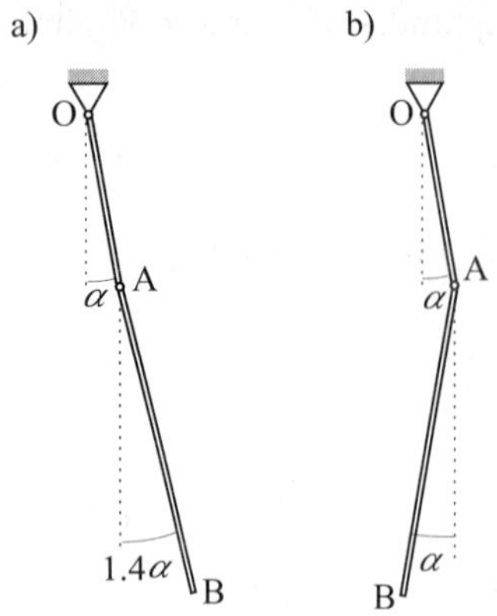

Figure 6.1.3

Figure 6.1.3a shows the first mode shape, while the second mode shape is presented in Figure 6.1.3b.

The solutions satisfying the initial conditions are

$$\varphi(t) = 0.05833\cos 6.772t + 0.04167\cos 2.256t, \tag{6.1.21}$$

$$\psi(t) = -0.05833\cos 6.772t + 0.05833\cos 2.256t. \tag{6.1.22}$$

These solutions are plotted in Figure 6.1.4.

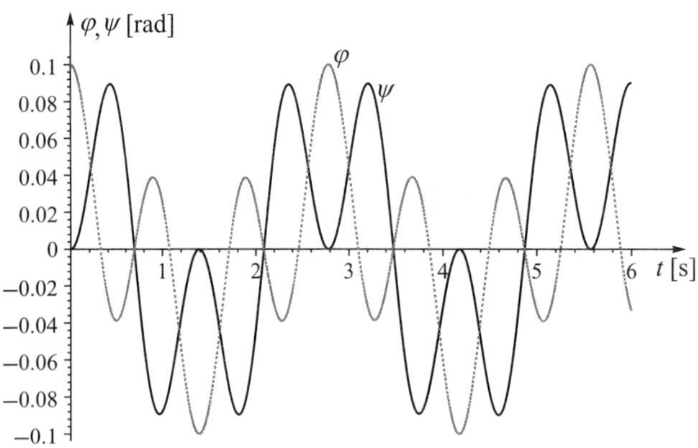

Figure 6.1.4

Example 6.2 A spring of stiffness k is attached to two equal sliders, each of which has mass m. The sliders can slide along two fixed smooth circular wires of radius R, so that the system performs small oscillations around the equilibrium position shown in Figure 6.2.1 (the spring is horizontal then). Assuming that the angles φ and ψ are given as the generalised coordinates, determine and present graphically the mode shapes, assuming $k = 10\,mg/R$. For the system parameters $R = 0.5$ m, $g = 9.81$ m/s^2, $\varphi(0) = 0.1$, $\psi(0) = 0$, $\dot{\varphi}(0) = 0$, $\dot{\psi}(0) = 0$, obtain also the solutions for motion and present them graphically.

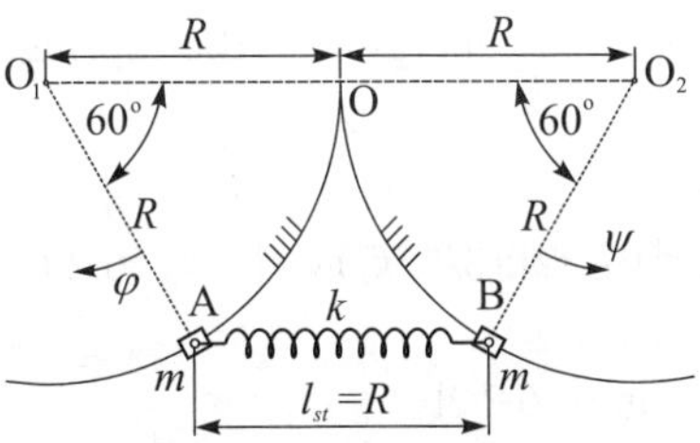

Figure 6.2.1

Solution

The static deflection f_{st} of the spring can be obtained by considering the forces acting on the slider A in the equilibrium position (Figure 6.2.2a).

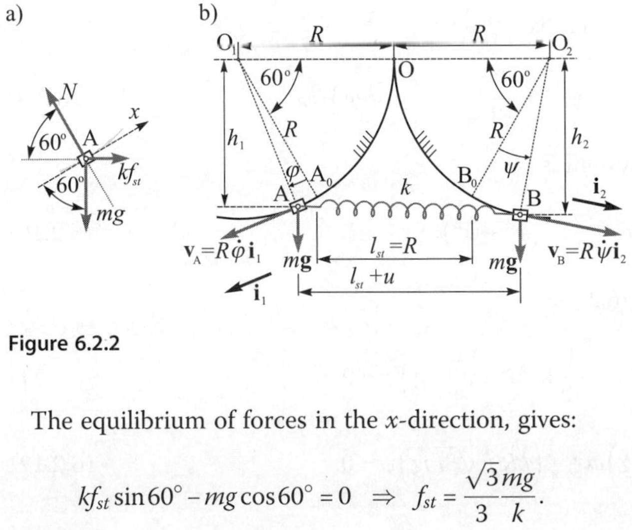

Figure 6.2.2

The equilibrium of forces in the x-direction, gives:

$$kf_{st}\sin 60^\circ - mg\cos 60^\circ = 0 \;\Rightarrow\; f_{st} = \frac{\sqrt{3}}{3}\frac{mg}{k}. \tag{6.2.1}$$

For the additional deflection of the spring $\lambda = u + v^2/(2l_{st})$, with $l_{st} = R$, one needs to determine the deflection in the u and v directions. Based on Figure 6.2.2b, one can express:

$$u = R(\cos 60^\circ - \cos(60^\circ + \varphi)) + R(\cos 60^\circ - \cos(60^\circ + \psi)), \tag{6.2.2}$$

$$v = R(\sin(60^\circ + \psi) - \sin(60^\circ + \varphi)). \tag{6.2.3}$$

Developing these into series, one can obtain:

$$u = \frac{R\sqrt{3}}{2}(\varphi + \psi) + \frac{R}{4}(\varphi^2 + \psi^2), \quad v = \frac{R}{2}(\psi - \varphi), \tag{6.2.4}$$

which leads to

$$\lambda = \frac{R\sqrt{3}}{2}(\varphi + \psi) + \frac{3R}{8}(\varphi^2 + \psi^2) - \frac{R}{4}\varphi\psi. \tag{6.2.5}$$

By using the following expression for the potential energy of the spring, $V_1 = k(f_{st} + \lambda)^2/2$, one can derive

$$V_1 = \frac{k}{128}\left[64f_{st}R\sqrt{3}(\varphi + \psi) + 48R(R + f_{st})(\varphi^2 + \psi^2) + 32R(3R - f_{st})\varphi\psi\right]. \tag{6.2.6}$$

The gravitational potential energy of the sliders is (Figure 6.2.2b)

$$V_2 = -mgh_1 - mgh_2 = -mgR\sin(60^\circ + \varphi) - mgR\sin(60^\circ + \psi). \tag{6.2.7}$$

After developing the sine functions into series, one can obtain

$$V_2 = -mg\frac{R}{2}(\varphi + \psi) + \frac{\sqrt{3}}{4}mgR(\varphi^2 + \psi^2). \tag{6.2.8}$$

Thus, the overall potential energy is

$$V = \frac{9R}{24}\left(kR + \sqrt{3}mg\right)(\varphi^2 + \psi^2) + \frac{R}{12}\left(9kR - \sqrt{3}mg\right)\varphi\psi. \tag{6.2.9}$$

The kinetic energy of the system is

$$T = \frac{m}{2}v_A^2 + \frac{m}{2}v_B^2 \Rightarrow T = \frac{m}{2}R^2(\dot{\varphi}^2 + \dot{\psi}^2), \tag{6.2.10}$$

so that Lagrange's equations (6.2) give

$$mR\ddot{\varphi} + \frac{3}{4}\left(kR + \sqrt{3}mg\right)\varphi + \frac{1}{12}\left(9kR - \sqrt{3}mg\right)\psi = 0, \tag{6.2.11}$$

$$mR\ddot{\psi} + \frac{1}{12}\left(9kR - \sqrt{3}mg\right)\varphi + \frac{3}{4}\left(kR + \sqrt{3}mg\right)\psi = 0. \tag{6.2.12}$$

The solutions are assumed in the form

$$\varphi(t) = \varphi_0\sin(\omega t + \beta), \quad \psi(t) = \psi_0\sin(\omega t + \beta), \tag{6.2.13}$$

where ω and β are constants, while φ_0 and ψ_0 are the amplitudes of φ and ψ. Finding their second time derivative

$$\ddot{\varphi}(t) = -\varphi_0\omega^2\sin(\omega t + \beta), \quad \ddot{\psi}(t) = -\psi_0\omega^2\sin(\omega t + \beta), \tag{6.2.14}$$

the differential equations of motion (6.2.11) and (6.2.12) give

$$\left(\frac{3}{4}\left(kR + \sqrt{3}mg\right) - mR\omega^2\right)\varphi_0 + \left(\frac{1}{12}\left(9kR - \sqrt{3}mg\right)\right)\psi_0 = 0, \tag{6.2.15}$$

$$\left(\frac{1}{12}\left(9kR-\sqrt{3}mg\right)\right)\varphi_0+\left(\frac{3}{4}\left(kR+\sqrt{3}mg\right)-mR\omega^2\right)\psi_0=0. \tag{6.2.16}$$

Equating the characteristic determinant with zero, the following characteristic equation is obtained:

$$\left(\frac{3}{4}\left(kR+\sqrt{3}mg\right)-mR\omega^2\right)^2=\left(\frac{1}{12}\left(9kR-\sqrt{3}mg\right)\right)^2. \tag{6.2.17}$$

Its roots give the squares of the natural frequencies

$$\omega_1^2=\frac{5\sqrt{3}}{6}\frac{g}{R}, \tag{6.2.18}$$

$$\omega_2^2=\frac{1}{6}\frac{9kR+4\sqrt{3}mg}{mR}. \tag{6.2.19}$$

Equation (6.2.15) gives the following mode shape ratios:

$$b_{1/2}=\frac{\psi_0}{\varphi_0}=\frac{12mR\,\omega_{1/2}^2-9\left(kR+\sqrt{3}mg\right)}{9kR-\sqrt{3}mg} \tag{6.2.20}$$

Introducing Equation (6.2.18) into Equation (6.2.20), one derives $b_1=-1$ and this case is plotted in Figure (6.2.3a). Substituting Equation (6.2.19) into Equation (6.2.20), one has $b_2=1$, and the corresponding mode shape is presented in Figure (6.2.3b).

For the given value $k=10mg/R$, the natural frequencies are:

$$\omega_1=1.201405707\sqrt{\frac{g}{R}},\quad \omega_2=4.019291049\sqrt{\frac{g}{R}}. \tag{6.2.21}$$

The solutions for motion satisfying the initial conditions are

$$\varphi(t)=0.05\cos 17.803t+0.05\cos 5.322t. \tag{6.2.22}$$

$$\psi(t)=0.05\cos 17.803t-0.05\cos 5.322t. \tag{6.2.23}$$

These are both shown in Figure 6.2.4, illustrating the aperiodic character of the response.

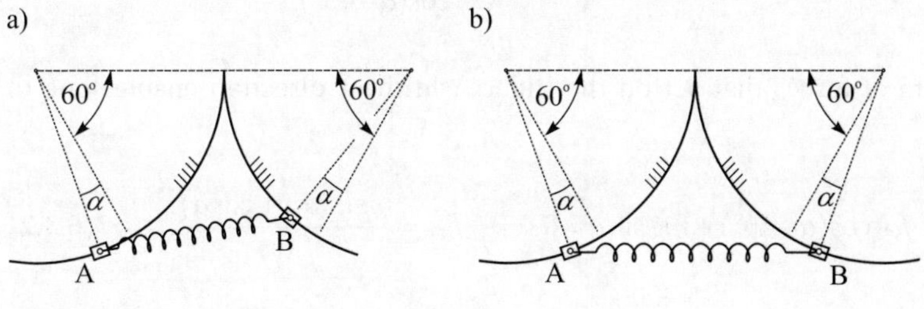

Figure 6.2.3

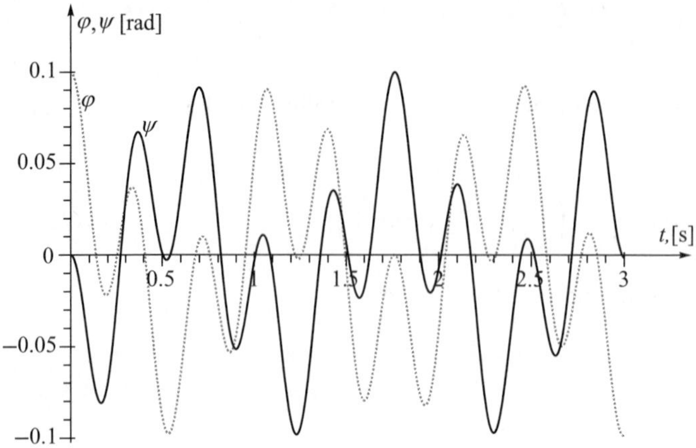

Figure 6.2.4

Example 6.3 Two equal sliders of mass m can move along two frictionless guides and are attached to two springs of stiffness k_1 and k_2, as shown in Figure 6.3.1. The directions of the guides are defined by the angle α. The length of the spring AB in the static equilibrium position is l. Determine the differential equations of motion for small oscillations for the system parameters: $k_1 = 8mg/l$ and $k_2 = 12mg/l$ and for two cases: $\alpha = 90^{\circ}$ (the guides are orthogonal) and $\alpha = 45^{\circ}$ (the guides form a straight line). Determine also the natural frequencies and the corresponding mode shapes.

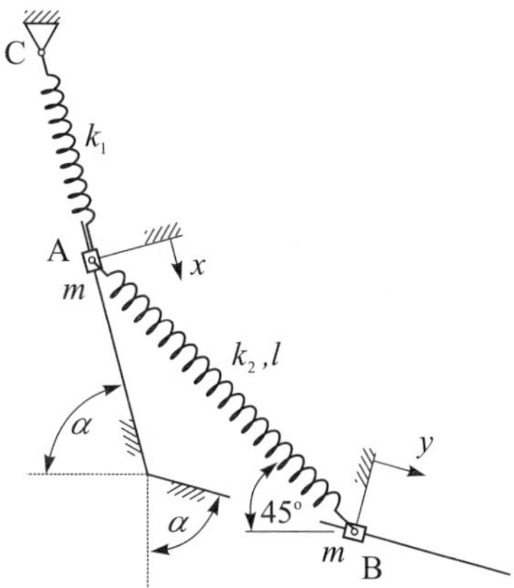

Figure 6.3.1

Solution

Fist, the static deflections of the springs f_{st1} and f_{st2} will be determined. The free-body diagrams are formed and shown in Figure 6.3.2a. The equilibrium of forces acting on the slider B in the y directions gives f_{st2}:

$$-k_2 f_{st2}\cos(\alpha - 45^{\circ}) + mg\cos\alpha = 0 \;\Rightarrow\; f_{st2} = \frac{mg\cos\alpha}{k_2\cos(\alpha - 45^{\circ})}. \tag{6.3.1}$$

The equilibrium of forces that act on the slider A in the x-direction enables one to determine f_{st1}:

$$-k_1 f_{st1} + k_2 f_{st2}\cos(\alpha - 45^{\circ}) + mg\sin\alpha = 0 \;\Rightarrow\; f_{st1} = \frac{mg(\sin\alpha + \cos\alpha)}{k_1}. \tag{6.3.2}$$

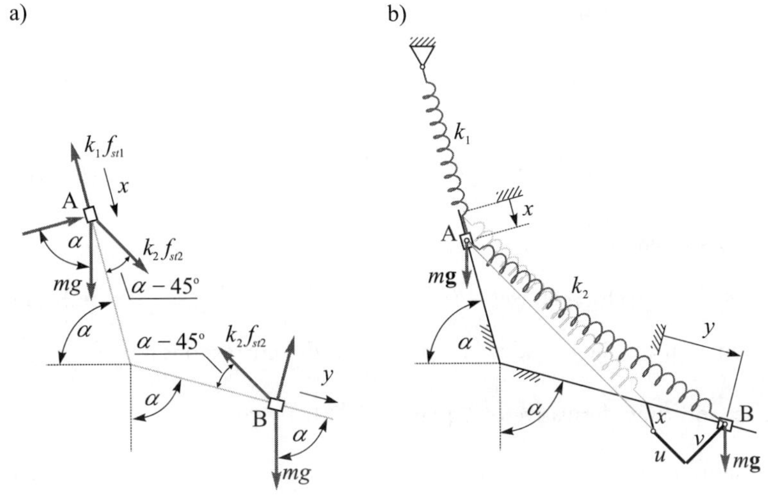

Figure 6.3.2

The overall potential energy takes the form

$$V = -mgx\sin\alpha - mgy\cos\alpha + \frac{k_1}{2}(f_{st1}+x)^2 + \frac{k_2}{2}(f_{st2}+\lambda)^2, \tag{6.3.3}$$

where the additional deflection λ is defined by

$$\lambda = u + \frac{1}{2l}v^2. \tag{6.3.4}$$

The deflections in the u and v directions for the spring AB are depicted in Figure 6.3.2b and are given by:

$$u = (y-x)\cos(\alpha - 45^\circ), \tag{6.3.5}$$

$$v = (y+x)\sin(\alpha - 45^\circ). \tag{6.3.6}$$

Thus, the potential energy (6.3.3) becomes

$$V = \frac{k_1}{2}x^2 + \frac{k_2}{2}\left((y-x)^2\cos^2(\alpha-45^\circ) + \frac{mg}{k_2 l}(y+x)^2\frac{\cos\alpha\sin^2(\alpha-45^\circ)}{\cos(\alpha-45^\circ)}\right). \tag{6.3.7}$$

The kinetic energy is

$$T = \frac{m}{2}\left(\dot{x}^2 + \dot{y}^2\right). \tag{6.3.8}$$

Lagrange's equations (6.2) become

$$m\ddot{x} + k_1 x - k_2(y-x)\cos^2(\alpha-45^\circ) + \frac{mg}{l}(y+x)\frac{\cos\alpha\sin^2(\alpha-45^\circ)}{\cos(\alpha-45^\circ)} = 0, \tag{6.3.9}$$

$$m\ddot{y} + k_2(y-x)\cos^2(\alpha-45^\circ) + \frac{mg}{l}(y+x)\frac{\cos\alpha\sin^2(\alpha-45^\circ)}{\cos(\alpha-45^\circ)} = 0. \tag{6.3.10}$$

The solutions are assumed as

$$x(t) = x_0\sin(\omega t+\beta),\quad y(t) = y_0\sin(\omega t+\beta), \tag{6.3.11}$$

where ω and β are constants, while x_0 and y_0 are the amplitudes corresponding to x and y, respectively.

Calculating the second time derivative of Equation (6.3.11) gives

$$\ddot{x}(t) = -x_0\omega^2\sin(\omega t+\beta),\quad \ddot{y}(t) = -y_0\omega^2\sin(\omega t+\beta). \tag{6.3.12}$$

Substituting (6.3.11) and (6.3.12) into (6.3.9) and (6.3.10) yields

$$\left(k_1 + k_2\cos^2(\alpha-45^\circ) + c - m\omega^2\right)x_0 + \left(-k_2\cos^2(\alpha-45^\circ) + c\right)y_0 = 0, \tag{6.3.13}$$

$$\left(-k_2\cos^2(\alpha-45^\circ) + c\right)x_0 + \left(k_2\cos^2(\alpha-45^\circ) + c - m\omega^2\right)y_0 = 0, \tag{6.3.14}$$

where $c = \dfrac{mg}{l}\dfrac{\cos\alpha\sin^2(\alpha-45^\circ)}{\cos(\alpha-45^\circ)}$.

Let us consider first the case $\alpha = 90^\circ$. The characteristic equation is obtained from (6.3.13) and (6.3.14) in the form

$$l^2\omega^4 - 20gl\omega^2 + 48g^2 = 0. \tag{6.3.15}$$

Its roots are given by

$$\omega_1 = \sqrt{\left(10-2\sqrt{13}\right)\frac{g}{l}} \approx 1.67\sqrt{\frac{g}{l}},\quad \omega_2 = \sqrt{\left(10+2\sqrt{13}\right)\frac{g}{l}} \approx 4.15\sqrt{\frac{g}{l}}. \tag{6.3.16}$$

One can also calculate the ratios $b_1 = y_0/x_0 = 1.868517092$, which corresponds to the first natural frequency, and $b_2 = y_0/x_0 = -0.5351837585$, which corresponds to the second natural frequency. The resulting mode shapes are presented in Figure 6.3.3.

If $\alpha = 45^\circ$, the characteristic equation stemming from (6.3.13) and (6.3.14) is obtained in the form

$$l^2\omega^4 - 32gl\omega^2 + 96g^2 = 0. \tag{6.3.17}$$

The two natural frequencies are

$$\omega_1 = 2\sqrt{\left(4-\sqrt{10}\right)\frac{g}{l}} \approx 1.83\sqrt{\frac{g}{l}},\quad \omega_2 = 2\sqrt{\left(4+\sqrt{10}\right)\frac{g}{l}} \approx 5.35\sqrt{\frac{g}{l}}. \tag{6.3.18}$$

The modal ratios are found to be: $b_1 = y_0/x_0 = 1.387425887$, and $b_2 = y_0/x_0 = -0.7207592201$. The corresponding mode shapes are shown in Figure 6.3.4.

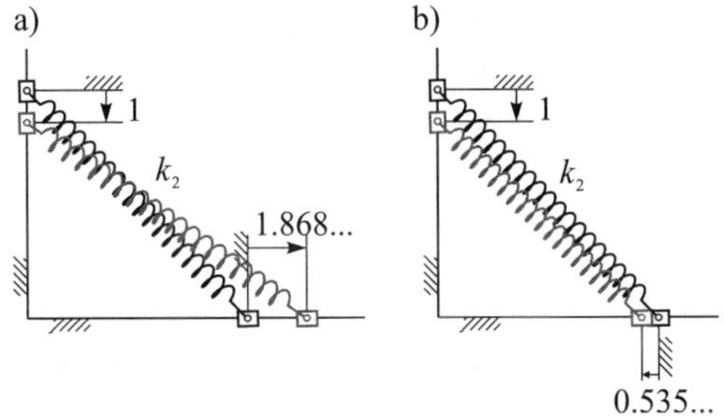

Figure 6.3.3

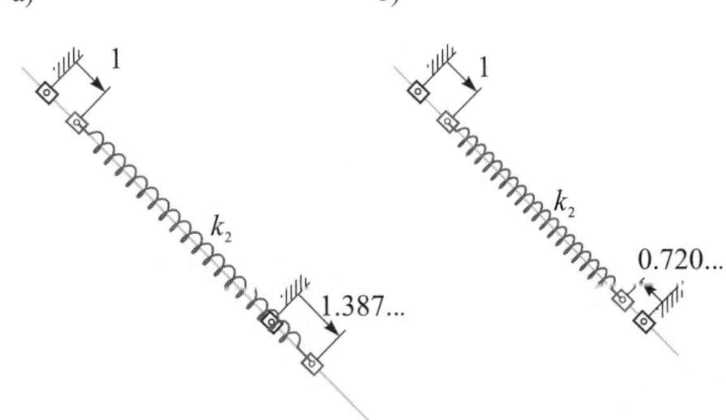

Figure 6.3.4

Example 6.4 A weightless rope is wrapped around two discs of radius R. The upper disc is fixed, while the lower disc is moveable and has mass m and radius R. The points A and B of the cable are attached to each other by a spring of stiffness k that has length l in the static equilibrium position shown in Figure 6.4.1a. There is no slipping between the

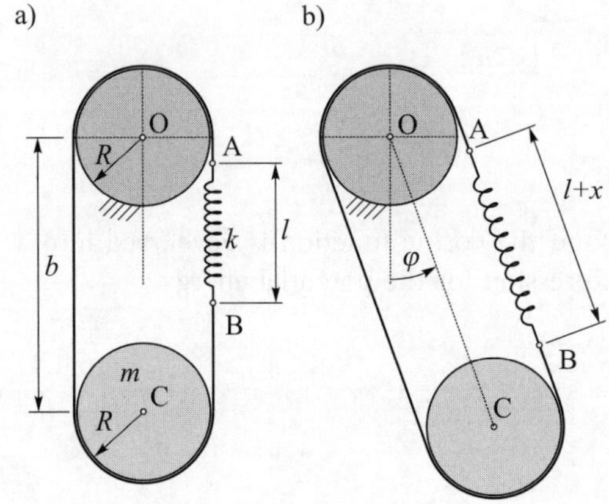

Figure 6.4.1

discs and the rope. Taking the coordinates φ and x (Figure 6.4.1b) as the generalised coordinates, determine the differential equations of motion for small oscillations. Assuming that $k = 9mg/b$, determine also the natural frequencies.

Solution

The system potential energy is defined by

$$V = -mg\overline{\mathrm{OC}}\cos\varphi + \frac{k}{2}(f_{st} + x)^2. \tag{6.4.1}$$

To determine $\overline{\mathrm{OC}}$ in an arbitrary position, one can consider the geometry presented in Figure 6.4.1.b and take into account that the length L of the cable is constant. Thus, one can write down

$$L + l = 2b + 2R\pi, \quad L + l + x = 2\overline{\mathrm{OC}} + 2R\pi \;\Rightarrow\; \overline{\mathrm{OC}} = b + \frac{x}{2}. \tag{6.4.2}$$

After forming the free-body diagram for the lower disc (Figure 6.4.2a), the zero sum of the moments about point K gives

$$kf_{st}2R - mgR = 0 \;\Rightarrow\; f_{st} = \frac{mg}{2k}. \tag{6.4.3}$$

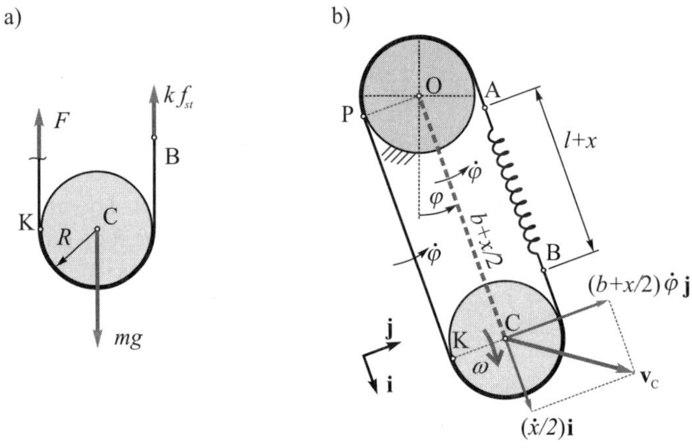

Figure 6.4.2

Equations (6.4.1)–(6.4.3) are now used, and the cosine function is developed into a Maclaurin series to derive the following expression for the potential energy

$$V \approx mgb\frac{\varphi^2}{2} + k\frac{x^2}{2}. \tag{6.4.4}$$

The kinetic energy is defined by

$$T = \frac{1}{2}mv_C{}^2 + \frac{1}{2}J_C\omega^2, \tag{6.4.5}$$

where

$$J_C = \frac{1}{2} m R^2. \tag{6.4.6}$$

To determine the velocity of the point C, an arbitrary position presented in Figure 6.4.2b is considered, yielding

$$\mathbf{v}_C = \frac{\dot{x}}{2}\mathbf{i} + \left(b + \frac{x}{2}\right)\dot{\varphi}\mathbf{j} \;\Rightarrow\; v_C{}^2 = \mathbf{v}_C \cdot \mathbf{v}_C = \frac{\dot{x}^2}{4} + \left(b + \frac{x}{2}\right)^2 \dot{\varphi}^2. \tag{6.4.7}$$

Given the fact that the system performs small oscillations around the equilibrium position, the kinetic energy should be calculated assuming that the system passes through this position. Thus, by using $x = 0$ and $\varphi = 0$, one has

$$v_C{}^2 = \frac{\dot{x}^2}{4} + b^2 \dot{\varphi}^2. \tag{6.4.8}$$

The point P is the instantaneous centre of zero velocity of the part of the rope in general plane motion. The velocity of the point K is then

$$\mathbf{v}_K = \mathrm{PK}\dot{\varphi}\mathbf{j}. \tag{6.4.9}$$

To determine the angular velocity of the lower disc, one can consider the relationship relating the velocities of the points K and C (Figure 6.4.2b), as follows

$$\mathbf{v}_C = \mathbf{v}_K + \mathbf{v}_{C/K} \;\Rightarrow\; \mathbf{v}_C \cdot \mathbf{i} = \mathbf{v}_K \cdot \mathbf{i} + \mathbf{v}_{C/K} \cdot \mathbf{i} \;\Rightarrow\; \frac{\dot{x}}{2} = 0 + R\omega \;\Rightarrow\; \omega = \frac{\dot{x}}{2R}. \tag{6.4.10}$$

Thus, the kinetic energy (6.4.5) becomes

$$T \approx \frac{1}{2} m b^2 \dot{\varphi}^2 + \frac{3}{16} m \dot{x}^2. \tag{6.4.11}$$

Using Equations (6.4.4) and (6.4.11) with Equation (6.2), the following differential equations of motion are derived:

$$\ddot{\varphi} + \frac{g}{b}\varphi = 0, \tag{6.4.12}$$

$$\ddot{x} + \frac{8k}{3m} x = 0. \tag{6.4.13}$$

As these equations are uncoupled, the squares of the natural frequencies are easily obtained as the coefficients in front of the coordinates:

$$\omega_1 = \sqrt{\frac{g}{b}}, \quad \omega_2 = \sqrt{\frac{8k}{3m}} = \sqrt{24}\sqrt{\frac{g}{b}}. \tag{6.4.14}$$

The fact that the equations of motion are uncoupled implies that the generalised coordinates represent the principal coordinates as well.

Example 6.5 A slender bar of mass m and length $2l$ is hinged to a disc of mass $M = 8m$ and radius $R = l$. The disc rolls without slipping along a horizontal surface. A spring of stiffness $k = 9mg/l$ is attached to the disc at one end and to the wall at the other end (Figure 6.5.1). The spring is horizontal in the static equilibrium position shown and its length is $2l$. Determine the natural frequencies and mode shapes. Assuming $l = 0.1$ m, $g = 9.81$ m/s^2, $\varphi(0) = 0$, $\dot{\varphi}(0) = 0$, $\psi(0) = 0.4$, $\dot{\psi}(0) = 0$, determine the analytic solution for motion and present it graphically.

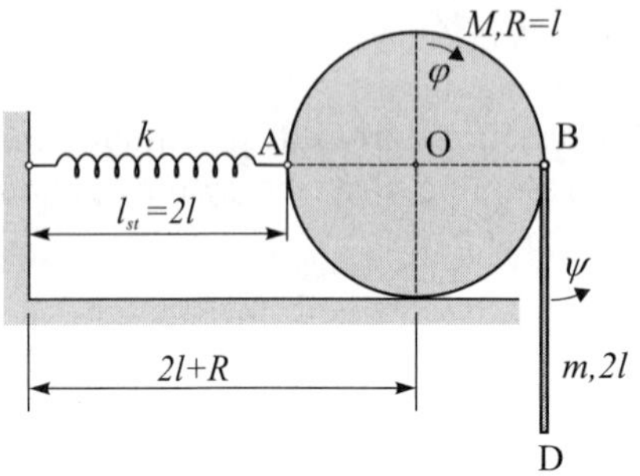

Figure 6.5.1

Solution

The overall potential energy can be expressed as (see Figure 6.5.2a)

$$V = mgy_C + \frac{k}{2}(f_{st} + \lambda)^2, \tag{6.5.1}$$

where

$$y_C = -R\sin\varphi - l\cos\psi \approx -l\varphi - l\left(1 - \frac{\psi^2}{2}\right), \quad \lambda = u + \frac{1}{2 \cdot 2l}v^2 \approx l\varphi + \frac{3}{4}l\varphi^2. \tag{6.5.2}$$

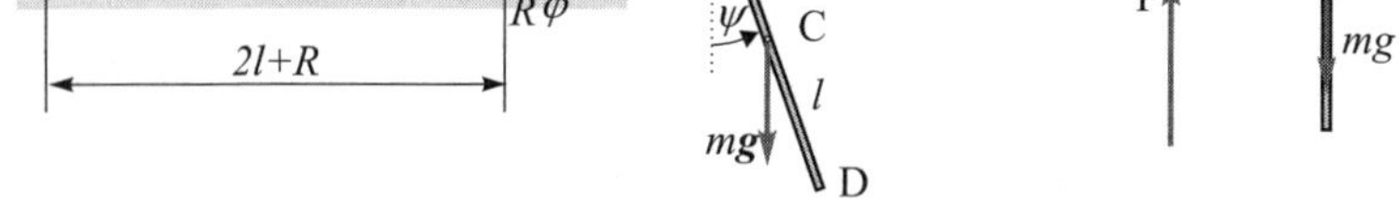

Figure 6.5.2

The deflections in the u and v directions can be represented as

$$u = x_A(\varphi) - x_A(0) = 2l + R + R\varphi - R\cos\varphi - 2l \approx R\varphi + R\frac{\varphi^2}{2}, \quad v = y_A(\varphi) = R\sin\varphi \approx R\varphi. \tag{6.5.3}$$

Forming the free-body diagram for the disc as shown in Figure 6.5.2b, one can write down the zero sum of the moments about the point P and obtain the static deflection of the spring f_{st} as follows:

$$kf_{st}R - mgR = 0 \;\Rightarrow\; f_{st} = \frac{mg}{k}. \tag{6.5.4}$$

Thus, the potential energy becomes

$$V \approx -mgl\varphi + mgl\frac{\psi^2}{2} + \frac{k}{2}\left(\frac{mg}{k} + l\varphi + \frac{3}{4}l\varphi^2\right)^2 \approx \frac{1}{2}mgl\psi^2 + \frac{21}{4}mgl\varphi^2. \tag{6.5.5}$$

The kinetic energy stems from the disc and the bar, both being in general plane motion:

$$T = \frac{1}{2}Mv_O{}^2 + \frac{1}{2}J_O\dot{\varphi}^2 + \frac{1}{2}mv_C{}^2 + \frac{1}{2}J_C\dot{\psi}^2, \tag{6.5.6}$$

where

$$J_O = \frac{1}{2}MR^2,\; J_C = \frac{1}{12}m(2l)^2. \tag{6.5.7}$$

The kinematics of the system passing through the equilibrium position is presented in Figure 6.5.3, and is used to determine the velocities of the points O and C as follows:

$$\begin{aligned}
&\mathbf{v}_O = R\dot{\varphi}\mathbf{i},\\
&\mathbf{v}_B = \mathbf{v}_O + \mathbf{v}_{B/O},\; \mathbf{v}_C = \mathbf{v}_B + \mathbf{v}_{C/B} = \mathbf{v}_O + \mathbf{v}_{B/O} + \mathbf{v}_{C/B},\; \mathbf{v}_{B/O} = -R\dot{\varphi}\mathbf{j},\; \mathbf{v}_{C/B} = l\dot{\psi}\mathbf{i},\; \Rightarrow\\
&v_O^2 = \mathbf{v}_O\cdot\mathbf{v}_O = l^2\dot{\varphi}^2,\; v_C^2 = (\mathbf{v}_O + \mathbf{v}_{B/O} + \mathbf{v}_{C/B})\cdot(\mathbf{v}_O + \mathbf{v}_{B/O} + \mathbf{v}_{C/B}) = l^2(\dot{\varphi} + \dot{\psi})^2 + l^2\dot{\varphi}^2.
\end{aligned} \tag{6.5.8}$$

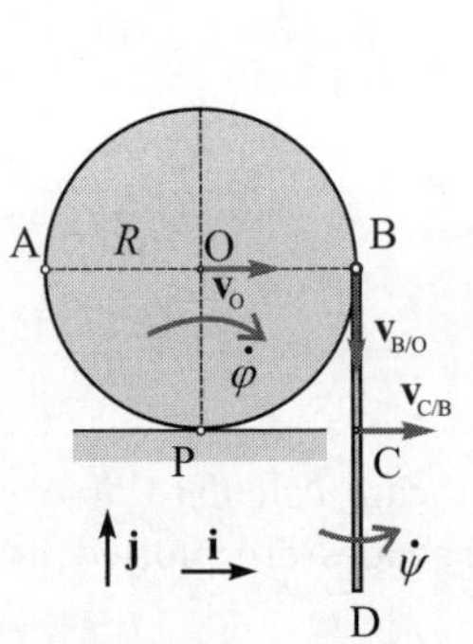

Figure 6.5.3

The expression for the kinetic energy (6.5.6)–(6.5.8) now turns into

$$T \approx \frac{3}{4}Ml^2\dot{\varphi}^2 + \frac{1}{2}ml^2\left((\dot{\varphi} + \dot{\psi})^2 + \dot{\varphi}^2\right) + \frac{1}{6}ml^2\dot{\psi}^2. \tag{6.5.9}$$

Using (6.5.5) and (6.5.9), Lagrange's equations (6.2) give

$$14l\ddot{\varphi} + l\ddot{\psi} + \frac{21}{2}g\varphi = 0, \tag{6.5.10}$$

$$l\ddot{\varphi} + \frac{4}{3}l\ddot{\psi} + g\psi = 0. \tag{6.5.11}$$

Assuming their solutions as

$$\varphi(t) = \varphi_0 \sin(\omega t + \beta), \quad \psi(t) = \psi_0 \sin(\omega t + \beta), \tag{6.5.12}$$

and finding its second time derivative, the following equations for small oscillations are derived:

$$\left(\frac{21}{2}g - 14l\omega^2\right)\varphi_0 + \left(-l\omega^2\right)\psi_0 = 0, \tag{6.5.13}$$

$$\left(-l\omega^2\right)\varphi_0 + \left(g - \frac{4}{3}l\omega^2\right)\psi_0 = 0. \tag{6.5.14}$$

By considering the characteristic determinant, the characteristic equation is obtained

$$106l^2\omega^4 - 168gl\omega^2 + 63g^2 = 0. \tag{6.5.15}$$

Its roots give the natural frequencies:

$$\omega_1 = 0.7804073556\sqrt{\frac{g}{l}}, \quad \omega_2 = 0.9878613363\sqrt{\frac{g}{l}}. \tag{6.5.16}$$

The two associated mode shape ratios are defined by

$$b_{1/2} = \frac{\psi_0}{\varphi_0} = \frac{3l\,\omega_{1/2}^2}{3g - 4l\,\omega_{1/2}^2}. \tag{6.5.17}$$

Substituting Equation (6.5.16) into Equation (6.5.17), one can calculate: $b_1 = 3.240370349$ and $b_2 = -3.240370349$. The corresponding mode shapes are plotted in Figure 6.5.4a and b.

For the given parameter values and the initial conditions, the solutions for motion are found to be

$$\varphi(t) = -0.06172\cos 9.784t + 0.06172\cos 7.730t, \tag{6.5.18}$$

$$\psi(t) = 0.2\cos 9.784t + 0.2\cos 7.730t. \tag{6.5.19}$$

These solutions are plotted in Figure 6.5.5.

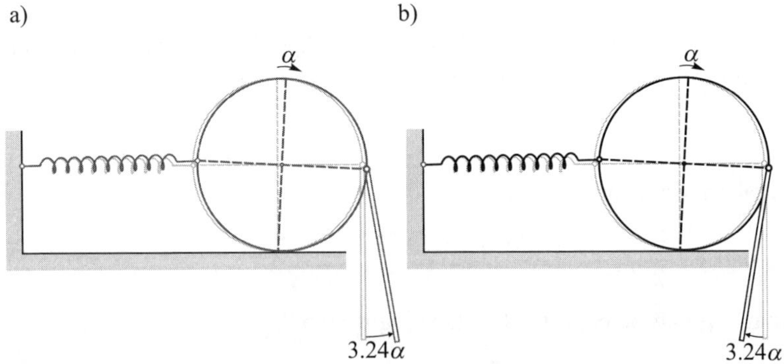

Figure 6.5.4

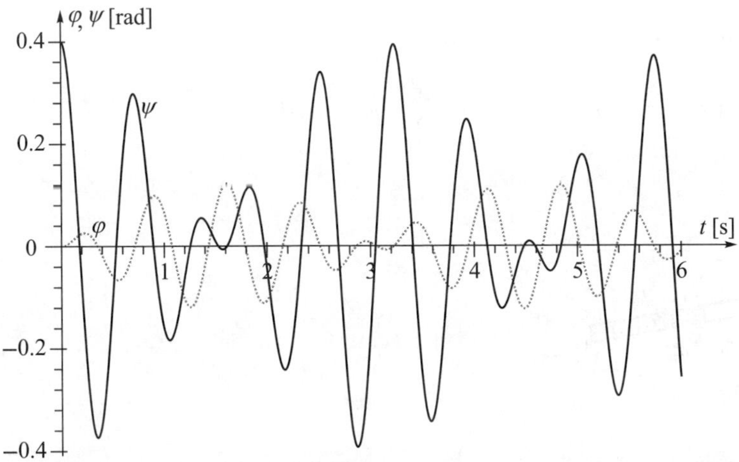

Figure 6.5.5

Example 6.6 A slider D of mass m moves without friction inside a straight pipe of mass $M = 2m$ and length $l = 4R$. The pipe rolls without slipping along the fixed cylindrical surface of radius R. A spring OA of stiffness $k_1 = 12mg/R$ is attached to the left end of the pipe as shown in Figure 6.6.1. Another spring is attached to the slider and the right end of

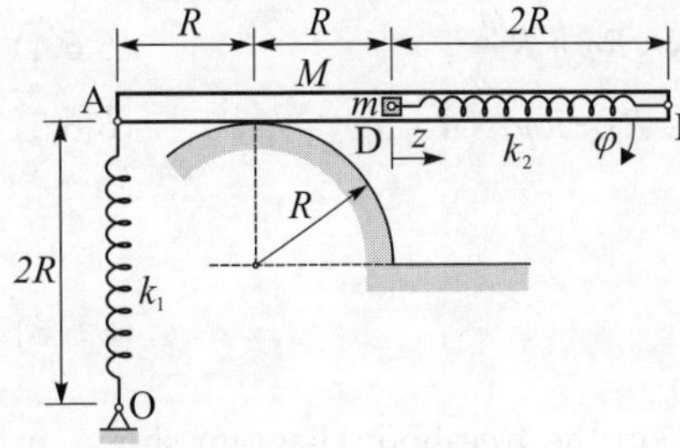

Figure 6.6.1

the pipe; its stiffness coefficient is $k_2 = 4mg/R$. In the static equilibrium shown, the spring OA is vertical and it has length $l_{st} = 2R$, while the pipe takes the horizontal position. Determine the natural frequencies and the corresponding mode shapes.

Solution

The overall potential energy is defined by

$$V = Mgy_{\mathrm{C}} + mgy_{\mathrm{D}} + \frac{k_1}{2}(f_{st} + \lambda)^2 + \frac{k_2}{2}z^2, \tag{6.6.1}$$

where the coordinate y_{C} can be expressed as (see Figure 6.6.2)

$$y_{\mathrm{C}} = R\cos\varphi - (R - R\varphi)\sin\varphi \approx R\left(1 - \frac{\varphi^2}{2}\right) - R\varphi + R\varphi^2 = -R\varphi + R\frac{\varphi^2}{2} + const. \tag{6.6.2}$$

while the coordinate y_{D} is

$$y_{\mathrm{D}} = R\cos\varphi - (R - R\varphi + z)\sin\varphi \approx -R\varphi + R\frac{\varphi^2}{2} - \varphi z + const. \tag{6.6.3}$$

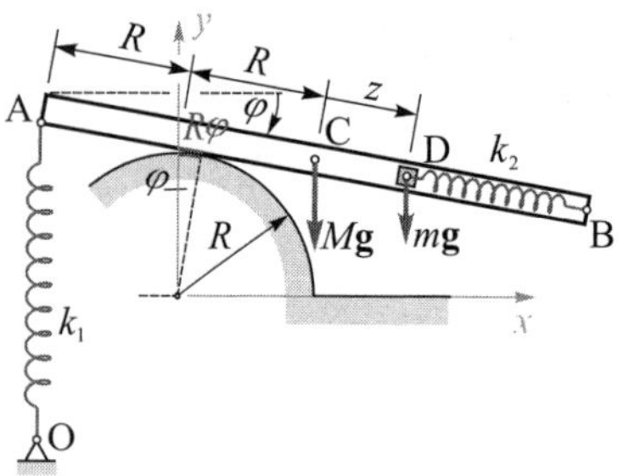

Figure 6.6.2

Given the fact that the deflections of the spring OA in the u and v directions can be represented as

$$u = y_{\mathrm{A}}(\varphi) - y_{\mathrm{A}}(0) = R\cos\varphi + (R + R\varphi)\sin\varphi - R \approx R\varphi + R\frac{\varphi^2}{2}, \tag{6.6.4}$$

$$v = x_{\mathrm{A}}(\varphi) - x_{\mathrm{A}}(0) = R\sin\varphi - (R + R\varphi)\cos\varphi - (-R) \approx R\varphi - (R + R\varphi) + R \approx 0, \tag{6.6.5}$$

its additional deflection is found to be

$$\lambda = u + \frac{1}{2 \cdot 2R}v^2 \approx R\varphi + \frac{1}{2}R\varphi^2. \tag{6.6.6}$$

Its static deflection f_{st} can be determined based on the free-body diagram shown in Figure 6.6.3.

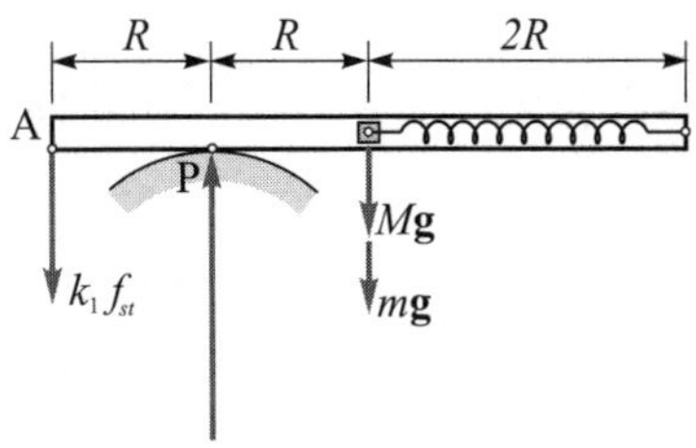

Figure 6.6.3

The zero sum of the moments about the point P gives:

$$k_1 f_{st} R - mgR - MgR = 0 \;\Rightarrow\; f_{st} = \frac{(M+m)g}{k_1} = \frac{R}{4}. \tag{6.6.7}$$

The potential energy now simplifies to

$$V \approx 9mgR\varphi^2 - mg\varphi z + \frac{2mg}{R} z^2. \tag{6.6.8}$$

The overall kinetic energy is defined by

$$T = \frac{1}{2} M v_C{}^2 + \frac{1}{2} J_C \dot{\varphi}^2 + \frac{1}{2} m v_D{}^2, \tag{6.6.9}$$

where

$$J_C = \frac{1}{12} M l^2 = \frac{1}{12} 2m(4R)^2 = \frac{8}{3} mR^2. \tag{6.6.10}$$

To determine the absolute velocity of the slider, one needs to take into account that its actually exhibits compound motion: it moves relatively with respect to the pipe (the relative velocity is labelled by $\mathbf{v}_{rel}$ in Figure 6.6.4), while the pipe itself performs general plane motion so that the slider has also the so-called transportation velocity (labelled by $\mathbf{v}_{tr}$ in Figure 6.6.4). This transportation velocity can be interpreted as the velocity of the point below the slider but on the pipe, which 'transfers' its velocity to the slider. In the position when the system passes through the equilibrium, the point below the slider corresponds to the point C, so that $\mathbf{v}_C = \mathbf{v}_{tr}$. Noting that the point P represents the instantaneous centre of zero velocity, the following kinematic relationships can be written down

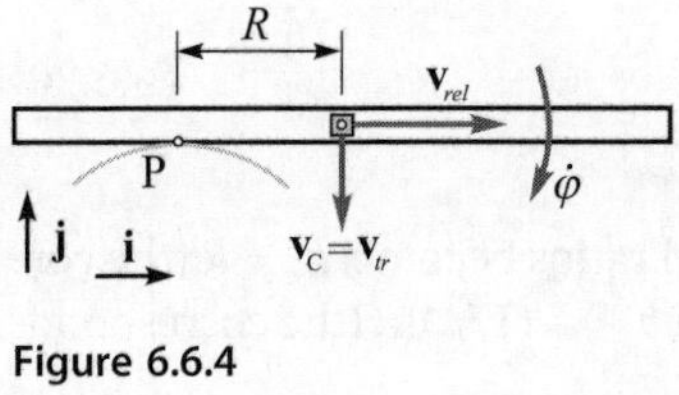

Figure 6.6.4

$$\mathbf{v}_{\mathrm{C}} = -R\dot{\varphi}\mathbf{j}, \quad \mathbf{v}_{\mathrm{D}} = \mathbf{v}_{tr} + \mathbf{v}_{rel} = -R\dot{\varphi}\mathbf{j} + \dot{z}\mathbf{i} \;\Rightarrow$$
$$v_{\mathrm{C}}^2 = \mathbf{v}_{\mathrm{C}}\cdot\mathbf{v}_{\mathrm{C}} = R^2\dot{\varphi}^2, \; v_{\mathrm{D}}^2 = (-R\dot{\varphi}\mathbf{j} + \dot{z}\mathbf{i})\cdot(-R\dot{\varphi}\mathbf{j} + \dot{z}\mathbf{i}) = R^2\dot{\varphi}^2 + \dot{z}^2. \tag{6.6.11}$$

Thus, the kinetic energy becomes

$$T \approx \frac{17}{6} mR^2\dot{\varphi}^2 + \frac{1}{2} m\dot{z}^2. \tag{6.6.12}$$

Lagrange's equations (6.2) now give

$$\frac{17}{3} R^2\ddot{\varphi} + 18gR\varphi - gz = 0, \tag{6.6.13}$$

$$\ddot{z} - g\varphi + \frac{4g}{R} z = 0. \tag{6.6.14}$$

Their solutions are assumed as

$$\varphi(t) = \varphi_0 \sin(\omega t + \beta), \quad z(t) = z_0 \sin(\omega t + \beta), \tag{6.6.15}$$

and yield the following system of algebraic equations:

$$\left(18gR - \frac{17}{3} R^2\omega^2\right)\varphi_0 + (-g)z_0 = 0, \tag{6.6.16}$$

$$(-g)\varphi_0 + \left(\frac{4g}{R} - \omega^2\right) z_0 = 0. \tag{6.6.17}$$

A non-trivial solution exists provided that the characteristic determinant is equal to zero, which leads to the following characteristic equation:

$$17R^2\omega^4 - 122gR\omega^2 + 213g^2 = 0. \tag{6.6.18}$$

The roots correspond to two natural frequencies

$$\omega_1 = \sqrt{3}\sqrt{\frac{g}{R}}, \quad \omega_2 = \sqrt{\frac{71}{17}}\sqrt{\frac{g}{R}}. \tag{6.6.19}$$

The modal ratios are obtained as

$$b_{1/2} = \frac{z_0}{\varphi_0} = \frac{gR}{4g - R\,\omega_{1/2}^2}. \tag{6.6.20}$$

For the two values of the natural frequencies, the modal ratios become: $b_1 = R$ (the corresponding mode shape is presented in Figure 6.6.5a) and $b_2 = -(17/3)R$ (the corresponding mode shape is shown in Figure 6.6.5b).

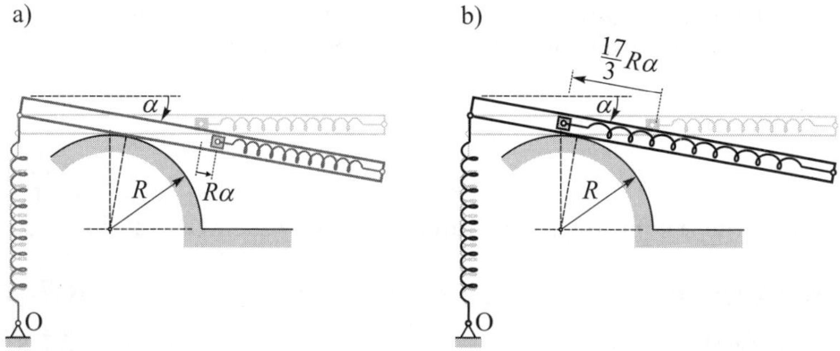

Figure 6.6.5

Example 6.7 A particle of mass m is attached to two springs of stiffness k_1 and k_2. The system is placed in the horizontal plane. In the static equilibrium position shown in Figure 6.7.1, these two springs are orthogonal. Assume first that the generalised coordinates are defined by x and y, and determine the solutions for motion corresponding to $x(0) = x_0$, $y(0) = y_0$, $\dot{x}(0) = 0$, $\dot{y}(0) = 0$. In addition, consider the reference frame rotated for the angle α, defined by the axis ξ and η. Assuming $m = 1$ kg, $x_0 = 5$ cm and $y_0 = 3$ cm, plot the solutions $x(t)$, $y(t)$, $y(x)$, $\xi(t)$, $\eta(t)$, $\eta(\xi)$ during the first 10 seconds, taking different combinations of the stiffness coefficients as follows: (1) $k_1 = 5$ N/cm, $k_2 = 35$ N/cm; (2) $k_1 = 10$ N/cm, $k_2 = 30$ N/cm; (3) $k_1 = 18$ N/cm, $k_2 = 22$ N/cm; (4) $k_1 = 19.5$ N/cm, $k_2 = 20.5$ N/cm.

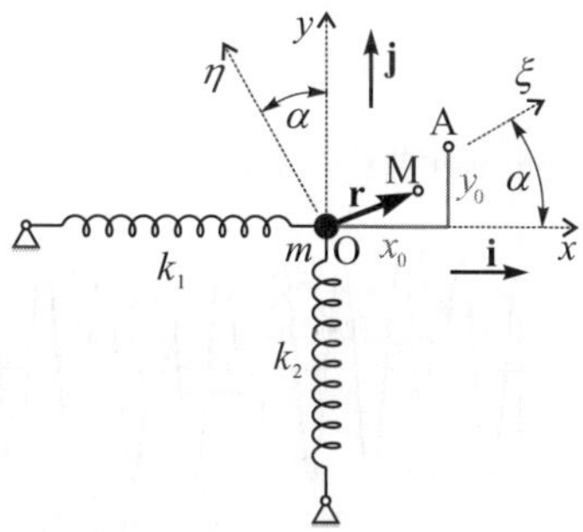

Figure 6.7.1

Solution

The potential energy of the particle is

$$V = \frac{1}{2}k_1x^2 + \frac{1}{2}k_2y^2, \tag{6.7.1}$$

while the kinetic energy has the form

$$T = \frac{m}{2}\left(\dot{x}^2 + \dot{y}^2\right). \tag{6.7.2}$$

Lagrange's equations with respect to generalised coordinate x and y are

$$\ddot{x} + \omega_1^2 x = 0, \quad \omega_1 = \sqrt{\frac{k_1}{m}}, \tag{6.7.3}$$

$$\ddot{y} + \omega_2^2 y = 0, \quad \omega_2 = \sqrt{\frac{k_2}{m}}. \tag{6.7.4}$$

They are obviously uncoupled, implying that x and y are principal coordinates. The solutions of these equations satisfying the initial conditions given are

$$x(t) = x_0 \cos\omega_1 t, \tag{6.7.5}$$

$$y(t) = y_0 \cos\omega_2 t. \tag{6.7.6}$$

Their projections onto the second reference frame defined by the axis ξ and η are:

$$\xi(t) = x(t)\cos\alpha + y(t)\sin\alpha, \tag{6.7.7}$$

$$\eta(t) = -x(t)\sin\alpha + y(t)\cos\alpha. \tag{6.7.8}$$

Time-history diagrams and trajectories corresponding to four cases defined are shown in Figures 6.7.2–6.7.5.

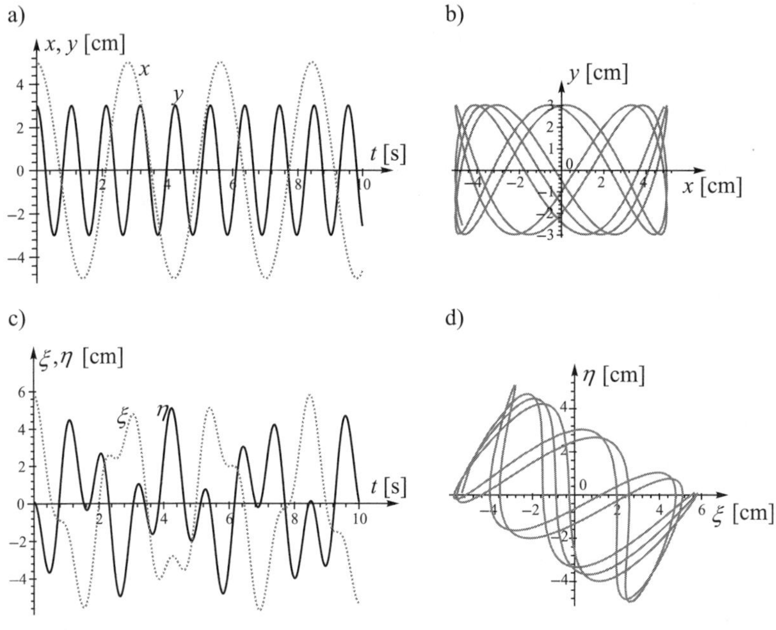

Figure 6.7.2

The trajectories shown in parts (b) and (d) represent the so-called Lissajous figures. The axes of symmetry of a rectangular in which they are placed are two principal axes and they are orthogonal to each other. Note that these four cases have been chosen in a way that ω_1 and ω_2 first differ considerably and then their mutual difference becomes smaller and smaller. When this difference is small, beating oscillations occur, which are seen in Figure 6.7.4c and Figure 6.7.5c, as well as in Figure 6.7.6, which corresponds to the one plotted in Figure 6.7.4c, but the time interval considered is longer so that the beating period can easily be recognised.

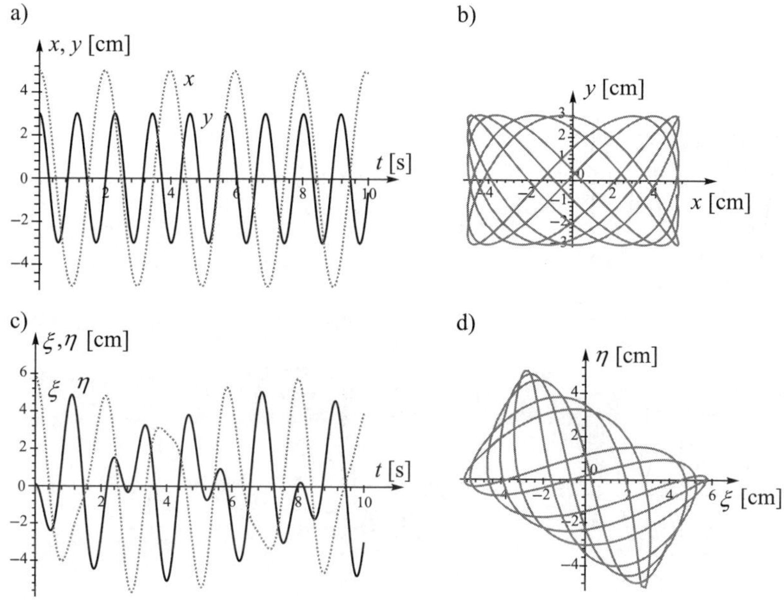

Figure 6.7.3

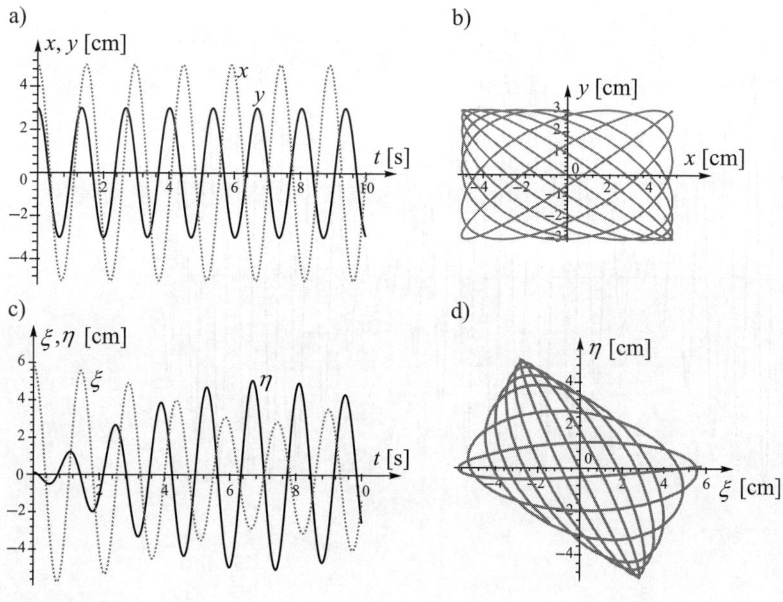

Figure 6.7.4

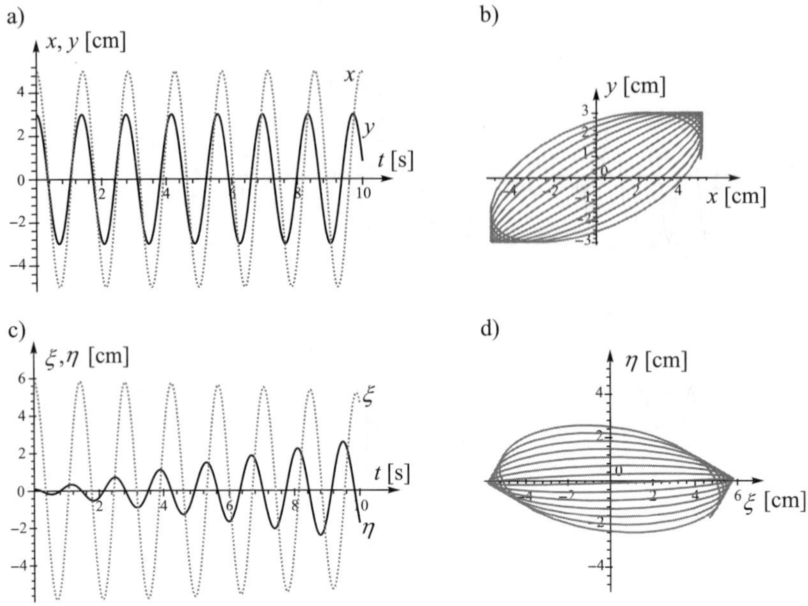

Figure 6.7.5

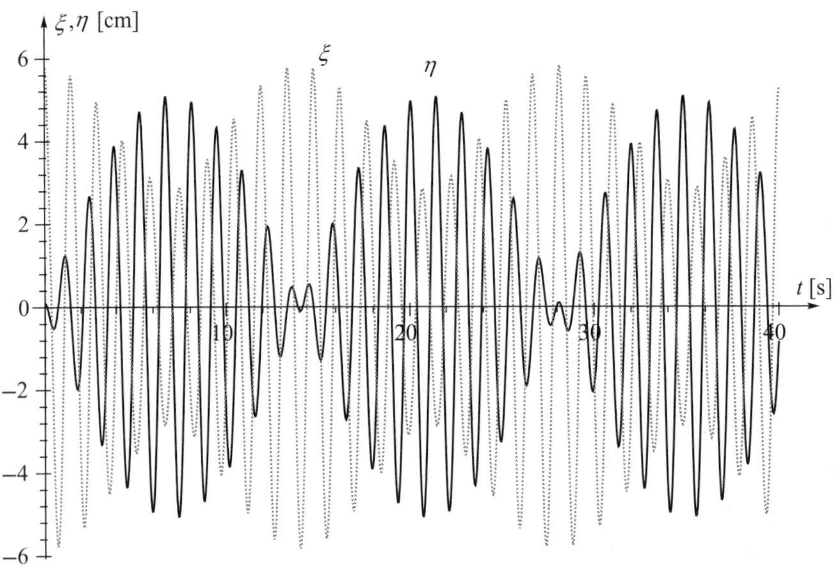

Figure 6.7.6

7

Forced Vibration of Two-Degree-of-Freedom Systems

Chapter Outline

This chapter is concerned with the use of Lagrange's equations to obtain the differential equation of motion of forced undamped and damped two-degree-of-freedom oscillatory systems that perform small vibrations. Six examples with harmonic excitations are presented and solved in detail, illustrating how the equations of motion can be derived and further used to get an insight into a system's behaviour. This includes dealing with the concept of dynamic absorbers and certain aspects of stability of motion.

Chapter Objectives

- To present Lagrange's equations of the second kind for forced undamped and damped two-degree-of-freedom systems
- To illustrate a variety of examples in which they yield the differential equations of motion for the external harmonic force or harmonic base/kinematic excitation
- To demonstrate how the Routh–Hurwitz stability criterion can be used to determine the parameters yielding asymptotic stability

Theoretical Introduction

To derive the differential equations of motion for two-degree-of-freedom forced non-conservative systems, we will start from the following Lagrange's equations:

$$\frac{d}{dt}\frac{\partial T}{\partial \dot{q}_i} - \frac{\partial T}{\partial q_i} + \frac{\partial D}{\partial \dot{q}_i} + \frac{\partial V}{\partial q_i} = Q_{q_i}, \quad i = 1, 2. \tag{7.1}$$

If the system performs small oscillations around the equilibrium position $q_1 = 0$ and $q_2 = 0$, these Lagrange's equations can be simplified to

$$\frac{d}{dt}\frac{\partial T}{\partial \dot{q}_i} + \frac{\partial D}{\partial \dot{q}_i} + \frac{\partial V}{\partial q_i} = Q_{q_i}, \quad i = 1, 2. \tag{7.2}$$

Mechanical Vibration: Fundamentals with Solved Examples, First Edition.
Ivana Kovačić and Dragi Radomirović.

When there is no damping, they have the form

$$\frac{d}{dt}\frac{\partial T}{\partial \dot{q}_i} + \frac{\partial V}{\partial q_i} = Q_{q_i}, \quad i = 1,2. \tag{7.3}$$

The kinetic energy has the form given by Equation (6.3) and the potential energy takes the form given by Equation (6.6). If the excitation is assumed to be harmonic with respect to both generalised coordinates, the resulting equations are

$$\begin{aligned} \widetilde{T}_{11}\ddot{q}_1 + \widetilde{T}_{12}\ddot{q}_2 + \widetilde{V}_{11}q_1 + \widetilde{V}_{12}q_2 &= h_1\cos(\Omega t + \beta), \\ \widetilde{T}_{12}\ddot{q}_1 + \widetilde{T}_{22}\ddot{q}_2 + \widetilde{V}_{12}q_1 + \widetilde{V}_{22}q_2 &= h_2\cos(\Omega t + \beta). \end{aligned} \tag{7.4}$$

Their particular integrals (forced solutions) can be sought in the form

$$\begin{aligned} q_{1p} &= A_1\cos(\Omega t + \beta), \\ q_{2p} &= A_2\cos(\Omega t + \beta). \end{aligned} \tag{7.5}$$

By substituting (7.5) into (7.4), one can derive

$$\begin{aligned} A_1\left(\widetilde{V}_{11} - \Omega^2\widetilde{T}_{11}\right) + A_2\left(\widetilde{V}_{12} - \Omega^2\widetilde{T}_{12}\right) &= h_1, \\ A_1\left(\widetilde{V}_{12} - \Omega^2\widetilde{T}_{12}\right) + A_2\left(\widetilde{V}_{22} - \Omega^2\widetilde{T}_{22}\right) &= h_2. \end{aligned} \tag{7.6}$$

The amplitudes of forced oscillations can be calculated as

$$\begin{aligned} A_1 &= \frac{h_1\left(\widetilde{V}_{22} - \Omega^2\widetilde{T}_{22}\right) - h_2\left(\widetilde{V}_{12} - \Omega^2\widetilde{T}_{12}\right)}{\left(\widetilde{V}_{11} - \Omega^2\widetilde{T}_{11}\right)\left(\widetilde{V}_{22} - \Omega^2\widetilde{T}_{22}\right) - \left(\widetilde{V}_{12} - \Omega^2\widetilde{T}_{12}\right)^2}, \\ A_2 &= \frac{-h_1\left(\widetilde{V}_{12} - \Omega^2\widetilde{T}_{12}\right) + h_2\left(\widetilde{V}_{11} - \Omega^2\widetilde{T}_{11}\right)}{\left(\widetilde{V}_{11} - \Omega^2\widetilde{T}_{11}\right)\left(\widetilde{V}_{22} - \Omega^2\widetilde{T}_{22}\right) - \left(\widetilde{V}_{12} - \Omega^2\widetilde{T}_{12}\right)^2}. \end{aligned} \tag{7.7}$$

The general solutions consist of the particular integrals (7.5) and the homogenous solutions, which are obtained by putting $h_1 = h_2 = 0$ into Equation (7.4). Note that the natural frequencies $\omega_{1,2}$ existing in the homogenous solutions are calculated from the characteristic (frequency) equation, as explained and illustrated in Chapter 6. It should also be pointed out that the denominators in the expressions (7.7) have the same form as the left-hand side of the characteristic equation. This implies that the case when the forcing frequency is equal to any of them ($\Omega = \omega_1$ and $\Omega = \omega_2$), the resonance phenomenon occurs.

An interesting case occurs if one of the generalised forces is zero, for instance, $h_2 = 0$. The solutions for the amplitudes of forced oscillations (7.7) become

$$\begin{aligned} A_1 &= \frac{h_1\left(\widetilde{V}_{22} - \Omega^2\widetilde{T}_{22}\right)}{\left(\widetilde{V}_{11} - \Omega^2\widetilde{T}_{11}\right)\left(\widetilde{V}_{22} - \Omega^2\widetilde{T}_{22}\right) - \left(\widetilde{V}_{12} - \Omega^2\widetilde{T}_{12}\right)^2}, \\ A_2 &= \frac{-h_1\left(\widetilde{V}_{12} - \Omega^2\widetilde{T}_{12}\right)}{\left(\widetilde{V}_{11} - \Omega^2\widetilde{T}_{11}\right)\left(\widetilde{V}_{22} - \Omega^2\widetilde{T}_{22}\right) - \left(\widetilde{V}_{12} - \Omega^2\widetilde{T}_{12}\right)^2}. \end{aligned} \tag{7.8}$$

If

$$\Omega^2 = \frac{\widetilde{V}_{22}}{\widetilde{T}_{22}}, \tag{7.9}$$

one has

$$A_1 = 0, \;\; A_2 = \frac{h_1}{\left(\widetilde{V}_{12} - \Omega^2 \widetilde{T}_{12}\right)}. \tag{7.10}$$

So, when the condition (7.9) is fulfilled, the oscillations corresponding to q_1 are absorbed. This is, in a nutshell, the concept of dynamic absorbers, which will be illustrated in some of the following examples.

Example 7.1 An oscillatory mechanical system consists of two physical pendula and a spring AB of stiffness $k = 12\ mg/a$, which is attached to both pendula. The first pendulum is a right-angle link composed of two rigidly connected bars O_1T and TA. The other pendulum comprises two orthogonal bars O_2E and BD. The masses and lengths of these bars are given in Figure 7.1.1, which shows the system in the equilibrium position around which it performs small oscillations. First, assuming that the system oscillates freely, determine the natural frequencies and modal shapes ratios. Present the modal shapes graphically taking $g = 9.81\ \text{m/s}^2$ and $a = 0.4$ m. Plot the solutions $\varphi(t)$ and $\psi(t)$ for the initial conditions assuming that the system oscillates in the first and second mode. Then, find and plot the solutions $\varphi(t)$ and $\psi(t)$ that correspond to the same initial conditions while an external torque acts on the left pendulum: $\mathfrak{M}(t) = 2mga\sin\Omega t$, with $\Omega = 10\ \text{s}^{-1}$. Finally, determine the value of Ω for which the right pendulum represents the dynamic absorber for the left one, and plot the frequency–amplitude diagram.

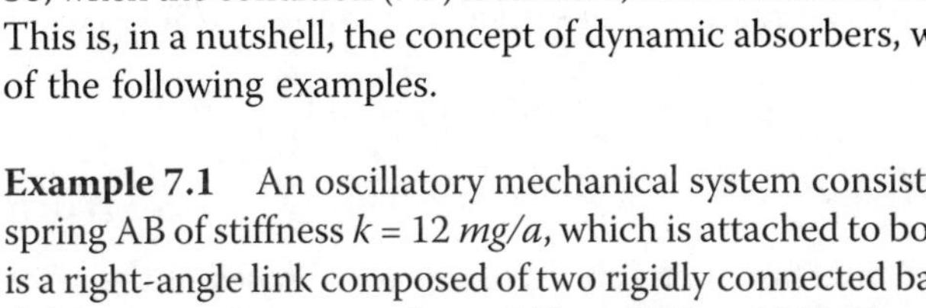

Figure 7.1.1

Solution

The overall potential energy is the sum of the gravitational potential energy and the spring's potential energy:

$$V = -mgy_{C_1} - \sqrt{3}mgy_{C_2} - mgy_{C_3} - 3mgy_{C_4} + \frac{1}{2}k(f_{st} + \lambda)^2, \tag{7.1.1}$$

where the y-coordinates can be found based on Figure 7.1.2, and then approximated as follows

$$y_{C_1} = \frac{a}{2}\cos\varphi \approx \frac{a}{2}\left(1 - \frac{\varphi^2}{2}\right),\;\; y_{C_2} = a\cos\varphi - \frac{a\sqrt{3}}{2}\sin\varphi \approx a\left(1 - \frac{\varphi^2}{2}\right) - \frac{a\sqrt{3}}{2}\varphi,$$

$$y_{C_3} = \frac{a}{2}\cos\psi \approx \frac{a}{2}\left(1 - \frac{\psi^2}{2}\right),\;\; y_{C_4} = a\cos\psi + \frac{a}{2}\sin\psi \approx a\left(1 - \frac{\psi^2}{2}\right) + \frac{a}{2}\psi. \tag{7.1.2}$$

The additional deflection of the spring can be obtained based on the deflection in the horizontal direction u and the deflection in the vertical direction v, as follows

$$\lambda \approx u + \frac{1}{2l_{st}} v^2, \tag{7.1.3}$$

where the length of the spring the static equilibrium position is $l_{st} = 2a$ (see Figure 7.1.1). To find the deflection in the horizontal direction u, one can consider the x-coordinates of the points A and B (see Figure 7.1.2):

$$u = (x_B(\psi) - x_B(0)) - (x_A(\varphi) - x_A(0)) = (a\sin\psi + 2a(1-\cos\psi)) - (a\sin\varphi - a\sqrt{3}(1-\cos\varphi))$$
$$\Rightarrow \quad u \approx (a\psi + a\psi^2) - \left(a\varphi - a\frac{\sqrt{3}}{2}\varphi^2\right), \tag{7.1.4}$$

while for the deflection in the v-direction, one can use the y-coordinates of the same points:

$$v = y_B(\psi) - y_A(\varphi) = (a\cos\psi + 2a\sin\psi) - (a\cos\varphi - a\sqrt{3}\sin\varphi)$$
$$\Rightarrow \quad v \approx 2a\psi + a\sqrt{3}\varphi. \tag{7.1.5}$$

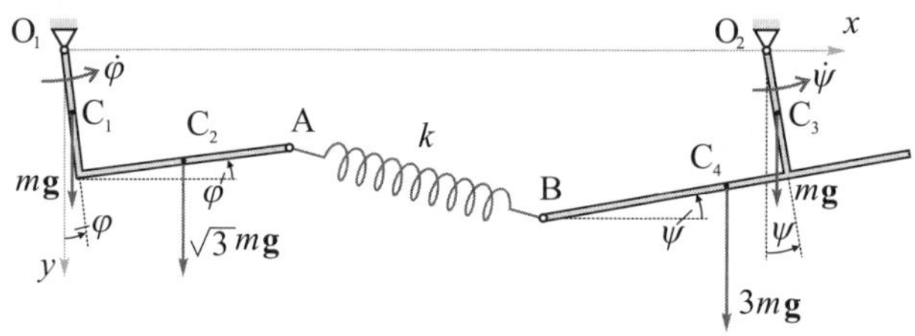

Figure 7.1.2

Substituting Equations (7.1.4) and (7.1.5) into Equation (7.1.3) leads to

$$\lambda \approx a(\psi - \varphi) + a\psi^2 + a\frac{\sqrt{3}}{2}\varphi^2 + \frac{a}{4}\left(2\psi + \sqrt{3}\varphi\right)^2. \tag{7.1.6}$$

The static deflection of the spring f_{st} can be calculated by equating to zero the moment sum about O_1 (Figure 7.1.3a) or O_2 (Figure 7.1.3b):

$$\Sigma M_{O_1} = 0 \;\Rightarrow\; kf_{st}a - \sqrt{3}mga\frac{\sqrt{3}}{2} = 0 \;\Rightarrow\; f_{st} = \frac{3mg}{2k} = \frac{a}{8}. \tag{7.1.7}$$

Using (7.1.1), (7.1.2) and (7.1.4)–(7.1.7), the potential energy of small oscillations is found to be

$$V \approx \frac{1+5\sqrt{3}}{4}mga\varphi^2 + \frac{13mga}{4}\psi^2 + 6mga(\psi-\varphi)^2 + \frac{3mga}{8}\left(2\psi + \sqrt{3}\varphi\right)^2. \tag{7.1.8}$$

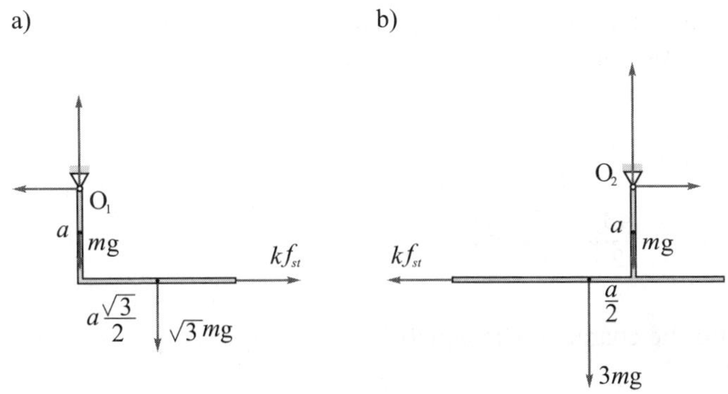

Figure 7.1.3

Both pendula are in rotational motion around the axes that pass through O_1 and O_2, and the system kinetic energy is

$$T = \frac{1}{2} J_{O_1} \dot{\varphi}^2 + \frac{1}{2} J_{O_2} \dot{\psi}^2, \tag{7.1.9}$$

where the mass moments of inertia can be found from the parallel-axis theorem as follows:

$$J_{O_1} = \frac{1}{12} ma^2 + m\overline{O_1C_1}^2 + \frac{1}{12}\sqrt{3}m\left(\sqrt{3}a\right)^2 + \sqrt{3}m\overline{O_1C_2}^2 = \frac{\left(1+6\sqrt{3}\right)}{3} ma^2, \tag{7.1.10}$$

$$J_{O_2} = \frac{1}{12} ma^2 + m\overline{O_2C_3}^2 + \frac{1}{12} 3m(3a)^2 + 3m\overline{O_2C_4}^2 = \frac{19}{3} ma^2. \tag{7.1.11}$$

Thus, the kinetic energy (7.1.9) becomes

$$T = \frac{\left(1+6\sqrt{3}\right)}{6} ma^2 \dot{\varphi}^2 + \frac{19}{6} ma^2 \dot{\psi}^2. \tag{7.1.12}$$

When the torque acts, the following expression for the virtual work holds:

$$\delta A = \delta A^{\mathfrak{M}} = (2mga\sin\Omega t)\,\delta\varphi + (0)\,\delta\psi, \tag{7.1.13}$$

so that the generalised forces can be recognised as

$$Q_\varphi = 2mga\sin\Omega t, \quad Q_\psi = 0. \tag{7.1.14}$$

Using (7.1.8), (7.1.12) and (7.1.14), Lagrange's equations (7.3) give the following equations of motion when the torque acts:

$$\frac{1+6\sqrt{3}}{3} a\ddot{\varphi} + \frac{59+10\sqrt{3}}{4} g\varphi + \frac{3\sqrt{3}-24}{2} g\psi = 2g\sin\Omega t, \tag{7.1.15}$$

$$\frac{19}{3} a\ddot{\psi} + \frac{-24+3\sqrt{3}}{2} g\varphi + \frac{43}{2} g\psi = 0. \tag{7.1.16}$$

When there is no torque, the right-hand side of Equation (7.1.15) is equal to zero. Then, the following characteristic determinant can be derived:

$$\begin{vmatrix} \frac{59+10\sqrt{3}}{4}g-\frac{1+6\sqrt{3}}{3}a\omega^2 & \frac{3\sqrt{3}-24}{2}g \\ \frac{3\sqrt{3}-24}{2}g & \frac{43}{2}g-\frac{19}{3}a\omega^2 \end{vmatrix}=0. \tag{7.1.17}$$

This can be developed into the characteristic equation

$$\left(\frac{59+10\sqrt{3}}{4}g-\frac{1+6\sqrt{3}}{3}a\omega^2\right)\left(\frac{43}{2}g-\frac{19}{3}a\omega^2\right)-\left(\frac{3\sqrt{3}-24}{2}g\right)^2=0. \tag{7.1.18}$$

Its solutions are:

$$\omega_1\approx 1.458242\sqrt{\frac{g}{a}}\approx 7.22161\ \mathrm{s}^{-1},\quad \omega_2\approx 2.508535\sqrt{\frac{g}{a}}\approx 12.42295\ \mathrm{s}^{-1}. \tag{7.1.19}$$

Given Equation (6.12), the following mode shape ratios are obtained

$$b_{1/2}=\frac{3\left(24-3\sqrt{3}\right)g}{129g-38a\,\omega_{1/2}^2}. \tag{7.1.20}$$

After using Eqs. (7.1.19), one obtains $b_1 = 1.170505142$ and $b_2 = -0.5122531891$.

Now, the particular solution for forced vibrations will be found. To that end, the solutions are assumed as follows

$$\varphi_p(t)=A_\varphi \sin\Omega t,\quad \psi_p(t)=A_\psi \sin\Omega t, \tag{7.1.21}$$

where A_φ and A_ψ stand for the respective amplitudes of φ and ψ. Introducing Equation (7.1.21) into Equations (7.1.15) and (7.1.16), the following system of algebraic equations with respect to A_φ and A_ψ is obtained:

$$\left(\frac{59+10\sqrt{3}}{4}g-\frac{1+6\sqrt{3}}{3}a\Omega^2\right)A_\varphi+\left(\frac{3\sqrt{3}-24}{2}g\right)A_\psi=2g, \tag{7.1.22}$$

$$\left(\frac{3\sqrt{3}-24}{2}g\right)A_\varphi+\left(\frac{43}{2}g-\frac{19}{3}a\Omega^2\right)A_\psi=0. \tag{7.1.23}$$

Their solutions are

$$A_\varphi=\frac{2g\left(\frac{43}{2}g-\frac{19}{3}a\Omega^2\right)}{\left(\frac{59+10\sqrt{3}}{4}g-\frac{1+6\sqrt{3}}{3}a\Omega^2\right)\left(\frac{43}{2}g-\frac{19}{3}a\Omega^2\right)-\left(\frac{3\sqrt{3}-24}{2}g\right)^2}, \tag{7.1.24}$$

$$A_{\psi} = \frac{-2g\left(\frac{3\sqrt{3}-24}{2}g\right)}{\left(\frac{59+10\sqrt{3}}{4}g - \frac{1+6\sqrt{3}}{3}a\Omega^2\right)\left(\frac{43}{2}g - \frac{19}{3}a\Omega^2\right) - \left(\frac{3\sqrt{3}-24}{2}g\right)^2}. \tag{7.1.25}$$

The solutions $\varphi(t)$ and $\psi(t)$ corresponding to the first and second mode are plotted in Figures 7.1.4a and 7.1.4b, while the related solutions corresponding to the case when the torque acts are presented in Figure 7.1.4c and 7.1.4d.

The purpose of the dynamic absorber is to make the amplitude A_{φ} equal to zero, as a result of which the left pendulum does not oscillate while the right one does. Given Equation (7.1.24), this amplitude is equal to zero if

$$\frac{43}{2}g - \frac{19}{3}a\Omega_{da}^2 = 0 \;\Rightarrow\; \Omega_{da} = \sqrt{\frac{129g}{38a}} \approx 9.1245\ \mathrm{s}^{-1}. \tag{7.1.26}$$

The frequency–amplitude diagrams for both generalised coordinates are shown in Figure 7.1.5. The value of the frequency corresponding to the dynamic absorber is also labelled.

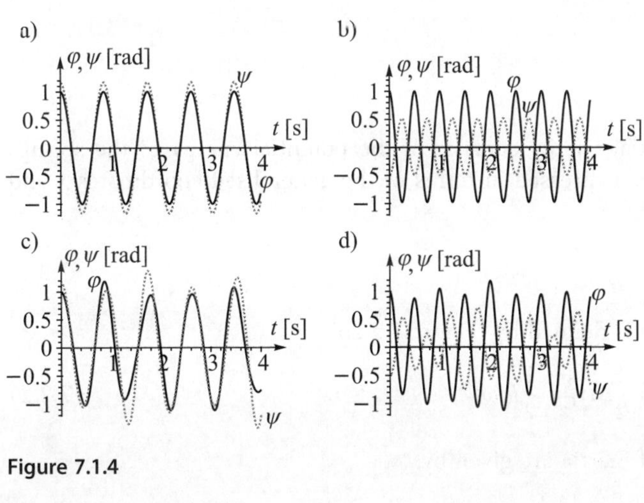

Figure 7.1.4

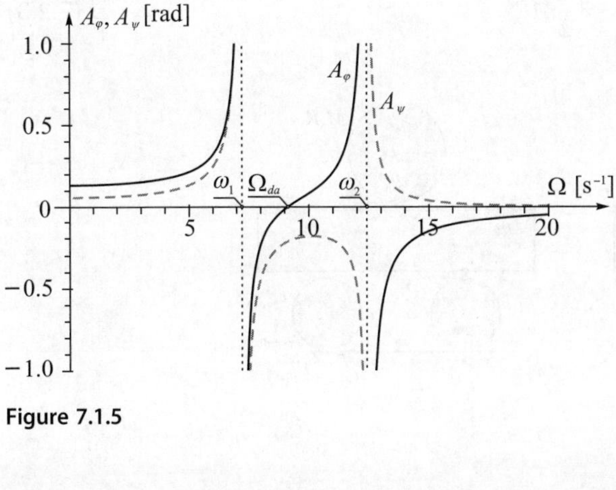

Figure 7.1.5

Example 7.2 A platform of mass $4m$ is subjected to the horizontal external harmonic excitation $\mathbf{F}(t) = F_0\cos\Omega t\mathbf{i}$, and is also attached to a spring AB of stiffness k_1, and can move in the horizontal direction. Two identical discs of mass m and radius r are placed below the platform and roll without slipping along a fixed horizontal surface. There is no slipping between the platform and the discs. In order to dynamically absorb the motion of the platform, a spring of stiffness k_2 is added, which is attached to a disc of mass M and radius R, but in two different ways, as shown in Figure 7.2.1a and 7.2.1b. In the first case, this disc rolls without slipping along the platform, and in the second case, along the horizontal surface. Assuming $m = 2.8$ kg, $F_0 = 100$ N, $\Omega = 18.732$ s^{-1}, $k_1 = 4000$ N/m and $k_2 = 3000$ N/m, determine M in both cases, and show the frequency–amplitude diagrams. The generalised coordinates are labelled in Figures 7.2.1a and 7.2.1b.

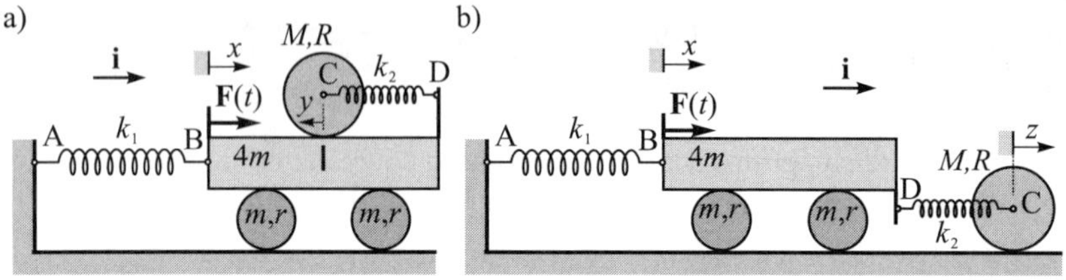

Figure 7.2.1

Solution

In the first case, the potential energy consists of the potential energy of the springs (Figure 7.2.2a), which can be expressed in terms of the generalised coordinates x and y as follows

$$V = \frac{1}{2}k_1x^2 + \frac{1}{2}k_2y^2. \tag{7.2.1}$$

The overall kinetic energy is defined by (Figure 7.2.2b)

$$T = \frac{1}{2}4m\dot{x}^2 + 2\frac{1}{2}J_P\omega_I^2 + \frac{1}{2}Mv_C^2 + \frac{1}{2}J_C\omega_{II}^2, \tag{7.2.2}$$

where the mass moments of inertia are given by:

$$J_P = \frac{1}{2}mr^2 + mr^2 = \frac{3}{2}mr^2, \quad J_C = \frac{1}{2}MR^2. \tag{7.2.3}$$

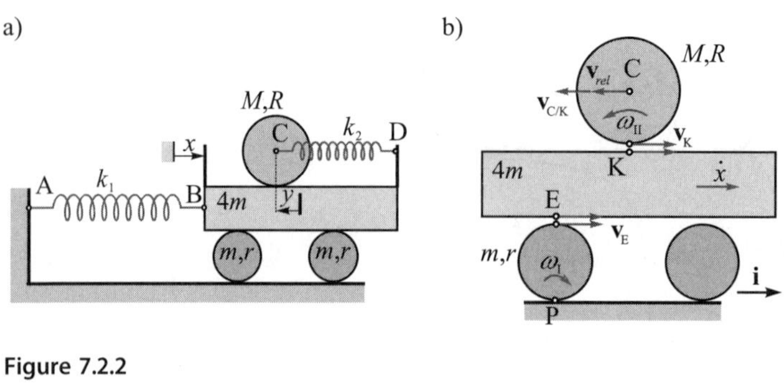

Figure 7.2.2

Using the equality of the velocities of the contact points during rolling without slipping and the basic kinematic equations for general plane motion and compound motion for the point C, one can obtain

$$\omega_{\mathrm{I}} = \frac{v_{\mathrm{E}}}{2r} = \frac{\dot{x}}{2r}, \tag{7.2.4}$$

$$\mathbf{v}_{\mathrm{C}} = \mathbf{v}_{tr} + \mathbf{v}_{rel} = \dot{x}\mathbf{i} - \dot{y}\mathbf{i}, \ \ \mathbf{v}_{\mathrm{C}} = \mathbf{v}_{\mathrm{K}} + \mathbf{v}_{\mathrm{C/K}} \ \Rightarrow \ \dot{x}\mathbf{i} - \dot{y}\mathbf{i} = \dot{x}\mathbf{i} - \omega_{\mathrm{II}} R\mathbf{i} \ \Rightarrow \ \omega_{\mathrm{II}} = \frac{\dot{y}}{R}. \tag{7.2.5}$$

Finally, substituting (7.2.3)–(7.2.5) into Equation (7.2.2), the kinetic energy becomes

$$T = \frac{19}{8} m\dot{x}^2 + \frac{1}{2} M(\dot{x} - \dot{y})^2 + \frac{1}{4} M\dot{y}^2. \tag{7.2.6}$$

The virtual work stems from the external excitation $\mathbf{F}(t)$ and is given by

$$\delta A = \delta A^{\mathbf{F}} = (F_0 \cos\Omega t)\,\delta x + (0)\,\delta y. \tag{7.2.7}$$

The coefficients in front of the variations of the generalised coordinates give the generalised forces as follows

$$Q_x = F_0 \cos\Omega t, \ \ Q_y = 0. \tag{7.2.8}$$

Given Equations (7.2.1), (7.2.6) and (7.2.8), Lagrange's equations (7.3) yield the following differential equations of motion:

$$\frac{19}{4} m\ddot{x} + M(\ddot{x} - \ddot{y}) + k_1 x = F_0 \cos\Omega t, \tag{7.2.9}$$

$$-M(\ddot{x} - \ddot{y}) + \frac{M}{2}\ddot{y} + k_2 y = 0. \tag{7.2.10}$$

The particular solutions have the form

$$x_p(t) = A_x \cos\Omega t, \ \ y_p(t) = A_y \cos\Omega t, \tag{7.2.11}$$

where A_x and A_y stand for the amplitudes in x and y, respectively.

Substituting (7.2.11) into (7.2.9)–(7.2.10), the following system of algebraic equations with respect to A_x and A_y is derived:

$$\left(k_1 - \left(\frac{19}{4} m + M\right)\Omega^2\right) A_x + (M\Omega^2) A_y = F_0, \tag{7.2.12}$$

$$(M\Omega^2) A_x + \left(k_2 - \frac{3}{2} M\Omega^2\right) A_y = 0. \tag{7.2.13}$$

Its solutions are

$$A_x = \frac{F_0\left(k_2 - \frac{3}{2} M\Omega^2\right)}{\left(k_1 - \left(\frac{19}{4} m + M\right)\Omega^2\right)\left(k_2 - \frac{3}{2} M\Omega^2\right) - (M\Omega^2)^2}, \tag{7.2.14}$$

$$A_y = \frac{-F_0 M\Omega^2}{\left(k_1 - \left(\frac{19}{4}m + M\right)\Omega^2\right)\left(k_2 - \frac{3}{2}M\Omega^2\right) - \left(M\Omega^2\right)^2}. \tag{7.2.15}$$

Given the requirement that the amplitude A_x is equal to zero, Equation (7.2.14) yields

$$k_2 - \frac{3}{2}M\Omega^2 = 0 \Rightarrow M = \frac{2k_2}{3\Omega^2} \approx 5.7\,\text{kg}. \tag{7.2.16}$$

The corresponding frequency–amplitude diagram is plotted in Figure 7.2.3, with the natural frequencies $\omega_1 \approx 13.245\ \text{s}^{-1}$, $\omega_2 \approx 22.941\ \text{s}^{-1}$. These frequencies as well as the one corresponding to the dynamic absorber are clearly labelled in this figure.

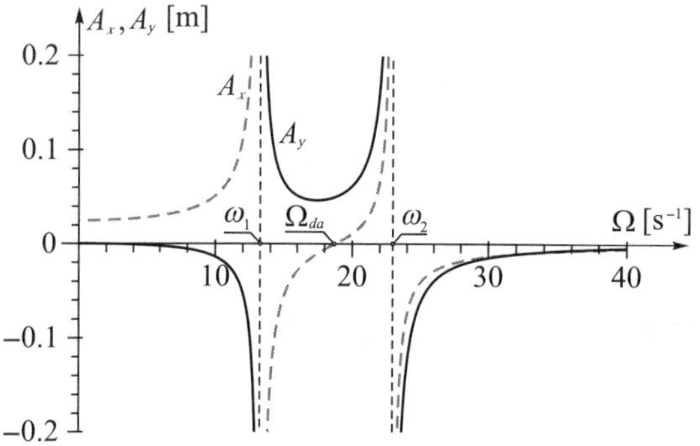

Figure 7.2.3

In the second case, the overall potential energy can be expressed in terms of the generalised coordinates x and z (Figure 7.2.4a) as follows

$$V = \frac{1}{2}k_1 x^2 + \frac{1}{2}k_2 (z - x)^2. \tag{7.2.17}$$

The kinetic energy is defined by Equation (7.2.2), where J_P, J_C and ω_I are the same as in Equation (7.2.3) and Equation (7.2.4). The instantaneous centre of zero velocity is located in the point Q (Figure 7.2.4b), so that:

$$\mathbf{v}_C = \dot{z}\mathbf{i}, \quad \omega_{II} = \frac{\dot{z}}{R}. \tag{7.2.18}$$

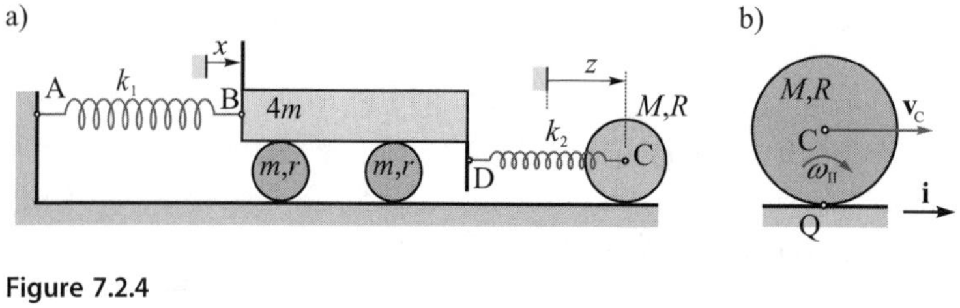

Figure 7.2.4

Substituting (7.2.3), (7.2.4) and (7.2.18) into Equation (7.2.2), the kinetic energy takes the form

$$T = \frac{19}{8}m\dot{x}^2 + \frac{3}{4}M\dot{z}^2. \tag{7.2.19}$$

The virtual work of the non-conservative force $\mathbf{F}(t)$ is the same as defined previously, so that the generalised forces for the generalised coordinates x and z are

$$Q_x = F_0 \cos\Omega t, \quad Q_z = 0. \tag{7.2.20}$$

Using (7.2.17), (7.2.19) and (7.2.20), Lagrange's equations (7.3) give the following differential equations of motion:

$$\frac{19}{4}m\ddot{x} + k_1 x - k_2(z-x) = F_0\cos\Omega t, \tag{7.2.21}$$

$$\frac{3}{2}M\ddot{z} + k_2(z-x) = 0. \tag{7.2.22}$$

The particular solutions are taken in the form

$$x_p(t) = A_x \cos\Omega t, \quad z_p(t) = A_z \cos\Omega t, \tag{7.2.23}$$

where A_x and A_z stand for the respective amplitudes of motion.

Introducing Equation (7.2.23) into Equations (7.2.21) and (7.2.22), one can derive:

$$\left(k_1 + k_2 - \frac{19}{4}m\Omega^2\right)A_x + (-k_2)A_z = F_0, \tag{7.2.24}$$

$$(-k_2)A_x + \left(k_2 - \frac{3}{2}M\Omega^2\right)A_z = 0. \tag{7.2.25}$$

The corresponding solutions are

$$A_x = \frac{F_0\left(k_2 - \frac{3}{2}M\Omega^2\right)}{\left(k_1 + k_2 - \frac{19}{4}m\Omega^2\right)\left(k_2 - \frac{3}{2}M\Omega^2\right) - k_2^2}, \tag{7.2.26}$$

$$A_z = \frac{F_0 k_2}{\left(k_1 + k_2 - \frac{19}{4}m\Omega^2\right)\left(k_2 - \frac{3}{2}M\Omega^2\right) - k_2^2}. \tag{7.2.27}$$

The disc will be the dynamic absorber for the platform when the amplitude A_x is equal to zero, which is satisfied when

$$k_2 - \frac{3}{2}M\Omega^2 = 0 \;\Rightarrow\; M = \frac{2k_2}{3\Omega^2} \approx 5.7\,\text{kg}. \tag{7.2.28}$$

The frequency–amplitude diagram is plotted in Figure 7.2.5. The natural frequencies $\omega_1 \approx 11.996\ \text{s}^{-1}$, $\omega_2 \approx 27.079\ \text{s}^{-1}$ are labelled as well as the frequency corresponding to the dynamic absorber.

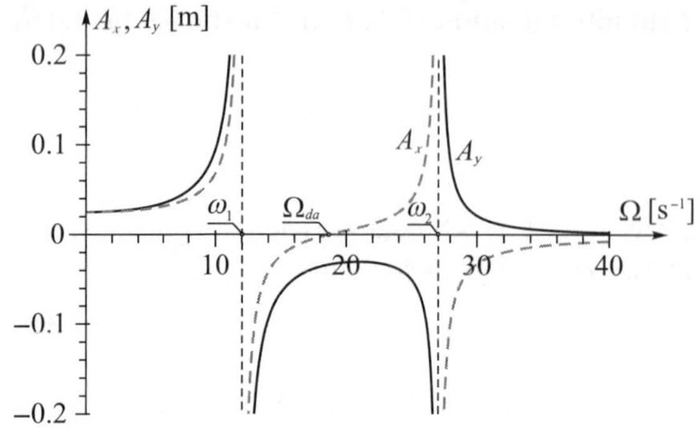

Figure 7.2.5

It is interesting to note that in both cases considered, the mass of the disc should be tuned to $M \approx 5.7$ kg, but the natural frequencies change and differ more with respect to each other in the second case.

Example 7.3 A disc of mass m and radius R rolls without slipping along a block that moves in accordance with $s(t) = s_0 \sin \Omega t$. The centre C of the disc is attached to this block with two springs AC and CB of stiffness k_1 (Figure 7.3.1). A body of mass M is placed on the disc and can move in the horizontal direction so that there is no slipping between these two elements. The body is attached to a block of mass M_a with a spring PT of stiffness k_2. The block moves along a horizontal guide. The oblique springs have length $R/\sin\alpha$ in the static equilibrium position and are extended for length δ in this position when the lower block does not move. The generalised coordinates are: y, which represents the relative displacement of the centre C of the disc with respect to the lower block, and x, which defines the motion of the upper block. The parameters of the system are: $m = 0.4$ kg, $M = 0.6$ kg, $M_a = 0.3$ kg, $R = 20$ cm, $s_0 = 1$ cm, $\Omega = 4.5\ \text{s}^{-1}$, $k_1 = 14$ N/cm, $k_2 = 9$ N/cm, $\alpha = 30°$ and $\delta = 2$ cm. Plot the solutions $y(t)$ and $x(t)$ for the case when the kinematic excitation $s(t)$ exists, as well as without it with three different sets of the initial conditions:

i) $y(0) = 1$ cm, $x(0) = b_1$ cm, $\dot{y}(0) = 0$, $\dot{x}(0) = 0$, which correspond to the first mode;
ii) $y(0) = 1$ cm, $x(0) = b_2$ cm, $\dot{y}(0) = 0$, $\dot{x}(0) = 0$, which correspond to the second mode;

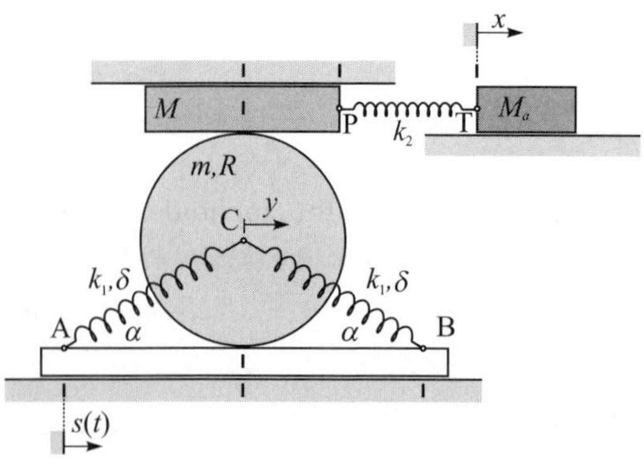

Figure 7.3.1

iii) $y(0) = 1$ cm, $x(0) = (b_1 + b_2)/2$ cm, $\dot{y}(0) = 0$, $\dot{x}(0) = 0$, which correspond to the case containing both angular frequencies.

Solution

The main part of the kinetic energy is given by Equation (5.6.8i), with the addition of the kinetic energy of the upper block (Figure 7.3.2). As this block performs translatory motion and its position is defined by the absolute coordinate x, the overall kinetic energy takes the form

$$T = \frac{1}{2}m(\dot{s}(t) + \dot{y})^2 + \frac{1}{4}m\dot{y}^2 + \frac{1}{2}M(\dot{s}(t) + 2\dot{y})^2 + \frac{1}{2}M_a\dot{x}^2. \tag{7.3.1}$$

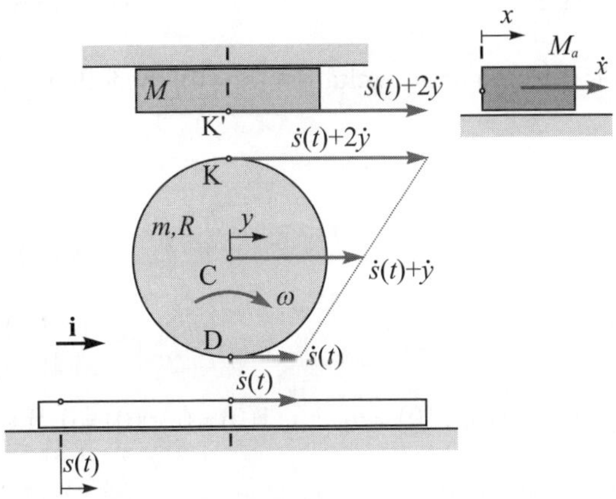

Figure 7.3.2

The potential energy for this system with two degrees of freedom (Figure 7.3.3) can be obtained in a similar way to Equation (5.6.11i), and has the form

$$V = \frac{1}{2}k_1\left(2\cos^2\alpha + \frac{2\delta\sin^3\alpha}{R}\right)y^2 + \frac{1}{2}k_2(x - s(t) - 2y)^2. \tag{7.3.2}$$

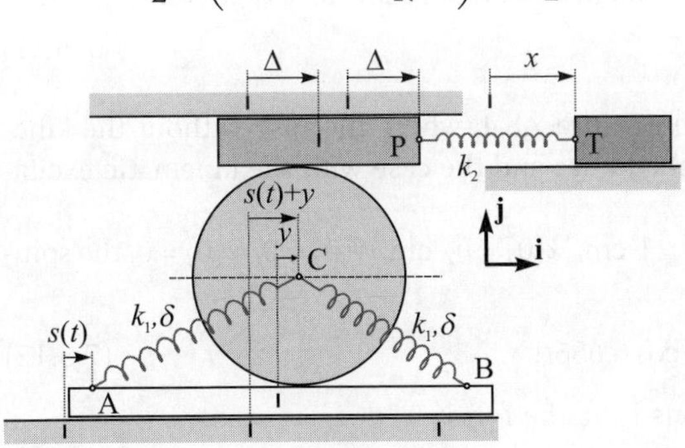

Figure 7.3.3

The derivatives needed for Lagrange's equations (7.3) are

$$\frac{d}{dt}\frac{\partial T}{\partial \dot{y}} = \left(\frac{3}{2}m + 4M\right)\ddot{y} + (m + 2M)\ddot{s}(t), \quad \frac{d}{dt}\frac{\partial T}{\partial \dot{x}} = M_a\ddot{x}, \tag{7.3.3}$$

$$\frac{\partial V}{\partial y} = \left(k_1\left(2\cos^2\alpha + \frac{2\delta\sin^3\alpha}{R}\right) + 4k_2\right)y - 2k_2(x - s(t)),\quad \frac{\partial V}{\partial x} = k_2(x - s(t) - 2y). \tag{7.3.4}$$

Thus, Lagrange's equations for the generalised coordinates y and x are:

$$\left(\frac{3}{2}m + 4M\right)\ddot{y} + \left(k_1\left(2\cos^2\alpha + \frac{2\delta\sin^3\alpha}{R}\right) + 4k_2\right)y - 2k_2x \\ = \left((m + 2M)\Omega^2 - 2k_2\right)s_0\sin\Omega t, \tag{7.3.5}$$

$$M_a\ddot{x} + k_2(x - 2y) = k_2s_0\sin\Omega t. \tag{7.3.6}$$

By considering these equations without the kinematic excitation ($s_0 = 0$), the following natural frequencies are obtained:

$$\omega_1 \approx 2.195\ \mathrm{s}^{-1},\quad \omega_2 \approx 6.656\ \mathrm{s}^{-1}, \tag{7.3.7}$$

while the modal shapes ratios are

$$b_1 = \frac{A_x}{A_y} \approx 2.383,\quad b_2 = \frac{A_x}{A_y} \approx -4.197. \tag{7.3.8}$$

For the first set of initial conditions: $y(0) = 1$ cm, $x(0) = b_1$ cm, $\dot{y}(0) = 0$, $\dot{x}(0) = 0$, the solutions for free oscillations are

$$y(t) \approx \cos 2.195t,\quad x(t) \approx 2.383\cos 2.195t, \tag{7.3.9}$$

while for the case with kinematic excitation, these solutions are

$$y(t) \approx -0.61124\sin 4.5t + 0.07932\sin 6.656t + \cos 2.195t + 1.01240\sin 2.195t, \tag{7.3.10}$$

$$x(t) \approx -0.6845\sin 4.5t - 0.3329\sin 6.656t + 2.383\cos 2.195t + 2.4124\sin 2.195t. \tag{7.3.11}$$

These four solutions are presented in Figure 7.3.4, where the case without the kinematic excitation is depicted by the dashed lines and the case with the kinematic excitation by the solid line.

When the initial conditions are: $y(0) = 1$ cm, $x(0) = b_2$ cm, $\dot{y}(0) = 0$, $\dot{x}(0) = 0$, the solutions for free vibrations are

$$y(t) \approx \cos 6.656t,\quad x(t) \approx -4.197\cos 6.656t, \tag{7.3.12}$$

while the solutions for forced vibrations have the form

$$y(t) \approx -0.61124\sin 4.5t + \cos 6.656t + 0.07932\sin 6.656t + 1.01240\sin 2.195t, \tag{7.3.13}$$

$$x(t) \approx -0.6845\sin 4.5t - 4.197\cos 6.656t - 0.3329\sin 6.656t + 2.4124\sin 2.195t. \tag{7.3.14}$$

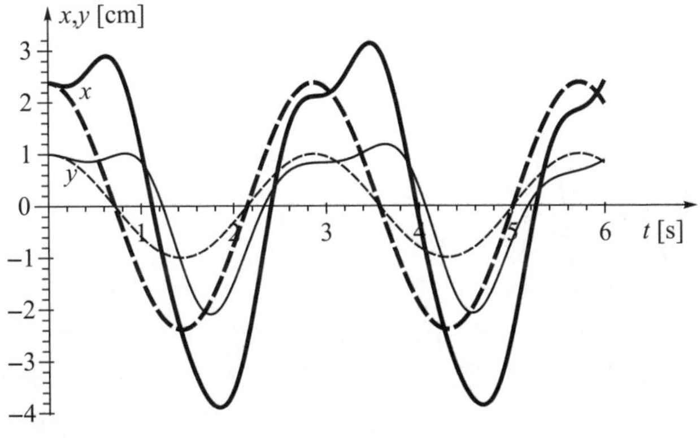

Figure 7.3.4

These four solutions are presented in Figure 7.3.5, where the case without the kinematic excitation is depicted by the dashed lines and the case with the kinematic excitation is plotted as the solid line.

In the third case with the initial conditions: $y(0) = 1$ cm, $x(0) = (b_1 + b_2)/2$ cm, $\dot{y}(0) = 0$, $\dot{x}(0) = 0$, the solutions for free vibrations are

$$y(t) \approx 0.5\cos 6.656t + 0.5\cos 2.195t, \quad x(t) \approx -2.098\cos 6.656t + 1.191\cos 2.195t, \tag{7.3.15}$$

while the forced responses are given by

$$\begin{aligned} y(t) \approx & -0.611\sin 4.5t + 0.5\cos 6.656t + 0.07932\sin 6.656t \\ & + 0.5\cos 2.195t + 1.0124\sin 2.195t, \end{aligned} \tag{7.3.16}$$

$$\begin{aligned} x(t) \approx & -0.685\sin 4.5t - 2.098\cos 6.656t - 0.3329\sin 6.656t \\ & + 1.191\cos 2.195t + 2.412\sin 2.195t. \end{aligned} \tag{7.3.17}$$

They are plotted in Figure 7.3.6, using the same legend as before.

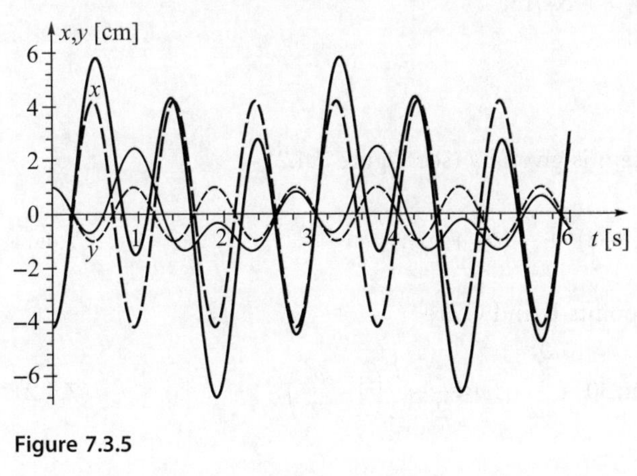

Figure 7.3.5

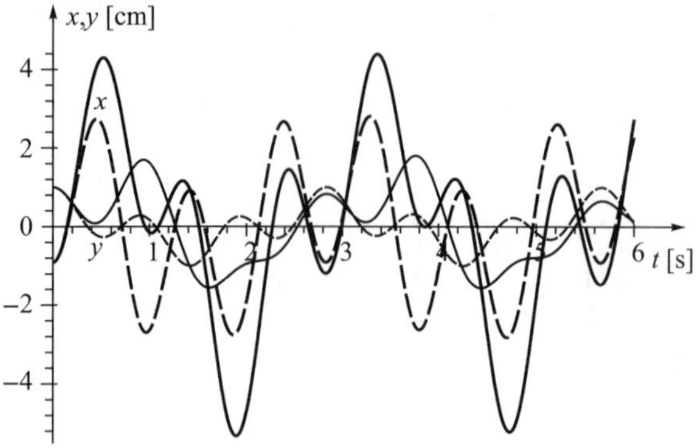

Figure 7.3.6

Example 7.4 A slider of mass m moves along a frictionless inclined guide of angle 30°. A spring AB of stiffness $k_1 = 12mg/l$ is attached to it. Its direction coincides with the guide, while its length in the static position is l (Figure 7.4.1). A bar BD of mass $M = 2\,m$ and length l is hinged to the slider and there is also a torsional spring of stiffness $k_2 = 6mgl$ between them. The system also contains a linear viscous damper AD of damping coefficient c. A force $\mathbf{F}(t) = F_0 \sin\Omega t \mathbf{i}_1$, with $F_0 = mg$ acts on the slider. The generalised coordinates are z and φ, and the static equilibrium position when the force does not act is shown in Figure 7.4.1. Determine the solutions for forced vibrations and also plot the frequency–amplitude diagram, assuming $m = 1$ kg, $l = 0.8$ m, $g = 9.81$ m/s^2 and for three values of c: (1) $c_1 = 1$ Ns/m, (2) $c_2 = 3$ Ns/m, (3) $c_3 = 5$ Ns/m.

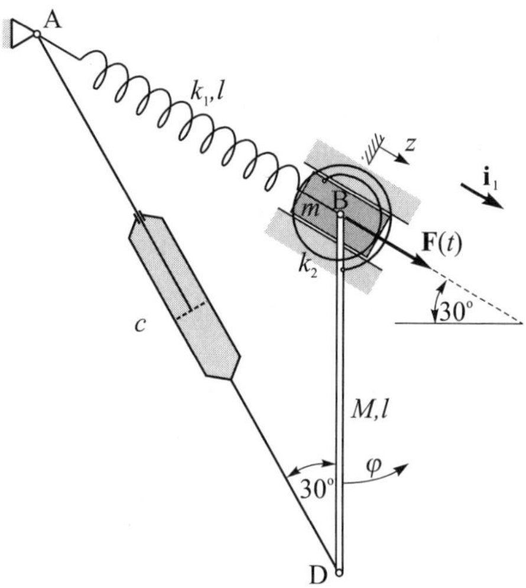

Figure 7.4.1

Solution

The potential energy of the system is given by (see Figure 7.4.2):

$$V = -mgy_B - Mgy_C + \frac{k_1}{2}(f_{st} + z)^2 + \frac{k_2}{2}(\alpha_{st} + \varphi)^2, \tag{7.4.1}$$

where the y-coordinates of the points B and C are

$$y_B = z\sin 30^\circ = \frac{z}{2}, \quad y_C = z\sin 30^\circ + \frac{l}{2}\cos\varphi \approx \frac{z}{2} + \frac{l}{2}\left(1 - \frac{\varphi^2}{2}\right). \tag{7.4.2}$$

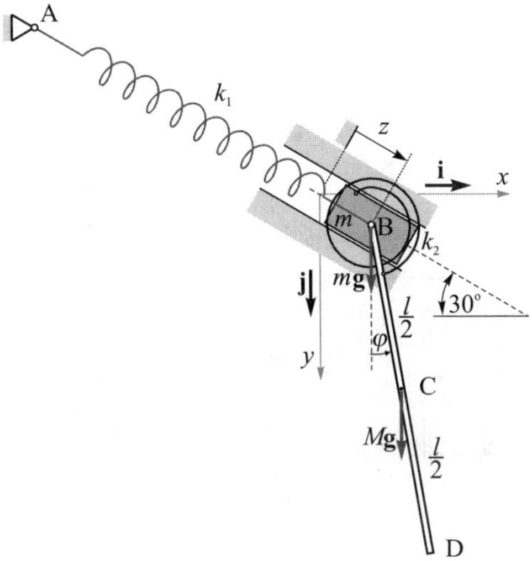

Figure 7.4.2

To determine the static deflection of the torsional spring, one can consider the moment sum about the point B (Figure 7.4.3a):

$$\sum M_B = 0 \;\Rightarrow\; -k_2 \alpha_{st} = 0 \;\Rightarrow\; \alpha_{st} = 0. \tag{7.4.3}$$

To obtain the static deflection of the spring AB, the equilibrium equation of the forces in the z-direction is utilised (Figure 7.4.3b):

$$\sum F_z = 0 \;\Rightarrow\; -k_1 f_{st} + mg\sin 30^\circ + Mg\sin 30^\circ = 0 \;\Rightarrow\; f_{st} = \frac{(m+M)g}{2k_1}. \tag{7.4.4}$$

Thus, the potential energy is found to be

$$V \approx \frac{k_1}{2} z^2 + \frac{1}{2}\left(k_2 + \frac{Mgl}{2}\right)\varphi^2. \tag{7.4.5}$$

The overall kinetic energy is defined by

$$T = \frac{1}{2} m v_B{}^2 + \frac{1}{2} M v_C{}^2 + \frac{1}{2} J_C \dot{\varphi}^2, \tag{7.4.6}$$

where

$$J_C = \frac{1}{12} M l^2. \tag{7.4.7}$$

To determine the velocities of the points B and C while the system is assumed to be passing through the equilibrium position (Figure 7.4.3c), the following kinematic relationships are used:

$$\mathbf{v}_C = \mathbf{v}_B + \mathbf{v}_{C/B}, \quad \mathbf{v}_B = \dot{z}\mathbf{i}_1 = \dot{z}\left(\cos 30^\circ \mathbf{i} + \sin 30^\circ \mathbf{j}\right), \quad \mathbf{v}_{C/B} = \frac{l}{2}\dot{\varphi}\mathbf{i} \;\Rightarrow \tag{7.4.8}$$

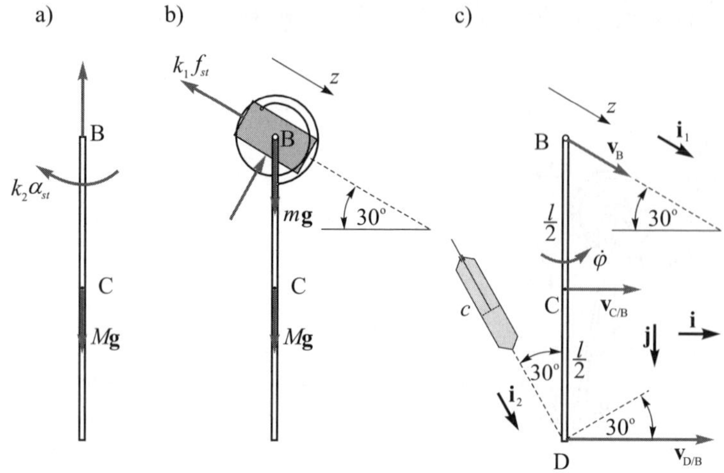

Figure 7.4.3

$$\mathbf{v}_C = \left(\dot{z}\frac{\sqrt{3}}{2} + \frac{l}{2}\dot{\varphi}\right)\mathbf{i} + \dot{z}\frac{1}{2}\mathbf{j}, \quad v_C{}^2 = \mathbf{v}_C \cdot \mathbf{v}_C = \left(\dot{z}\frac{\sqrt{3}}{2} + \frac{l}{2}\dot{\varphi}\right)^2 + \left(\frac{\dot{z}}{2}\right)^2. \tag{7.4.9}$$

Given Equations (7.4.6)–(7.4.9), the kinetic energy becomes

$$T \approx \frac{1}{2}\left(m + \frac{M}{4}\right)\dot{z}^2 + \frac{M}{2}\left(\dot{z}\frac{\sqrt{3}}{2} + \frac{l}{2}\dot{\varphi}\right)^2 + \frac{1}{2}\frac{1}{12}Ml^2\dot{\varphi}^2. \tag{7.4.10}$$

The dissipation function is defined by

$$D = \frac{1}{2}cv_{p/c}^2, \tag{7.4.11}$$

where the relative velocity of the piston with respect to the cylinder $v_{p/c}$ (Figure 7.4.3c) can be obtained as follows

$$\begin{gathered} v_{p/c} = \mathbf{v}_D \cdot \mathbf{i}_2, \quad \mathbf{v}_D = \mathbf{v}_B + \mathbf{v}_{D/B}, \quad \mathbf{v}_{D/B} = l\dot{\varphi}\mathbf{i}, \quad \mathbf{i}_2 = \sin 30^\circ \mathbf{i} + \cos 30^\circ \mathbf{j} \Rightarrow \\ v_{p/c} = \left(\dot{z}(\cos 30^\circ \mathbf{i} + \sin 30^\circ \mathbf{j}) + l\dot{\varphi}\mathbf{i}\right) \cdot \left(\sin 30^\circ \mathbf{i} + \cos 30^\circ \mathbf{j}\right) = \dot{z}\frac{\sqrt{3}}{2} + \frac{l}{2}\dot{\varphi}. \end{gathered} \tag{7.4.12}$$

Consequently, the dissipation function takes the form

$$D = \frac{1}{2}c\left(\dot{z}\frac{\sqrt{3}}{2} + \frac{l}{2}\dot{\varphi}\right)^2. \tag{7.4.13}$$

The virtual work of the non-conservative force $\mathbf{F}(t)$ is

$$\delta A = \delta A^{\mathbf{F}} = (F_0 \sin \Omega t)\,\delta z + (0)\,\delta\varphi, \tag{7.4.14}$$

where the generalised forces are recognised as

$$Q_z = F_0 \sin \Omega t, \quad Q_\varphi = 0. \tag{7.4.15}$$

Using (7.4.5), (7.4.10), (7.4.13) and (7.4.15), Lagrange's equations (7.2) give the following equations of motion:

$$(m+M)\ddot{z} + \frac{\sqrt{3}}{4}Ml\ddot{\varphi} + c\frac{\sqrt{3}}{2}\left(\frac{\sqrt{3}}{2}\dot{z} + \frac{l}{2}\dot{\varphi}\right) + k_1 z = F_0 \sin \Omega t, \tag{7.4.16}$$

$$\frac{\sqrt{3}}{4}Ml\ddot{z} + \frac{1}{3}Ml^2\ddot{\varphi} + c\frac{l}{2}\left(\frac{\sqrt{3}}{2}\dot{z} + \frac{l}{2}\dot{\varphi}\right) + \left(k_2 + \frac{Mgl}{2}\right)\varphi = 0. \tag{7.4.17}$$

By omitting the non-homogeneous and dissipation terms in these equations, the natural frequencies can be obtained (see Chapter 6), and they are calculated to be $\omega_1 \approx 6.450\ \mathrm{s}^{-1}$ and $\omega_2 \approx 15.585\ \mathrm{s}^{-1}$.

The particular solution of the system formed by Equations (7.4.16) and (7.4.17) is

$$z_p(t) = A_{z1}\sin \Omega t + A_{z2}\cos \Omega t, \quad \varphi_p(t) = A_{\varphi 1}\sin \Omega t + A_{\varphi 2}\cos \Omega t, \tag{7.4.18}$$

$$A_z = \sqrt{(A_{z1})^2 + (A_{z2})^2}, \quad A_\varphi = \sqrt{(A_{\varphi 1})^2 + (A_{\varphi 2})^2}. \tag{7.4.19}$$

Substituting these into (7.4.16) and (7.4.17), one derives:

$$\begin{aligned}&\left((k_1 - m\Omega^2 - M\Omega^2)A_{z1} - \frac{3}{4}c\Omega A_{z2} - \frac{\sqrt{3}}{4}Ml\Omega^2 A_{\varphi 1} - \frac{\sqrt{3}}{4}cl\Omega A_{\varphi 2}\right)\sin \Omega t\\ &+ \left(\frac{3}{4}c\Omega A_{z1} + (k_1 - m\Omega^2 - M\Omega^2)A_{z2} + \frac{\sqrt{3}}{4}cl\Omega A_{\varphi 1} - \frac{\sqrt{3}}{4}Ml\Omega^2 A_{\varphi 2}\right)\cos \Omega t\\ &= F_0 \sin \Omega t,\end{aligned} \tag{7.4.20}$$

$$\begin{aligned}&\left(-\frac{\sqrt{3}}{4}Ml\Omega^2 A_{z1} - \frac{\sqrt{3}}{4}cl\Omega A_{z2} + \left(k_2 + \frac{1}{2}Mgl - \frac{1}{3}Ml^2\Omega^2\right)A_{\varphi 1} - \frac{1}{4}cl^2\Omega A_{\varphi 2}\right)\sin \Omega t\\ &+ \left(\frac{\sqrt{3}}{4}cl\Omega A_{z1} - \frac{\sqrt{3}}{4}Ml\Omega^2 A_{z2} + \frac{1}{4}cl^2\Omega A_{\varphi 1} + \left(k_2 + \frac{1}{2}Mgl - \frac{1}{3}Ml^2\Omega^2\right)A_{\varphi 2}\right)\cos \Omega t = 0.\end{aligned} \tag{7.4.21}$$

The system of resulting algebraic equations with respect to A_{z1}, A_{z2}, $A_{\varphi 1}$ and $A_{\varphi 2}$, is

$$(k_1 - m\Omega^2 - M\Omega^2)A_{z1} - \frac{3}{4}c\Omega A_{z2} - \frac{\sqrt{3}}{4}Ml\Omega^2 A_{\varphi 1} - \frac{\sqrt{3}}{4}cl\Omega A_{\varphi 2} = F_0, \tag{7.4.22}$$

$$\frac{3}{4}c\Omega A_{z1} + (k_1 - m\Omega^2 - M\Omega^2)A_{z2} + \frac{\sqrt{3}}{4}cl\Omega A_{\varphi 1} - \frac{\sqrt{3}}{4}Ml\Omega^2 A_{\varphi 2} = 0, \tag{7.4.23}$$

$$-\frac{\sqrt{3}}{4}Ml\Omega^2 A_{z1} - \frac{\sqrt{3}}{4}cl\Omega A_{z2} + \left(k_2 + \frac{1}{2}Mgl - \frac{1}{3}Ml^2\Omega^2\right)A_{\varphi 1} - \frac{1}{4}cl^2\Omega A_{\varphi 2} = 0, \tag{7.4.24}$$

$$\frac{\sqrt{3}}{4}cl\Omega A_{z1} - \frac{\sqrt{3}}{4}Ml\Omega^2 A_{z2} + \frac{1}{4}cl^2\Omega A_{\varphi 1} + \left(k_2 + \frac{1}{2}Mgl - \frac{1}{3}Ml^2\Omega^2\right)A_{\varphi 2} = 0. \tag{7.4.25}$$

Solving these for the given parameters, the frequency–amplitude diagrams shown in Figure 7.4.4 and 7.4.5 are obtained.

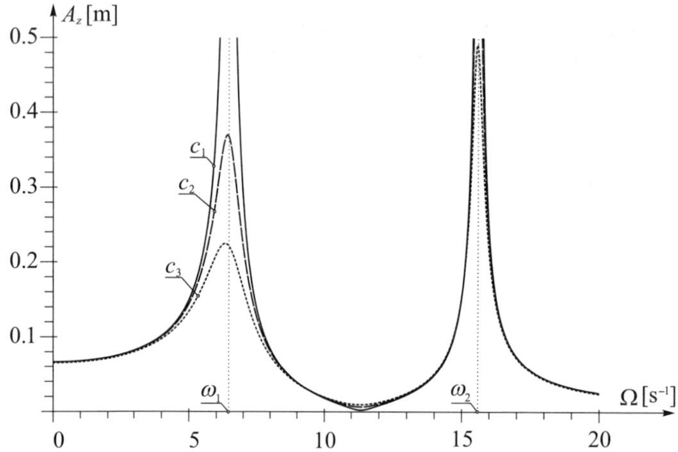

Figure 7.4.4

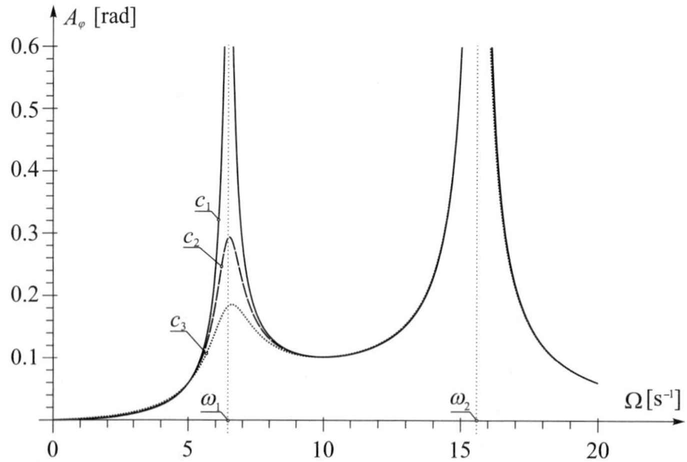

Figure 7.4.5

Example 7.5 A platform moves harmonically in the horizontal plane in the ξ-direction in accordance with the law $s(t) = s_0 \sin \Omega t$. A particle of mass m is attached to two orthogonal springs of stiffness k_1 and k_2, and moves along the platform. The position of one of the springs in the static equilibrium position when there is no kinematic excitation is shown in Figure 7.5.1 and is defined by the angle α. The generalised coordinates are the relative displacements of the particle x and y. Determine the forced vibrations for the generalised coordinates, and then find the forced vibrations in the ξ-directions. Plot them graphically using the following values of the system parameters: $m = 1$ kg, $k_1 = 9$ N/cm, $k_2 = 19$ N/cm, $\alpha = 20°$, $s_0 = 1$ cm and $\Omega = 7\ \mathrm{s}^{-1}$. In addition, determine and plot the relative trajectory caused by the kinematic excitation.

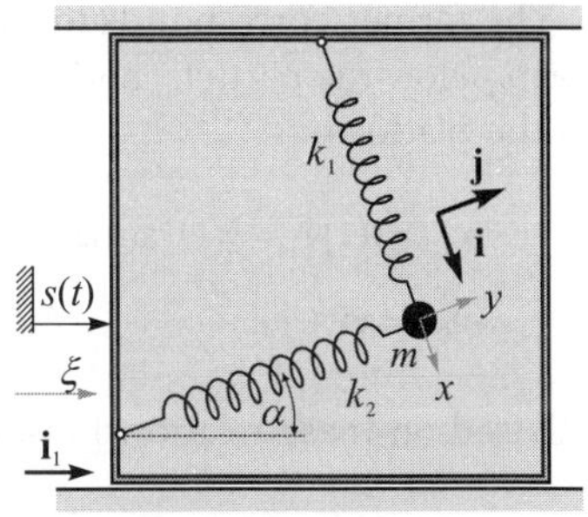

Figure 7.5.1

Solution

Given the fact that the system lies in the horizontal plane, the potential energy stems only from the springs (Figure 7.5.2a):

$$V = \frac{1}{2}k_1 x^2 + \frac{1}{2}k_2 y^2. \tag{7.5.1}$$

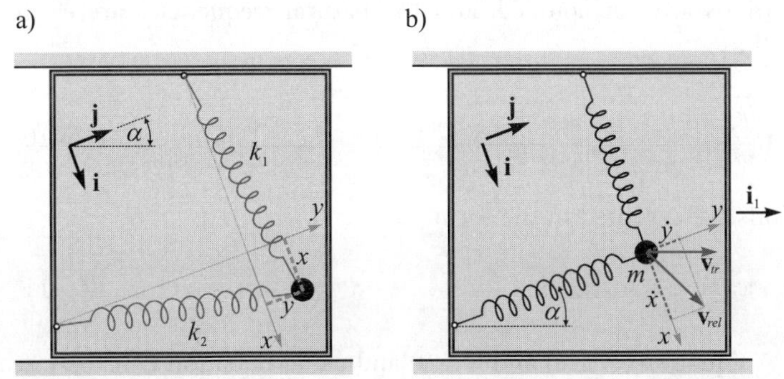

Figure 7.5.2

The kinetic energy of the system is

$$T = \frac{1}{2}mv^2, \tag{7.5.2}$$

where v stands for the absolute velocity of the particle. To obtain it, the vector sum of two velocities (transportational and relative) is considered (Figure 7.5.2b):

$$\mathbf{v} = \mathbf{v}_{tr} + \mathbf{v}_{rel}. \tag{7.5.3}$$

The former corresponds to the velocity of the platform, that is, the point on the platform below the particle, and the latter is the relative velocity of the particle with respect to the platform:

$$\mathbf{v}_{tr} = \dot{s}(t)\mathbf{i}_1 = \dot{s}(t)(\sin\alpha\mathbf{i} + \cos\alpha\mathbf{j}) = \dot{s}(t)\sin\alpha\mathbf{i} + \dot{s}(t)\cos\alpha\mathbf{j}, \tag{7.5.4}$$

$$\mathbf{v}_{rel} = \dot{x}\mathbf{i} + \dot{y}\mathbf{j}. \tag{7.5.5}$$

Based on these relationships, one can derive

$$\mathbf{v} = (\dot{x} + \dot{s}(t)\sin\alpha)\mathbf{i} + (\dot{y} + \dot{s}(t)\cos\alpha)\mathbf{j} \;\Rightarrow\; v^2 = (\dot{x} + \dot{s}(t)\sin\alpha)^2 + (\dot{y} + \dot{s}(t)\cos\alpha)^2. \tag{7.5.6}$$

Consequently, the kinetic energy (7.5.2) becomes

$$T = \frac{m}{2}\left((\dot{x} + \dot{s}(t)\sin\alpha)^2 + (\dot{y} + \dot{s}(t)\cos\alpha)^2\right). \tag{7.5.7}$$

Using (7.5.1) and (7.5.7), Lagrange's equations (7.3) give the following equations of motion:

$$m\ddot{x} + k_1 x = m\Omega^2 s_0 \sin\alpha \sin\Omega t, \tag{7.5.8}$$

$$m\ddot{y} + k_2 y = m\Omega^2 s_0 \cos\alpha \sin\Omega t. \tag{7.5.9}$$

Note that these equations are not coupled, and the natural frequencies are easily obtained (see Chapter 3) as

$$\omega_1 = \sqrt{\frac{k_1}{m}}, \quad \omega_2 = \sqrt{\frac{k_2}{m}}. \tag{7.5.10}$$

The particular solutions are assumed in the form

$$x_p(t) = A_x \sin\Omega t, \quad y_p(t) = A_y \sin\Omega t. \tag{7.5.11}$$

Substituting them into Equations (7.5.8) and (7.5.9) and using Equation (7.5.10), one derives:

$$\left(k_1 - m\Omega^2\right)A_x \sin\Omega t = s_0 m\Omega^2 \sin\alpha \sin\Omega t \;\Rightarrow\; A_x = \frac{s_0\Omega^2 \sin\alpha}{\omega_1^2 - \Omega^2}, \tag{7.5.12}$$

$$\left(k_2 - m\Omega^2\right)A_x \sin\Omega t = s_0 m\Omega^2 \cos\alpha \sin\Omega t \;\Rightarrow\; A_y = \frac{s_0\Omega^2 \cos\alpha}{\omega_2^2 - \Omega^2}. \tag{7.5.13}$$

Equations (7.5.11) now become

$$x_p(t) = \frac{s_0\Omega^2 \sin\alpha}{\omega_1^2 - \Omega^2}\sin\Omega t, \quad y_p(t) = \frac{s_0\Omega^2 \cos\alpha}{\omega_2^2 - \Omega^2}\sin\Omega t. \tag{7.5.14}$$

The forced vibrations in the ξ-direction are

$$\xi_p = x_p \sin\alpha + y_p \cos\alpha \;\Rightarrow\; \xi_p(t) = \left(\frac{\sin^2\alpha}{\omega_1^2-\Omega^2} + \frac{\cos^2\alpha}{\omega_2^2-\Omega^2}\right) s_0\Omega^2 \sin\Omega t. \qquad (7.5.15)$$

The expression for the trajectory is obtained by eliminating time from (7.5.14):

$$\frac{x_p}{y_p} = \tan\alpha \frac{\omega_2^2-\Omega^2}{\omega_1^2-\Omega^2} \quad \text{or} \quad y_p = \frac{\omega_1^2-\Omega^2}{\omega_2^2-\Omega^2}\cot\alpha x_p. \qquad (7.5.16)$$

The time–history diagrams for forced vibrations are plotted in Figure 7.5.3a, while the trajectory is plotted in Figure 7.5.3b.

It is interesting to note that the trajectory is a straight line.

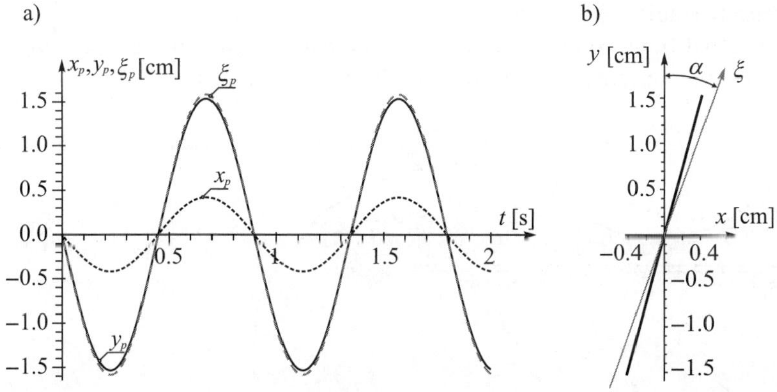

Figure 7.5.3

The frequency–amplitude diagrams are presented in Figure 7.5.4.

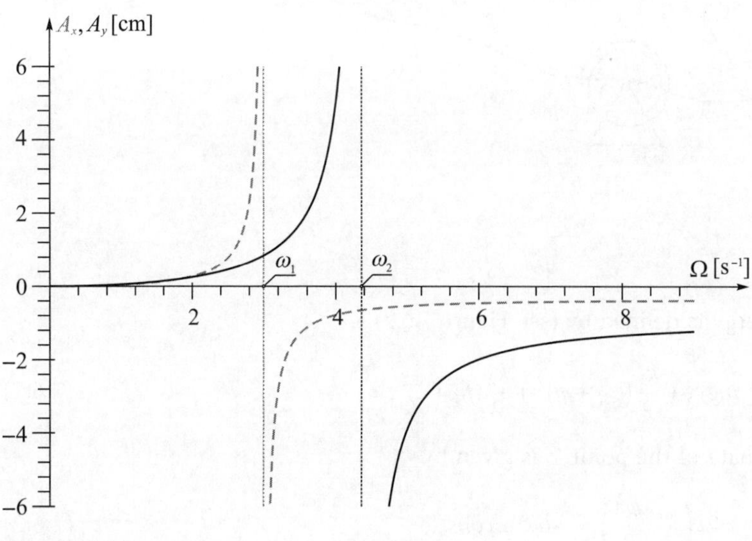

Figure 7.5.4

Example 7.6 A plate of mass M is hinged at its point O. It has a radius of gyration i_O about the axis that is perpendicular to the vertical plane in which the plate moves and passes through O. There is also a torsional spring of stiffness k_1 attached to the plate (Figure 7.6.1). The centre of mass C of the body is at a distance $2l$ from the point O. The body is subjected to a torque $\mathfrak{M}(t) = \mathfrak{M}_0 \sin\Omega t$. A bar AB of mass m and length $4\,l$ moves along the plate with weightless sliders at its ends, which move along two frictionless orthogonal guides O_1A and O_1B embedded into the plate. There is also a spring O_1A of stiffness k_2 and a linear viscous damper O_1B of damping coefficient c. Their positions in the static equilibrium when the torque does not act are shown in Figure 7.6.1. The parameters of the system are: $M = 4\,m$, $i_O = \sqrt{5}l$, $k_2 = 8\,mg/l$, $\mathfrak{M}_0 = mgl/10$ and $c = (m/10)\sqrt{g/l}$.

First, assuming that there is no torque, determine the stiffness k_1 so that the equilibrium is asymptotically stable.

Second, assuming that the torque acts, but there is no damper and taking $k_1 = 20mgl$, determine the frequency–amplitude diagram and plot it for $l = 0.2$ m.

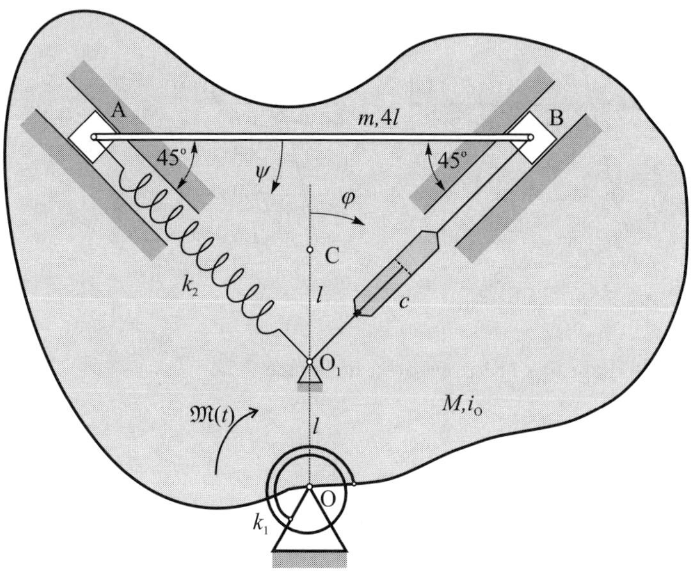

Figure 7.6.1

Solution

The potential energy is defined by (see Figure 7.6.2)

$$V = Mgy_C + mgy_S + \frac{k_1}{2}(\alpha_{st} + \varphi)^2 + \frac{k_2}{2}(f_{st} + \lambda)^2, \tag{7.6.1}$$

where the coordinate of the point C is given by

$$y_C = 2l\cos\varphi \approx 2l\left(1 - \frac{\varphi^2}{2}\right) = -l\varphi^2 + const., \tag{7.6.2}$$

while the coordinates for the point S can be found in exact form first, and then approximated as follows:

$$\begin{aligned} y_S &= l\cos\varphi + \overline{O_1A}\sin(45^\circ + \varphi) - \overline{AS}\sin(\varphi+\psi) \\ &= l\cos\varphi + 4l\cos(45^\circ - \psi)\sin(45^\circ + \varphi) - 2l\sin(\varphi+\psi) \\ &\approx 2l\left(-\frac{3\varphi^2}{4} - \frac{\psi^2}{2} + \varphi\psi\right) + const. \end{aligned} \quad (7.6.3)$$

To obtain the static deflection of the torsional spring, one can equate to zero the moment sum about O for the system as a whole (see Figure 7.6.3a)

$$\sum M_O = 0 \;\Rightarrow\; k_1\alpha_{st} = 0 \;\Rightarrow\; \alpha_{st} = 0. \quad (7.6.4)$$

To obtain the static deflection of the other spring, one can consider the bar only (Figure 7.6.3b) and equate to zero the moment sum about the point P:

$$\sum M_P = 0 \;\Rightarrow\; k_2 f_{st} = 0 \;\Rightarrow\; f_{st} = 0, \quad (7.6.5)$$

Thus, one concludes that neither of the springs is predeformed in the static equilibrium position.

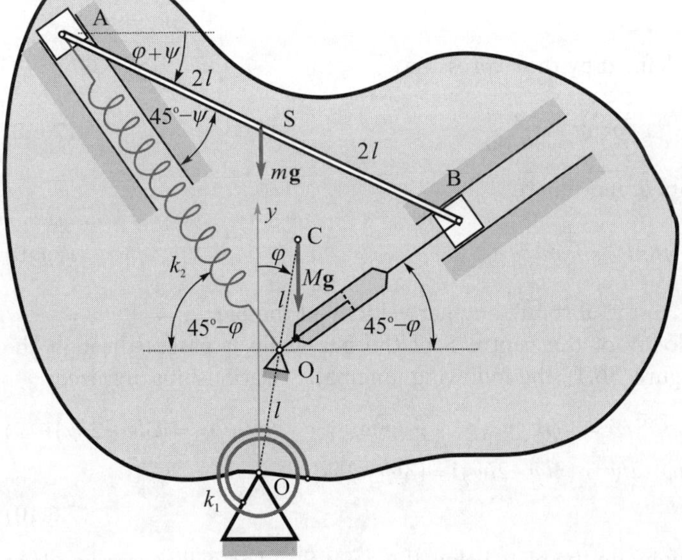

Figure 7.6.2

The additional deflection of the spring of stiffness k_2 is equal to the total deflection, which corresponds to relative displacement of the point A with respect to the plate (see Figure 7.6.2) and is given by

$$\lambda = \overline{O_1A}(\psi) - \overline{O_1A}(0) = 4l\cos(45^\circ - \psi) - 4l\cos 45^\circ \quad \Rightarrow$$

$$\lambda \approx 4l\cos 45^\circ + \psi 4l\sin 45^\circ - 4l\cos 45^\circ = 2l\sqrt{2}\psi. \tag{7.6.6}$$

Using Equations (7.6.1)–(7.6.6), the potential energy for small oscillations is found to be

$$V \approx \left(\frac{k_1}{2} - \left(M + \frac{3m}{2}\right)gl\right)\varphi^2 + \left(4k_2l^2 - mgl\right)\psi^2 + 2mgl\varphi\psi. \tag{7.6.7}$$

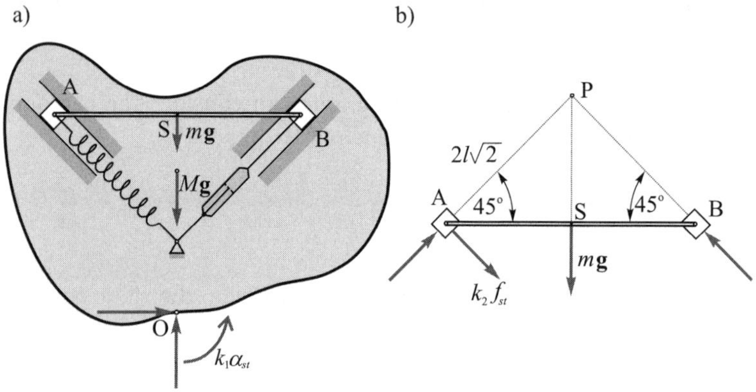

Figure 7.6.3

The kinetic energy is defined by the expression

$$T = \frac{1}{2}J_O\dot{\varphi}^2 + \frac{1}{2}mv_S{}^2 + \frac{1}{2}J_S(\dot{\varphi} + \dot{\psi})^2, \tag{7.6.8}$$

where the mass moments of inertia are

$$J_O = Mi_O^2, \quad J_S = \frac{1}{12}m(4l)^2 = \frac{4}{3}ml^2, \tag{7.6.9}$$

while $\dot{\varphi} + \dot{\psi}$ corresponds to the absolute angular velocity of the bar.

To determine the velocity of the centre S of the bar while it passes through the equilibrium position (Figure 7.6.4), the following kinematic relationships are used:

$$\mathbf{v}_S = \mathbf{v}_{S'} + \mathbf{v}_{Srel}, \ \mathbf{v}_{S'} = \overline{OS}\,\dot{\varphi}\mathbf{i} = 3l\dot{\varphi}\mathbf{i}, \ \mathbf{v}_{Srel} = -\overline{P_{rel}S}\omega_{rel}\mathbf{i} = -2l\dot{\psi}\mathbf{i}, \ \mathbf{v}_S = (3l\dot{\varphi} - 2l\dot{\psi})\mathbf{i} \ \Rightarrow$$
$$v_S^2 = \mathbf{v}_S\cdot\mathbf{v}_S = (3l\dot{\varphi} - 2l\dot{\psi})\mathbf{i}\cdot(3l\dot{\varphi} - 2l\dot{\psi})\mathbf{i} = (3l\dot{\varphi} - 2l\dot{\psi})^2, \tag{7.6.10}$$

where S' stands for the point on the plate below the point S and actually corresponds to the transportation velocity.

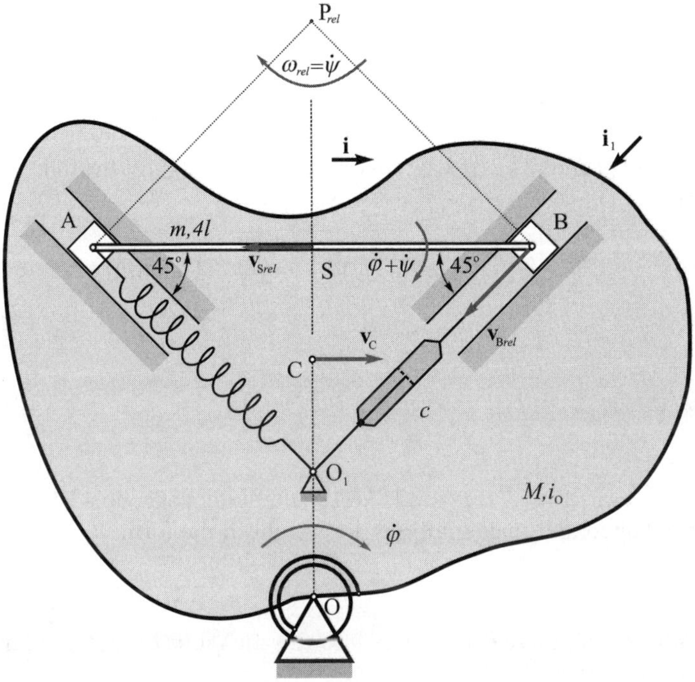

Figure 7.6.4

The kinetic energy is now calculated to be

$$T = \frac{1}{2} M i_O^2 \dot{\varphi}^2 + \frac{1}{2} m (3l\dot{\varphi} - 2l\dot{\psi})^2 + \frac{1}{2}\frac{4}{3} m l^2 (\dot{\varphi} + \dot{\psi})^2. \tag{7.6.11}$$

The dissipation function is defined by

$$D = \frac{1}{2} c v_{p/c}^2, \tag{7.6.12}$$

where the relative velocity of the piston with respect to the cylinder $v_{p/c}$ in the position when the system passes through the equilibrium position (Figure 7.6.4), can be obtained by projecting the relative velocity of the point B on the relative position of the damper

$$v_{p/c} = \mathbf{v}_{Brel} \cdot \mathbf{i}_1 \approx \overline{P_{rel}B}\,\dot{\psi}\,\mathbf{i}_1 \cdot \mathbf{i}_1 = 2l\sqrt{2}\dot{\psi} \;\Rightarrow\; v_{p/c}^2 = 8l^2\dot{\psi}^2. \tag{7.6.13}$$

The dissipative function now takes the form

$$D = 4cl^2\dot{\psi}^2. \tag{7.6.14}$$

The virtual work of the non-conservative torque $\mathfrak{M}(t)$ is

$$\delta A = \delta A^{\mathfrak{M}} = (\mathfrak{M}_0 \sin \Omega t)\delta\varphi + (0)\delta\psi, \tag{7.6.15}$$

where the generalised forces are recognised as the coefficients in front of $\delta\varphi$ and $\delta\psi$

$$Q_\varphi = \mathfrak{M}_0 \sin\Omega t, \quad Q_\psi = 0. \tag{7.6.16}$$

Using (7.6.7), (7.6.11), (7.6.14) and (7.6.16), Lagrange's equations (7.2) give the following equations of motion:

$$\left(Mi_O^2 + \frac{31}{3}ml^2\right)\ddot{\varphi} - \frac{14}{3}ml^2\ddot{\psi} + (k_1 - (2M+3m)gl)\varphi + 2mgl\psi = \mathfrak{M}_0 \sin\Omega t, \tag{7.6.17}$$

$$-\frac{14}{3}ml^2\ddot{\varphi} + \frac{16}{3}ml^2\ddot{\psi} + 8cl^2\dot{\psi} + \left(8k_2l^2 - 2mgl\right)\psi + 2mgl\varphi = 0. \tag{7.6.18}$$

Removing only the forcing term from (7.6.17) and (7.6.18), the system of coupled linear differential equations is obtained and their solutions are sought in the form

$$\varphi = Ae^{\lambda t}, \quad \psi = Be^{\lambda t}, \tag{7.6.19}$$

where λ is a constant. Substituting these expressions into the equations of motion and performing some transformations leads to:

$$\left(k_1 - 11mgl + \frac{91}{3}ml^2\lambda^2\right)A + \left(2mgl - \frac{14}{3}ml^2\lambda^2\right)B = 0, \tag{7.6.20}$$

$$\left(2g - \frac{14}{3}l\lambda^2\right)A + \left(62g + \frac{4}{5}\sqrt{gl}\lambda + \frac{16}{3}l\lambda^2\right)B = 0. \tag{7.6.21}$$

To achieve a non-trivial solution, the corresponding determinant should be equal to zero, which gives the following characteristic equation

$$a_0\lambda^4 + a_1\lambda^3 + a_2\lambda^2 + a_3\lambda + a_4 = 0, \tag{7.6.22}$$

where

$$\begin{aligned} a_0 &= 140ml^3, \quad a_1 = \frac{364}{15}ml^{5/2}\sqrt{g}, \quad a_2 = \frac{5522}{3}mgl^2 + \frac{16}{3}k_1l, \\ a_3 &= \frac{4}{5}k_1\sqrt{gl} - \frac{44}{5}m(gl)^{3/2}, \quad a_4 = 62k_1g - 686mg^2l. \end{aligned} \tag{7.6.23}$$

The Routh–Hurwitz criterion of asymptotic stability are fulfilled if all the coefficients a_0, a_1, a_2, a_3 and a_4 are positive and if the following inequality holds:

$$a_3(a_1a_2 - a_0a_3) - a_1^2a_4 > 0. \tag{7.6.24}$$

These conditions guarantee that the real parts of the roots of the characteristic equations are negative, yielding asymptotic stability.

In this particular case, the stiffness k_1 can be calculated from the conditions $a_3 > 0$ and $a_4 > 0$, as all other coefficients are already positive. The condition (7.6.24) can be transformed into

$$a_3(a_1a_2 - a_0a_3) - a_1^2a_4 = \frac{12544}{255}m^2l^5g^2(mgl + k_1) + \frac{3136}{255}mgl^4k_1^2 > 0. \tag{7.6.25}$$

For the condition $a_3 > 0$ to be fulfilled, it is required that

$$k_1 > 11mgl, \tag{7.6.26}$$

while $a_4 > 0$ imposes

$$k_1 > \frac{343}{31}mgl \approx 11.064mgl. \tag{7.6.27}$$

Given that the value in Equation (7.6.27) is higher, it represents the required value of the stiffness coefficient.

Let us now analyse the case when there is no damper, there is no torque and the stiffness coefficient is $k_1 = 20mgl$. Applying the procedure demonstrated in Chapter 6, the two natural frequencies are obtained to be

$$\omega_1 \approx 3.775\,\mathrm{s}^{-1}, \quad \omega_2 \approx 25.846\,\mathrm{s}^{-1}. \tag{7.6.28}$$

The particular solutions for the undamped forced case are assumed as

$$\varphi_p = A_\varphi \sin\Omega t, \quad \psi_p = A_\psi \sin\Omega t. \tag{7.6.29}$$

The equation of motion for undamped motion can be transformed into the following system of algebraic equations

$$\left(20g - 11g - \frac{91}{3}l\Omega^2\right)A_\varphi + \left(2g + \frac{14}{3}l\Omega^2\right)A_\psi = \frac{g}{10}, \tag{7.6.30}$$

$$\left(2g + \frac{14}{3}l\Omega^2\right)A_\varphi + \left(62g - \frac{16}{3}l\Omega^2\right)A_\psi = 0. \tag{7.6.31}$$

Its solutions are

$$A_\varphi = \frac{(g/10)(93g - 8l\Omega^2)}{831g^2 - 2921gl\Omega^2 + 210l^2\Omega^4}, \quad A_\psi = -\frac{(g/10)(3g + 7l\Omega^2)}{831g^2 - 2921gl\Omega^2 + 210l^2\Omega^4}. \tag{7.6.32}$$

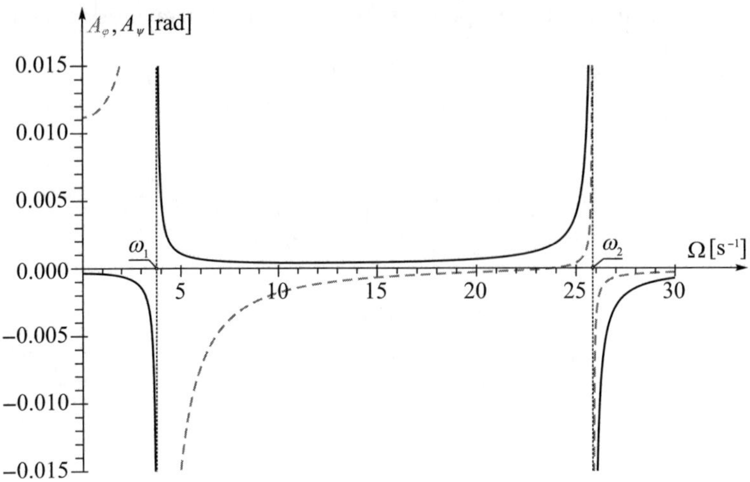

Figure 7.6.5

They are plotted in Figure 7.6.5 for $g = 9.81\ \mathrm{m/s^2}$ and $l = 0.2\ \mathrm{m}$.

8

Vibration of Systems with Infinite Number of Degrees of Freedom

Chapter Outline

This chapter is concerned with continuous oscillatory systems: bars that exhibit longitudinal (axial) vibration, shafts that perform torsional vibration and beams vibrating transversally. Their governing equations, which have the form of partial differential equations, are derived. Many examples are presented and solved to shown how natural frequencies and modes of vibration can be obtained. The influence of boundary conditions is explained in detail, while a few examples illustrate the influence of the initial conditions on the response in the time domain.

Chapter Objectives

- To present the governing equations for longitudinal vibration of bars, torsional vibration of shafts and transversal vibration of beams
- To illustrate a variety of examples with different boundary conditions and the way they can be used to derive the characteristic determinant, characteristic equation and natural frequencies
- To present mode shapes for certain typical problems
- To illustrate the influence of initial conditions on the response in the time domain.

8.1 Theoretical Introduction: Longitudinal Vibration of Bars

Previous chapters are concerned with *discrete* or *lumped-parameter* oscillatory systems. The corresponding equations of motion are given in the form of ordinary differential equations, with the displacements being dependent on time only and the resulting motion being an *initial-value problem*, that is, being dependent on the initial conditions. Their number of degrees of freedom defines the number of natural frequencies and mode shapes of vibration.

In this chapter, *continuous* or *distributed-parameter* oscillatory systems are dealt with: longitudinal (axial) vibration of bars, torsional vibration of shafts and transversal (lateral)

Mechanical Vibration: Fundamentals with Solved Examples, First Edition.
Ivana Kovačić and Dragi Radomirović.

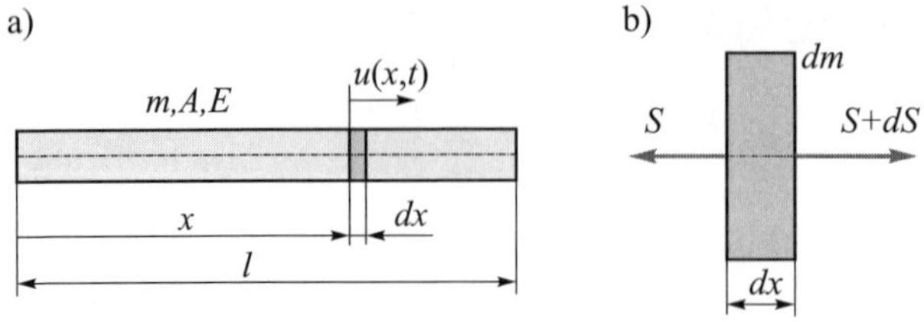

Figure 8.1.1

vibration of beams. It will be shown that unlike in discrete systems, their governing equations have the form of partial differential equations, since the displacement of a point or cross-section on these elastic elements depends both on its position and time. As these elastic elements have an infinite number of degrees of freedom, they also have an infinite number of natural frequencies and modes of vibration, which will be illustrated in this chapter.

Let us start with a uniform elastic bar of mass m, length l and cross-sectional area A shown in Figure 8.1.1a with a view to analysing its longitudinal vibration. It is assumed that the bar is made of Hooke's material whose Young's modulus is E. Let us focus now on an infinitesimally small element of the bar, placed at the location defined by x. Its mass is dm, and the corresponding free-body diagram is presented in Figure 8.1.1b. The forces acting on it are of magnitude S and $S + dS$, and the displacement is labelled by $u = u(x,t)$ as it depends both on x and time t. Newton's second law projected on the horizontal direction is

$$dm\frac{\partial^2 u(x,t)}{\partial t^2} = -S + S + dS. \tag{8.1.1}$$

The elementary mass can be obtained from the fact that the bar is uniform, so that:

$$\frac{dm}{dx} = \frac{m}{l} \quad \Rightarrow \quad dm = \frac{m}{l}dx. \tag{8.1.2}$$

To determine the expression for the force, one can turn to Hooke's law and relate the force to the axial stress σ_c and the axial strain $\partial u(x,t)/\partial x$ as follows:

$$S = \sigma A = E\frac{\partial u(x,t)}{\partial x}A. \tag{8.1.3}$$

Substituting (8.1.2) and (8.1.3) into (8.1.1), one derives

$$\frac{\partial^2 u(x,t)}{\partial t^2} = \frac{EAl}{m}\frac{\partial}{\partial x}\left(\frac{\partial u(x,t)}{\partial x}\right). \tag{8.1.4}$$

Recognising that the ratio m/Al corresponds to the mass density ρ of the bar, one can write the governing equation in the form

$$\frac{\partial^2 u(x,t)}{\partial t^2} = c^2\frac{\partial^2 u(x,t)}{\partial x^2}, \tag{8.1.5}$$

where $c^2 = E/\rho$ (note that the unit for c is m/s).

Equation (8.1.5) can be represented as

$$\ddot{u}(x,t) = c^2 u''(x,t), \tag{8.1.6}$$

where the dots in this Chapter stand for partial differentiation with respect to time t, and the primes for partial differentiation with respect to x. To solve this partial differential equation, the principle of separation of variables is applied, assuming the solutions to be the product of two functions, each of which depends on one independent variable only:

$$u(x,t) = X(x)T(t). \tag{8.1.7}$$

Substituting it into Equation (8.1.6) leads to

$$X\ddot{T} = c^2 X'' T. \tag{8.1.8}$$

This can be represented as the ratios

$$\frac{\ddot{T}}{T} = c^2 \frac{X''}{X} = -\omega_n^2, \tag{8.1.9}$$

which due to different independent variables should be equal to a constant. This constant is labelled here by ω_n, and needs to be determined. Now, the ratios in Equation (8.1.9) give two differential equations:

$$X'' + \frac{\omega_n^2}{c^2} X = 0, \tag{8.1.10}$$

$$\ddot{T} + \omega_n^2 T = 0. \tag{8.1.11}$$

Their solutions can be represented as (see Equations (3.6) and (3.8)):

$$X = B\cos\frac{\omega_n x}{c} + C\sin\frac{\omega_n x}{c}, \tag{8.1.12}$$

$$T = P\cos\omega_n t + Q\sin\omega_n t. \tag{8.1.13}$$

The constants B, C, P and Q should be calculated from the boundary and initial conditions. Several illustrative examples are given below. They also include the presentation of modal shapes that corresponds to different values of ω_n.

Example 8.1.1 Find the natural frequencies, mode shapes and a free vibration solution of a uniform bar of mass m, length l, cross-sectional area A, made of Hooke's material with Young's modulus E that is fixed at both ends (Figure 8.1.1.1a). Assume first the initial condition for displacement in a triangular form (Figure 8.1.1.1b) and then in a

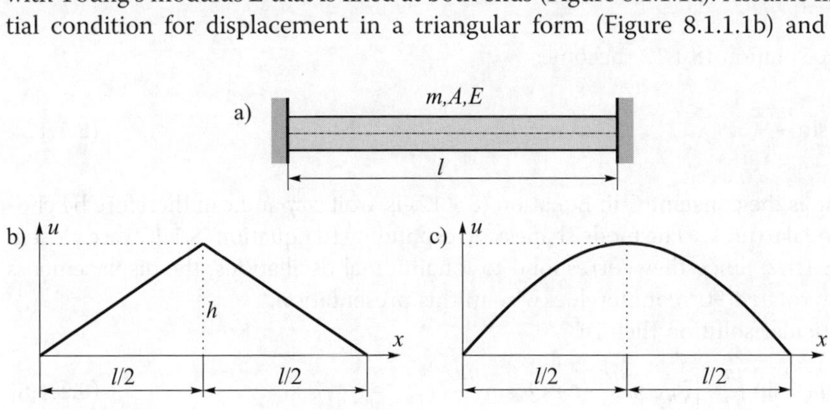

Figure 8.1.1.1

sinusoidal form (Figure 8.1.1.1c), while the other initial condition in both cases is $\partial u(x,t)/\partial t = 0$ for t = 0 (note that this and analogous conditions will be labelled further on as $\partial u(x,0)/\partial t = 0$).

Solution

The fact that the bar is fixed at both ends implies that these ends cannot move, that is, the displacement at each end $x = 0$ and $x = l$ is always zero, which can be written down in the form of the following boundary conditions:

$$u(0,t) = 0, \quad u(l,t) = 0. \tag{8.1.1.1}$$

Equation (8.1.7) transforms these two conditions into:

$$X(0) = 0, \quad X(l) = 0. \tag{8.1.1.2}$$

The solution (8.1.12) satisfying the first conditions from (8.1.1.2) gives $B = 0$. The second condition is satisfied when

$$\sin\frac{\omega_n l}{c} = 0, \tag{8.1.1.3}$$

which represents the so-called *characteristic (frequency) equation.* Its solutions are

$$\frac{\omega_n}{c} l = n\pi, \quad n = 1,2,3,\ldots \tag{8.1.1.4}$$

giving

$$\omega_n = \frac{n\pi c}{l}, \quad n = 1,2,3\ldots \tag{8.1.1.5}$$

It is seen that there is an infinite number of natural frequencies, the first three of which are

$$\omega_1 = \frac{\pi c}{l}, \quad \omega_2 = \frac{2\pi c}{l}, \quad \omega_3 = \frac{3\pi c}{l}. \tag{8.1.1.6}$$

Thus, the solution (8.1.12) becomes

$$X = \sin\frac{n\pi x}{l}. \tag{8.1.1.7}$$

Note that as the constant C in Equation (8.1.12) is arbitrary and can therefore be chosen to be equal to unity. The mode shapes corresponding to Equation (8.1.1.7) are plotted in Figure 8.1.1.2. Since they correspond to longitudinal oscillations, the displacements are actually rotated 90° counterclockwise in this presentation.

One particular solution then is:

$$u(x,t) = \sin\frac{n\pi x}{l}\left[P_n \cos\frac{n\pi c}{l}t + Q_n \sin\frac{n\pi c}{l}t\right], \tag{8.1.1.8}$$

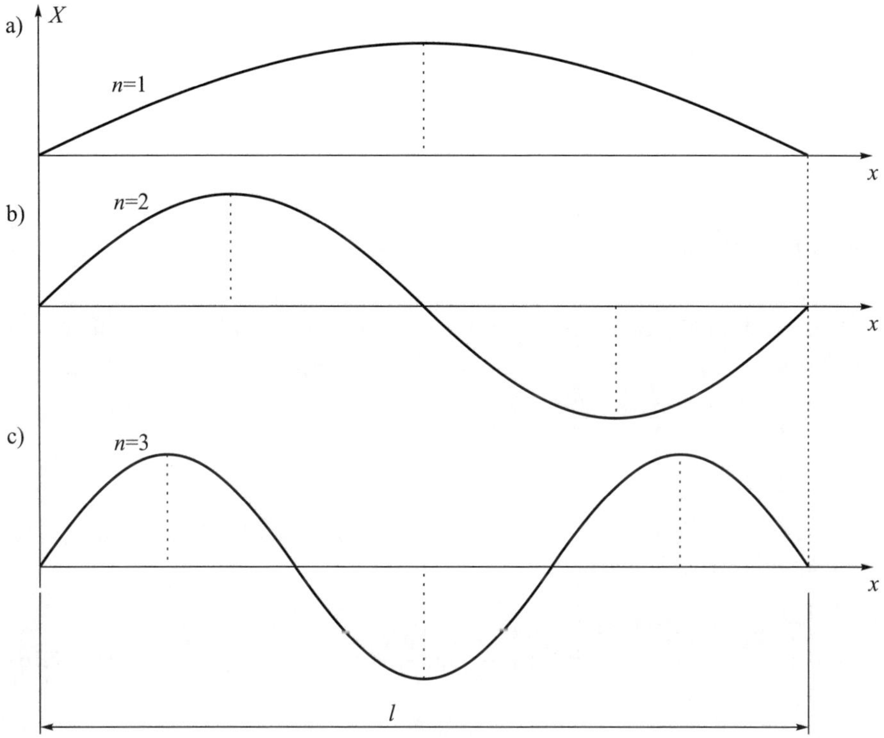

Figure 8.1.1.2

and the general solution for free vibration (8.1.7) can be presented by using the mode superposition method as

$$u(x,t) = \sum_{n=1}^{\infty} \sin\frac{n\pi x}{l}\left[P_n \cos\frac{n\pi c}{l}t + Q_n \sin\frac{n\pi c}{l}t\right]. \tag{8.1.1.9}$$

The initial condition $\partial u(x,0)/\partial t = 0$ implies that $Q_n = 0$. To analyse the influence of the initial condition for the displacement, one can use Equation (8.1.7) and consider the following:

$$u(x,0) \equiv u_0(x) = \sum_{n=1}^{\infty} P_n \sin\frac{n\pi x}{l}. \tag{8.1.1.10}$$

This can be expanded as

$$u_0(x) = P_1 \sin\frac{\pi x}{l} + P_2 \sin\frac{2\pi x}{l} + P_3 \sin\frac{3\pi x}{l} + \ldots + P_n \sin\frac{n\pi x}{l} + \ldots \tag{8.1.1.11}$$

To find the coefficients P_n, both sides are multiplied by $\sin\frac{n\pi x}{l}$ and then integrated from 0 to l. Knowing that

$$\int_0^l \sin\frac{N\pi x}{l}\sin\frac{n\pi x}{l}dx = 0, \quad \text{for } N \neq n, \tag{8.1.1.12}$$

$$\int_0^l \sin^2\frac{n\pi x}{l}dx = \frac{1}{2}\int_0^l \left(1-\cos\frac{2n\pi x}{l}\right)dx = \frac{l}{2}, \tag{8.1.1.13}$$

one has

$$P_n = \frac{2}{l}\int_0^l u_0(x)\sin\frac{n\pi x}{l}dx. \tag{8.1.1.14}$$

The initial conditions plotted in Figure 8.1.1.1b can be presented in the following analytical form:

$$u_0(x) = \begin{cases} \dfrac{2hx}{l}, & 0 \le x \le l/2 \\ 2h\left(1-\dfrac{x}{l}\right), & l/2 \le x \le l. \end{cases} \tag{8.1.1.15}$$

Thus, Equation (8.1.1.14) becomes:

$$P_n = \frac{2}{l}\int_0^{l/2} \frac{2hx}{l}\sin\frac{n\pi x}{l}dx + \frac{2}{l}\int_{l/2}^{l} 2h\left(1-\frac{x}{l}\right)\sin\frac{n\pi x}{l}dx, \tag{8.1.1.16}$$

or:

$$P_n = \frac{8h}{n^2\pi^2}\sin\frac{n\pi}{2}. \tag{8.1.1.17}$$

This enables one to present the solution (8.1.1.8) in the form:

$$u(x,t) = \frac{8h}{n^2\pi^2}\sum_{n=1}^{\infty}\sin\frac{n\pi x}{l}\sin\frac{n\pi}{2}\cos\frac{n\pi c}{l}t. \tag{8.1.1.18}$$

Recognising the natural frequencies (8.1.1.5), this can be written down as

$$u(x,t) = \frac{8h}{\pi^2}\left(\sin\frac{\pi x}{l}\cos\omega_1 t - \frac{1}{9}\sin\frac{3\pi x}{l}\cos\omega_3 t + \frac{1}{25}\sin\frac{5\pi x}{l}\cos\omega_5 t + \dots\right), \tag{8.1.1.19}$$

which gives an insight into the content with respect to the existence of modes plotted in Figure 8.1.1.2. and illustrates that the resulting oscillations are obtained as the superposition of the modes.

If the initial conditions are given by $u_0(x) = \sin\pi x/l$, which actually corresponds to the first mode shape, the coefficients (8.1.1.14) become:

$$P_n = \frac{2}{l}\int_0^l \sin\frac{\pi x}{l}\sin\frac{n\pi x}{l}dx. \tag{8.1.1.20}$$

By using (8.1.1.12) and (8.1.1.13), one obtains that only the first coefficient is different from zero, that is,

$$P_1 = 1. \tag{8.1.1.21}$$

The solution (8.1.1.8) simplifies to

$$u(x,t) = \sin\frac{\pi x}{l}\cos\omega_1 t. \tag{8.1.1.22}$$

So, as the initial displacements are equal to the first mode, the response has only one existing frequency that is associated with the first mode.

Example 8.1.2 Find the natural frequencies, modes shapes and a free vibration solution for a uniform bar of mass m, length l, cross-sectional area A, made of Hooke's material with Young's modulus E that is fixed at one end and free at the other (Figure 8.1.2.1a). Assume the initial condition for the displacement as having a straight-line form (Figure 8.1.2.1b), while the other initial condition is $\partial u(x,0)/\partial t = 0$. Determine the solutions for longitudinal oscillations of the point at the right end and in the middle of the bar, assuming $h = l/2 = 0.5$ m.

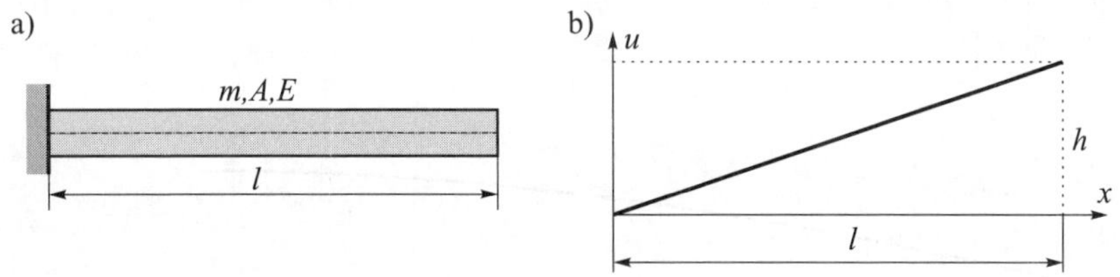

Figure 8.1.2.1

Solution

The fixed left end imposes the boundary condition:

$$u(0,t) = 0, \tag{8.1.2.1}$$

yielding $B = 0$ in Equation (8.1.12).

By focusing on the right end, one can plot the free-body diagram, representing it as an infinitesimally small element (Figure 8.1.2.2). The equilibrium of forces in the horizontal direction gives $S(l,t) = 0$, which due to Equation (8.1.3), implies

$$\frac{\partial u(l,t)}{\partial x} = 0 \Rightarrow X'(l) = 0. \tag{8.1.2.2}$$

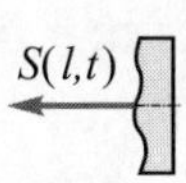

Figure 8.1.2.2

This leads to the following characteristic equation:

$$\cos\frac{\omega_n l}{c} = 0. \tag{8.1.2.3}$$

Its solutions are given by

$$\frac{\omega_n}{c} l = (2n-1)\frac{\pi}{2}, \quad n = 1,2,3,... \tag{8.1.2.4}$$

The natural frequencies are

$$\omega_n = (2n-1)\frac{\pi\ c}{2l}, \quad n = 1,2,3,..., \tag{8.1.2.5}$$

and the first three of them are

$$\omega_1 = \frac{\pi\ c}{2l}, \quad \omega_2 = \frac{3\pi\ c}{2l}, \quad \omega_3 = \frac{5\pi\ c}{2l}. \tag{8.1.2.6}$$

Thus, the solution (8.1.12) for c = 1 has the form

$$X = \sin\frac{(2n-1)\pi x}{2l}. \tag{8.1.2.7}$$

The first three mode shapes corresponding to Equation (8.1.2.7) are plotted in Figure 8.1.2.3.

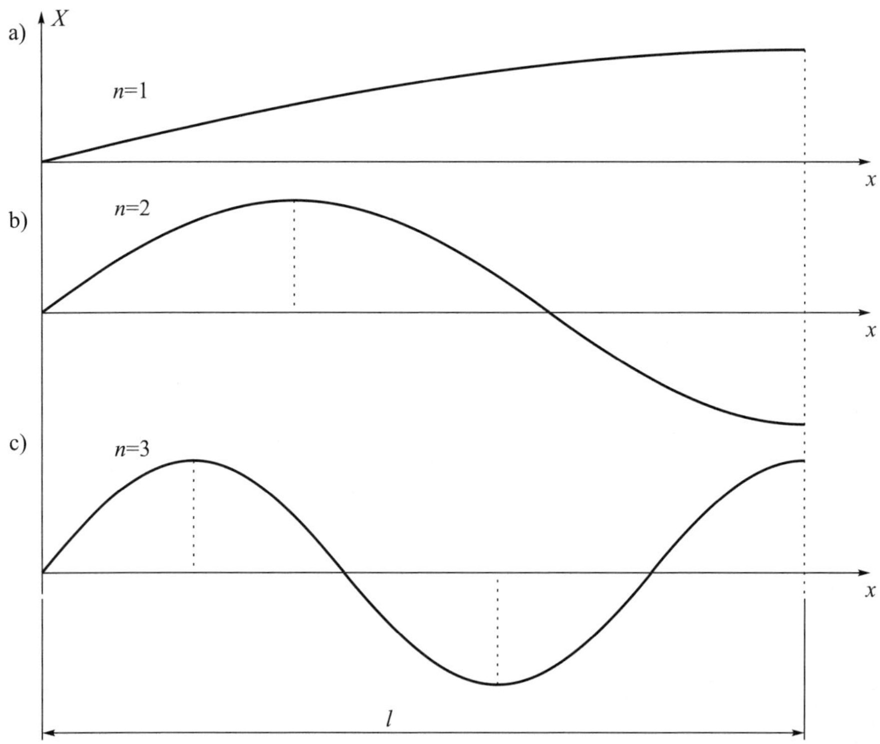

Figure 8.1.2.3

The general solution for free vibration (8.1.7) can be presented by using the mode superposition method as

$$u(x,t)=\sum_{n=1}^{\infty}\sin\frac{(2n-1)\pi\, x}{2l}\left[P_n\cos(2n-1)\frac{\pi\, c}{2l}t+Q_n\sin(2n-1)\frac{\pi\, c}{2l}t\right]. \tag{8.1.2.8}$$

The initial condition $\partial u(x,0)/\partial t=0$ gives that $Q_n=0$. To obtain the coefficients P_n, the initial conditions plotted in Figure 8.1.2.1b can be presented as:

$$u_0(x)=\frac{hx}{l},\quad 0\le x\le l. \tag{8.1.2.9}$$

Equation (8.1.1.14) gives:

$$P_n=\frac{2}{l}\int_0^l\frac{hx}{l}\sin\frac{n\pi x}{l}dx. \tag{8.1.2.10}$$

Solving it, one can obtain

$$P_n=-\frac{2h}{n\pi}\cos n\pi. \tag{8.1.2.11}$$

This transforms the solution (8.1.2.8) into:

$$u(x,t)=-\sum_{n=1}^{\infty}\frac{2h}{n\pi}\sin\frac{(2n-1)\pi\, x}{2l}\cos n\pi\cos(2n-1)\frac{\pi\, c}{2l}t. \tag{8.1.2.12}$$

Recognising the natural frequencies (8.1.2.5), this can be written as

$$u(x,t)=\frac{2h}{\pi}\left(\sin\frac{\pi\, x}{2l}\cos\omega_1 t-\frac{1}{2}\sin\frac{3\pi\, x}{2l}\cos\omega_2 t+\frac{1}{3}\sin\frac{5\pi\, x}{2l}\cos\omega_3 t+\ldots\right). \tag{8.1.2.13}$$

The solution for longitudinal oscillations of the point at the right end with $h=l/2=0.5$ m is:

$$u(l,t)=\frac{1}{\pi}\left(\cos\omega_1 t+\frac{1}{2}\cos\omega_2 t+\frac{1}{3}\cos\omega_3 t+\ldots\right). \tag{8.1.2.14}$$

The solution for longitudinal oscillations of the point in the middle of the bar is:

$$u(l/2,t)=\frac{1}{\pi}\left(\frac{\sqrt{2}}{2}\cos\omega_1 t-\frac{\sqrt{2}}{4}\cos\omega_2 t-\frac{\sqrt{2}}{6}\cos\omega_3 t+\ldots\right). \tag{8.1.2.15}$$

Example 8.1.3 The system consists of a uniform bar of mass m, length l, cross-sectional area A, made of Hooke's material with Young's modulus E that is free at one end and carries a mass M at the other end. Two parallel springs of equal stiffness k are attached to it as shown in Figure 8.1.3.1. Find the natural frequencies for the system described, assuming that $k = EA/(2l)$ and $M = m/2$. Find the natural frequencies also for the case when the springs are attached directly to the bar and there is no mass M.

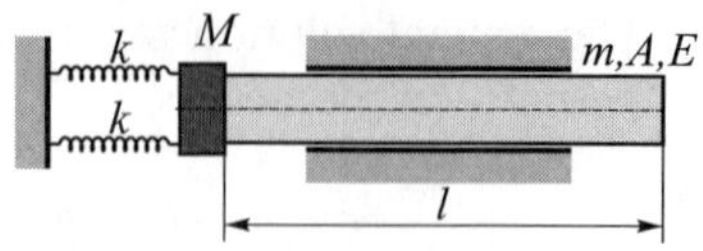

Figure 8.1.3.1

Solution

The free end of the bar performing longitudinal oscillations is considered Example 8.1.2 in with the corresponding boundary condition (see Equation (8.1.2.2)):

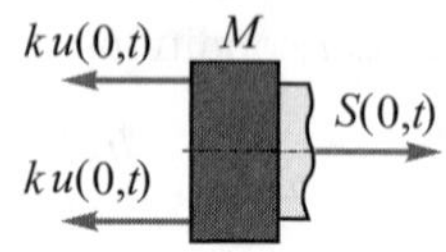

Figure 8.1.3.2

$$X'(l) = 0, \tag{8.1.3.1}$$

which, based on Equation (8.1.12), yields

$$-B\sin\frac{\omega_n l}{c} + C\cos\frac{\omega_n l}{c} = 0. \tag{8.1.3.2}$$

The free-body diagram for the mass M at the other end is shown in Figure 8.1.3.2.

The mass is subjected to two spring forces, which are proportional to the displacement at this end. In addition, the internal force of magnitude S also acts. Newton's second law projected on the horizontal axis is:

$$M\frac{\partial^2 u(0,t)}{\partial t^2} = -2ku(0,t) + S(0,t). \tag{8.1.3.3}$$

Using the expression for the internal force (8.1.3), one has

$$M\frac{\partial^2 u(0,t)}{\partial t^2} = -2ku(0,t) + EA\frac{\partial u(0,t)}{\partial x}, \tag{8.1.3.4}$$

which, due to $u(x,t) = X(x)T(t)$, can be represented as

$$MX(0)\ddot{T}(t) = -2kX(0)T(t) + EAX'(0)T(t). \tag{8.1.3.5}$$

Given Equation (8.1.8), this can be written as

$$Mc^2X''(0) = -2kX(0) + EAX'(0). \tag{8.1.3.6}$$

Using Equation (8.1.12), this becomes:

$$B\left(2k - M\omega_n^2\right) - EA\frac{\omega_n}{c}C = 0. \tag{8.1.3.7}$$

To solve the system defined by (8.1.3.2) and (8.1.3.7), one can form the characteristic determinant Δ and make it equal to zero for the sake of obtaining non-trivial solutions:

$$\Delta = \begin{vmatrix} -\sin\dfrac{\omega_n l}{c} & \cos\dfrac{\omega_n l}{c} \\ 2k - M\omega_n^2 & -EA\dfrac{\omega_n}{c} \end{vmatrix} = 0. \tag{8.1.3.8}$$

The following characteristic equation is derived from this:

$$\tan\frac{\omega_n l}{c} = \frac{2k - M\omega_n^2}{EA\dfrac{\omega_n}{c}}. \tag{8.1.3.9}$$

Let us solve first the simplest case when there is no mass. For $M = 0$, Equation (8.1.3.9) simplifies to

$$\tan\frac{\omega_n l}{c} = \frac{2k}{EA}\frac{c}{\omega_n}. \tag{8.1.3.10}$$

For $k = EA/(2l)$, this becomes

$$\tan y = \frac{1}{y}, \tag{8.1.3.11}$$

where $y = \omega_n l/c$. To solve this, one can find the intersection of the function on the left- and right-hand sides of Equation (8.1.3.11), as shown in Figure 8.1.3.3. The first three intersection points are

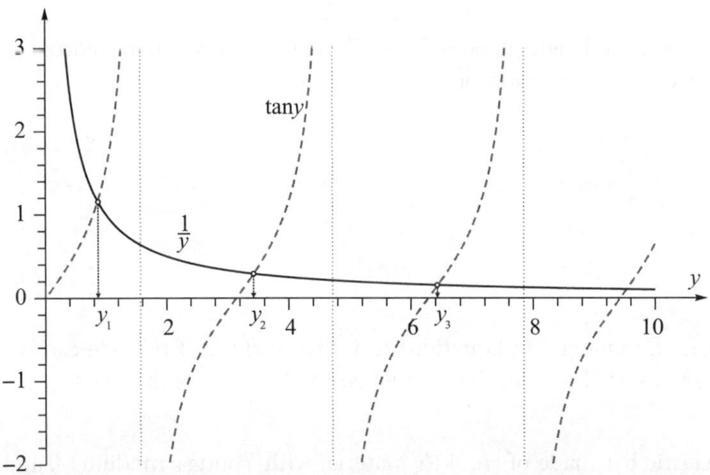

Figure 8.1.3.3

$$y_1 = 0.860334,\quad y_2 = 3.42562,\quad y_3 = 6.4373, \tag{8.1.3.12}$$

which yield

$$\omega_1 = 0.860334\frac{c}{l},\quad \omega_2 = 3.42562\frac{c}{l},\quad \omega_3 = 6.4373\frac{c}{l}. \tag{8.1.3.13}$$

Let us now solve Equation (8.1.3.9). For the parameters given $k = EA/(2l)$, $M = m/2$, it becomes

$$\tan y = \frac{1}{y} - \frac{1}{2}y. \tag{8.1.3.14}$$

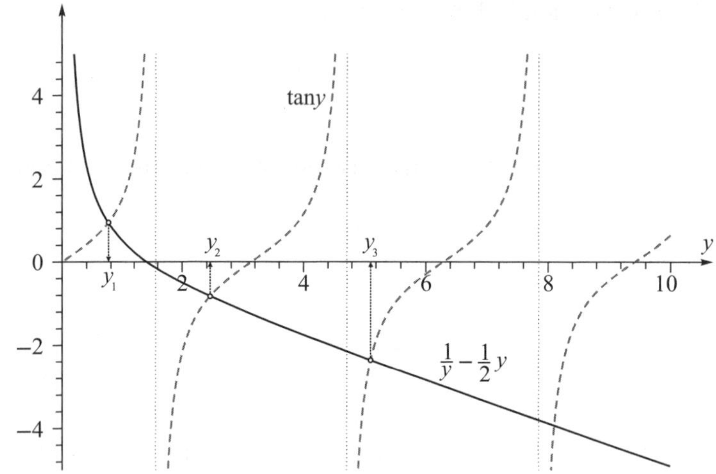

Figure 8.1.3.4

The functions on the left- and right-hand sides of Equation (8.1.3.14) are plotted in Figure 8.1.3.4. The first three intersection points are

$$y_1 = 0.756455, \quad y_2 = 2.45477, \quad y_3 = 5.11304, \tag{8.1.3.15}$$

which give

$$\omega_1 = 0.756455\frac{c}{l}, \quad \omega_2 = 2.45477\frac{c}{l}, \quad \omega_3 = 5.11304\frac{c}{l}. \tag{8.1.3.16}$$

Comparing the natural frequencies in Equations (8.1.3.13) and (8.1.3.16), one can see that the natural frequencies are lower in the case when the bar carries the mass.

Example 8.1.4 An elastic bar made of Hooke's material with Young's modulus E and mass density ρ has a changeable cross-section as shown in Figure 8.1.4.1, where the length of each part is l. A spring of stiffness k is attached to its free end and the middle section. Find the characteristic equation and the natural frequencies assuming $A_2 = A_1/2 = A/2$ and $k = EA/(2l)$.

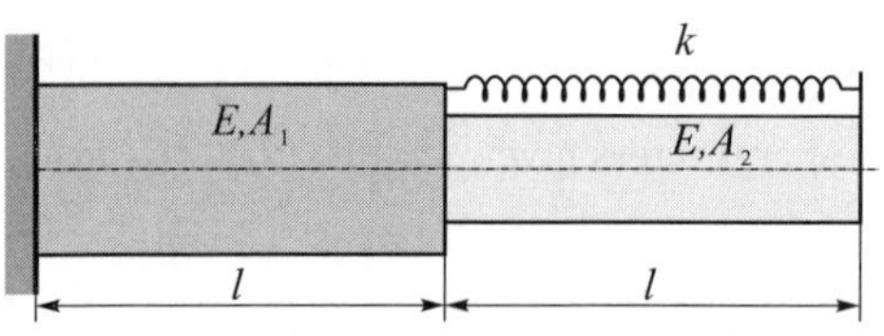

Figure 8.1.4.1

Solution

Due to the changeable cross-section, the bar will be treated as consisting of two bars as labelled in Figure 8.1.4.2, with the respective locations defined by x_1 and x_2, and the displacements u_1 and u_2. Thus, there are two governing equations:

$$\frac{\partial^2 u_1(x_1,t)}{\partial t^2} = c^2 \frac{\partial^2 u_1(x_1,t)}{\partial x_1^2}, \tag{8.1.4.1}$$

$$\frac{\partial^2 u_2(x_2,t)}{\partial t^2} = c^2 \frac{\partial^2 u_2(x_2,t)}{\partial x_2^2}, \tag{8.1.4.2}$$

for which we have

$$X_1 = B_1 \cos\frac{\omega_n x_1}{c} + C_1 \sin\frac{\omega_n x_1}{c}, \tag{8.1.4.3}$$

$$X_2 = B_2 \cos\frac{\omega_n x_2}{c} + C_2 \sin\frac{\omega_n x_2}{c}. \tag{8.1.4.4}$$

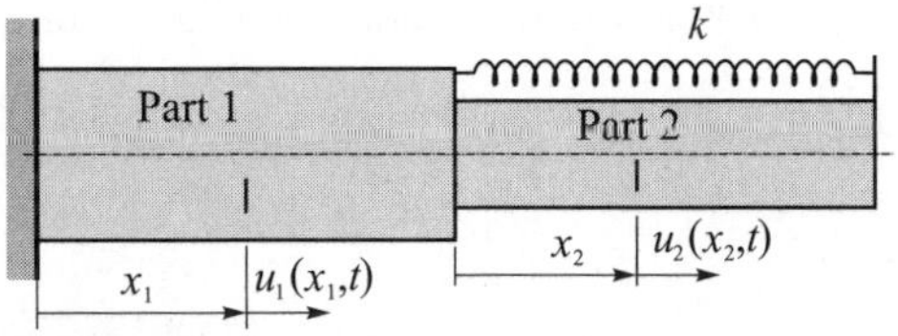

Figure 8.1.4.2

So, there are four unknown constants. The first boundary condition is easily recognised for Part 1, which is fixed at the left end:

$$u_1(0,t) = 0 \;\Rightarrow\; X_1(0) = 0, \tag{8.1.4.5}$$

which, based on Equation (8.1.4.3), yields $B_1 = 0$. Then, the fact that the ending section of Part 1 is the beginning section of Part 2, implies

$$u_1(l,t) = u_2(0,t) \;\Rightarrow\; X_1(l) = X_2(0), \tag{8.1.4.6}$$

resulting in

$$C_1 \sin\frac{\omega_n l}{c} - B_2 = 0. \tag{8.1.4.7}$$

Considering the infinitesimally small element in the middle (Figure 8.1.4.3a), one can write the following equilibrium equation for the forces in the horizontal direction:

$$-S_1(l,t) + S_2(0,t) + F_k = 0, \tag{8.1.4.8}$$

where

$$S_1(l,t) = EA_1 \frac{\partial u_1}{\partial x_1}(l,t), \;\; S_2(0,t) = EA_2 \frac{\partial u_2}{\partial x_2}(0,t), \;\; F_k = k[u_2(l,t) - u_2(0,t)]. \tag{8.1.4.9}$$

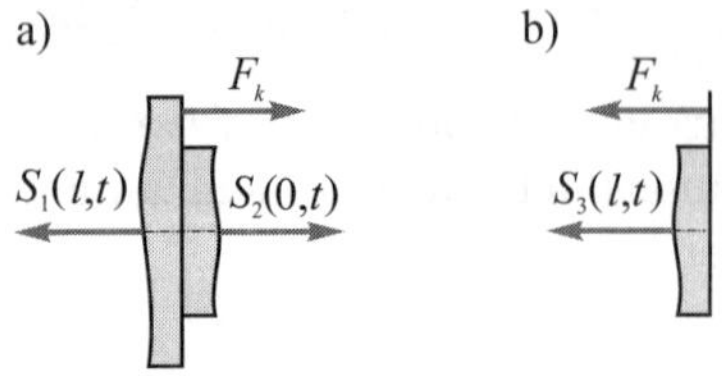

Figure 8.1.4.3

The equilibrium equation transforms into:

$$-EA_1X_1'(l)+EA_2X_2'(0)+k[X_2(l)-X_2(0)]=0, \tag{8.1.4.10}$$

and further also into

$$-EA_1C_1\frac{\omega_n}{c}\cos\frac{\omega_n l}{c}+EA_2C_2\frac{\omega_n}{c}+k\left(B_2\cos\frac{\omega_n l}{c}+C_2\sin\frac{\omega_n l}{c}-B_2\right)=0. \tag{8.1.4.11}$$

Considering the infinitesimally small element at the end of Part 2 (Figure 8.1.4.3b), the following equilibrium equation for the forces in the horizontal direction can be written:

$$-S_3(l,t)-F_k=0, \tag{8.1.4.12}$$

where

$$S_3(l,t)=EA_2\frac{\partial u_2}{\partial x_2}(l,t). \tag{8.1.4.13}$$

Equation (8.1.4.12) can be represented as

$$EA_2X_2'(l)+k[X_2(l)-X_2(0)]=0, \tag{8.1.4.14}$$

or as

$$EA_2\left(-\frac{\omega_n}{c}B_2\sin\frac{\omega_n l}{c}+\frac{\omega_n}{c}C_2\cos\frac{\omega_n l}{c}\right)+k\left(B_2\cos\frac{\omega_n l}{c}+C_2\sin\frac{\omega_n l}{c}-B_2\right)=0. \tag{8.1.4.15}$$

The characteristic determinant Δ with respect to C_1, B_2 and C_2 is:

$$\Delta=\begin{vmatrix}\sin\frac{\omega_n l}{c} & -1 & 0\\ -EA_1\frac{\omega_n}{c}\cos\frac{\omega_n l}{c} & k\left(\cos\frac{\omega_n l}{c}-1\right) & EA_2\frac{\omega_n}{c}+k\sin\frac{\omega_n l}{c}\\ 0 & -EA_2\frac{\omega_n}{c}\sin\frac{\omega_n l}{c}+k\left(\cos\frac{\omega_n l}{c}-1\right) & EA_2\frac{\omega_n}{c}\cos\frac{\omega_n l}{c}+k\sin\frac{\omega_n l}{c}\end{vmatrix}=0. \tag{8.1.4.16}$$

Using the given relationships between the cross-sectional areas and Equation (8.1.4.16) yields

$$-\frac{EA\,\omega_n^2}{2c^2}+\frac{k\omega_n}{c}\left(1-2\cos\frac{\omega_n l}{c}\right)\sin\frac{\omega_n l}{c}+\frac{3EA\,\omega_n^2}{4c^2}\sin^2\frac{\omega_n l}{c}=0. \tag{8.1.4.17}$$

Further, by using the given expression for k, one has

$$-\frac{\omega_n^2 l^2}{2c^2} + \frac{\omega_n l}{c}\left(\frac{1}{2} - \cos\frac{\omega_n l}{c}\right)\sin\frac{\omega_n l}{c} + \frac{3}{4}\frac{\omega_n^2 l^2}{c^2}\sin^2\frac{\omega_n l}{c} = 0. \tag{8.1.4.18}$$

Introducing the substitution $y = \omega_n l/c$, this can be solved numerically to obtain

$$\omega_1 = 1.002\frac{c}{l}, \quad \omega_2 = 2.57158\frac{c}{l}, \quad \omega_3 = 4.39278\frac{c}{l}. \tag{8.1.4.19}$$

It is also interesting to note that putting $k = 0$ into Equation (8.1.4.16) and using $A_1 = A_2 = A$, one can derive

$$\cos\frac{2\omega_n l}{c} = 0, \tag{8.1.4.20}$$

which is in agreement with the frequency equation obtained in Example 8.1.2 for a fixed–free bar, but here the length is $2l$ (see Equation (8.1.2.3)).

8.2 Theoretical Introduction: Torsional Vibration of Shafts

This section is concerned with torsional oscillations of shafts. The object of interest is a uniform elastic shaft of mass m, length l and polar moment of inertia I_o, as shown in Figure 8.2.1a. The shaft is made of material whose shear modulus is G. The infinitesimally small element of length dx, placed at the location x is considered, and the corresponding free-body diagram is presented in Figure 8.2.1b.

The (twisting) torques acting on this element are M and $M + dM$ and the angle of rotation (twist) is labelled by $\varphi = \varphi(x,t)$. Thus, this angle depends on the location x of the element and time t. The dynamic equation of rotation has the form

$$\rho I_o dx\frac{\partial^2\varphi(x,t)}{\partial t^2} = -M + M + dM, \tag{8.2.1}$$

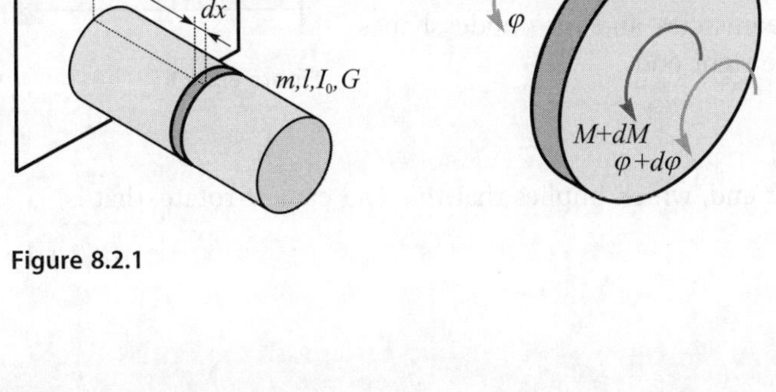

Figure 8.2.1

where ρ stands for the mass density and $\rho I_o dx$ for the moment of inertia of the small element of length dx. To define dM, we turn to Strength of Material and use the known expression between the torque and the twist angle, presenting it in the form:

$$M = GI_o \frac{\partial \varphi(x,t)}{\partial x}. \tag{8.2.2}$$

By expressing dM as $(\partial M/\partial x)dx$ and using Equation (8.2.2), one can transform Equation (8.2.1) into

$$\rho I_o \frac{\partial^2 \varphi(x,t)}{\partial t^2} = GI_o \frac{\partial^2 \varphi(x,t)}{\partial x^2}, \tag{8.2.3}$$

or, labelling $c^2 = G/\rho$, as

$$\frac{\partial^2 \varphi(x,t)}{\partial t^2} = c^2 \frac{\partial^2 \varphi(x,t)}{\partial x^2}. \tag{8.2.4}$$

It is obvious that this governing equation for torsional vibration of the shaft has the same form as the governing equation for longitudinal vibration of a bar, Equation (8.1.5). This further implies that the solution of this equation can also be obtained by the separation of variables

$$\varphi(x,t) = X(x)T(t). \tag{8.2.5}$$

where

$$\frac{\ddot{T}}{T} = c^2 \frac{X''}{X} = -\omega_n^2, \tag{8.2.6}$$

with:

$$X = B\cos\frac{\omega_n x}{c} + C\sin\frac{\omega_n x}{c}, \tag{8.2.7}$$

$$T = P\cos\omega_n t + Q\sin\omega_n t. \tag{8.2.8}$$

The constants B, C, P and Q should be calculated from the boundary and initial conditions. Only the former are demonstrated here because of different mechanical quantities, while the latter concern the time domain and use the same methodology as presented in the previous section.

Example 8.2.1 Find the natural frequencies of a shaft fixed at one end as shown in Figure 8.2.1.1. The shaft carries a rigid disc of mass moment of inertia J, where $J = \rho I_o l = GI_o l/c^2$.

Find also the natural frequencies and the mode shapes when there is no disc at the right end.

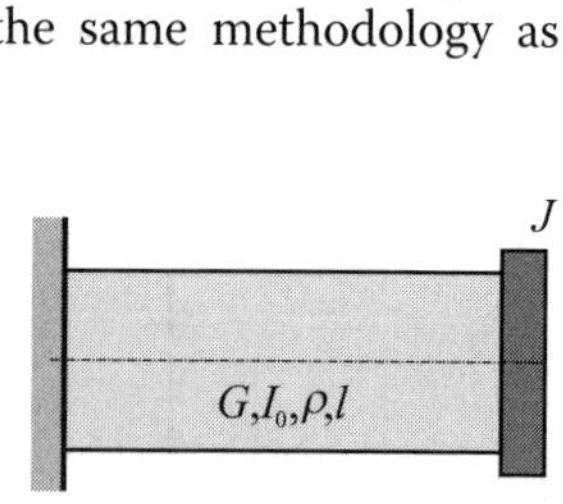

Figure 8.2.1.1

Solution

The shaft is fixed at the left end, which implies that this end cannot rotate, that is,

$$\varphi(0,t) = 0. \tag{8.2.1.1}$$

This condition can be transformed into:

$$X(0) = 0, \tag{8.2.1.2}$$

giving $B = 0$ in Equation (8.2.7).

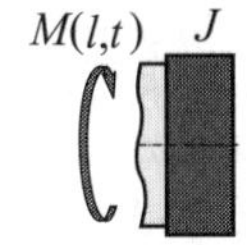

Figure 8.2.1.2

Focusing on the rigid disc on the right end, one can create its free-body diagram (Figure 8.2.1.2) and write down its dynamic equation of rotation:

$$J\frac{\partial^2\varphi(l,t)}{\partial t^2} = -M(l,t), \tag{8.2.1.3}$$

where the internal torque is defined by Equation (8.2.2) with $x = l$. Separating the variables as in Equation (8.2.5) and using (8.2.4), this becomes

$$Jc^2X''(l) = -GI_oX'(l). \tag{8.2.1.4}$$

Using the form given by Equation (8.2.7), Equation (8.2.1.4) transforms into

$$-Jc^2\left(\frac{\omega_n}{c}\right)^2 \sin\frac{\omega_n l}{c} + GI_o\frac{\omega_n}{c}\cos\frac{\omega_n l}{c} = 0, \tag{8.2.1.5}$$

that is,

$$\tan\frac{\omega_n l}{c} = \frac{GI_o l}{Jc^2}\frac{c}{\omega_n l}. \tag{8.2.1.6}$$

Given that $J = GI_o l/c^2$, this equation takes the same form as Equation (8.1.3.11), and the natural frequencies are then the same as given in Equation (8.1.3.13).

If there is no disc at the right end ($J = 0$), Equation (8.2.1.5) gives

$$\cos\frac{\omega_n l}{c} = 0, \tag{8.2.1.7}$$

which is the same frequency equation as the one obtained for the fixed–free bar, Equation (8.1.2.3), considered in Example (8.1.2). Thus, the mode shapes presented in Figure 8.1.2.3. are the same. However, they do not represent the displacement, but the twist angle.

Example 8.2.2 A shaft fixed at one end and free at the other has a changeable cross-section as shown in Figure 8.2.2.1, where the length of each part is l. There is a disc of moment of inertia J between these two parts and a torsional spring of stiffness k, which is

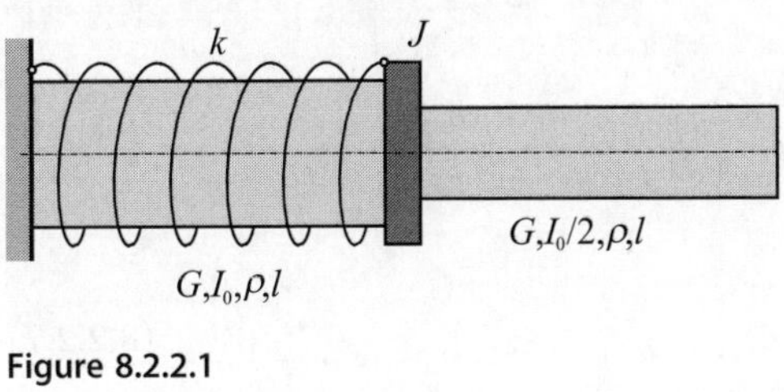

Figure 8.2.2.1

attached to the fixed wall and to the disc. For the physical parameters shown in Figure 8.2.2.1, find the characteristic equation and the natural frequencies assuming $k = GI_o/l$, $J = \rho I_o l$. Compare these natural frequencies with the cases when there is no disc, and when neither the spring nor the disc exists.

Solution

Due to the changeable cross-section, the shaft will be treated as consisting of two parts as labelled in Figure 8.2.2.2, with the respective positions defined by x_1 and x_2, and the twist angles φ_1 and φ_2. Thus, there are two governing equations

$$\frac{\partial^2 \varphi_1(x_1,t)}{\partial t^2} = c^2 \frac{\partial^2 \varphi_1(x_1,t)}{\partial x_1^2}, \tag{8.2.2.1}$$

$$\frac{\partial^2 \varphi_2(x_2,t)}{\partial t^2} = c^2 \frac{\partial^2 \varphi_2(x_2,t)}{\partial x_2^2}. \tag{8.2.2.2}$$

The separation of variables is introduced as $\varphi_1(x,t) = X_1(x)T(t)$ and $\varphi_2(x,t) = X_2(x)T(t)$, where

$$X_1 = B_1 \cos\frac{\omega_n x_1}{c} + C_1 \sin\frac{\omega_n x_1}{c}, \tag{8.2.2.3}$$

$$X_2 = B_2 \cos\frac{\omega_n x_2}{c} + C_2 \sin\frac{\omega_n x_2}{c}. \tag{8.2.2.4}$$

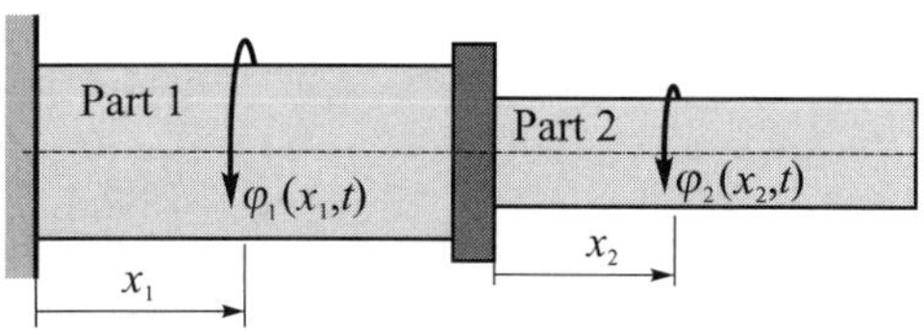

Figure 8.2.2.2

There are four existing constants and, thus, four boundary conditions are needed. The first boundary condition is easily recognised for Part 1, which is fixed at the left end:

$$\varphi_1(0,t) = 0 \Rightarrow X_1(0) = 0, \tag{8.2.2.5}$$

yielding $B_1 = 0$.

The fact that the ending section of Part 1 is the beginning section of Part 2 imposes the continuity condition

$$\varphi_1(l,t) = \varphi_2(0,t) \Rightarrow X_1(l) = X_2(0), \tag{8.2.2.6}$$

leading to

$$C_1 \sin\frac{\omega_n l}{c} - B_2 = 0. \tag{8.2.2.7}$$

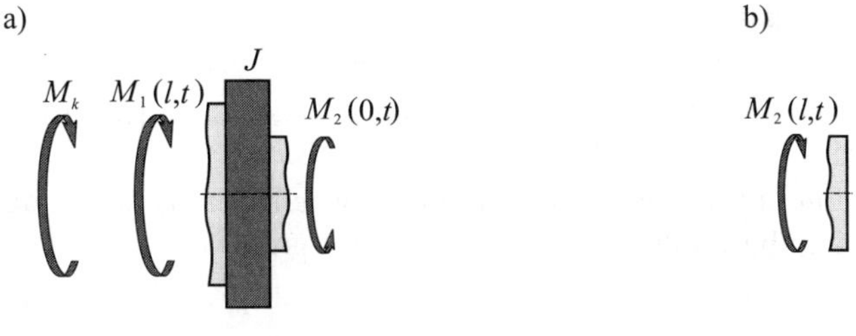

Figure 8.2.2.3

Considering the rotation of the disc in the middle (Figure 8.2.2.3a), one can write down the following dynamic equation of its rotation:

$$J\frac{\partial^2\varphi_2(0,t)}{\partial t^2} = -M_1(l,t) - M_k + M_2(0,t), \tag{8.2.2.8}$$

where

$$M_1(l,t) = GI_o\frac{\partial\varphi_1(l,t)}{\partial x_1}, \quad M_k = k\varphi_1(l,t), \quad M_2(0,t) = G\frac{I_o}{2}\frac{\partial\varphi_2(0,t)}{\partial x_2}. \tag{8.2.2.9}$$

Knowing that $\ddot{\varphi}_2(0,t) = c^2\varphi_2''(0,t)$, Equation (8.2.2.8) can be transformed into

$$Jc^2X_2''(0) = -GI_oX_1'(l) - kX_1(l) + G\frac{I_o}{2}X_2'(0), \tag{8.2.2.10}$$

and also into

$$-C_1\left(GI_o\frac{\omega_n}{c}\cos\frac{\omega_n l}{c} + k\sin\frac{\omega_n l}{c}\right) + Jc^2\left(\frac{\omega_n}{c}\right)^2 B_2 + G\frac{I_o}{2}\frac{\omega_n}{c}C_2 = 0. \tag{8.2.2.11}$$

The right end of the shaft, that is, the right end of Part 2 is free, and the free-body diagram for an infinitesimally small element at this location is given in Figure 8.2.2.3b. It is easy to see that the internal torque must be equal to zero at the end, which due to Equation (8.2.2), imposes:

$$\frac{\partial\varphi_2(l,t)}{\partial x_2} = 0 \;\Rightarrow\; X_2'(l,t) = 0. \tag{8.2.2.12}$$

This further gives:

$$-B_2\sin\frac{\omega_n l}{c} + C_2\cos\frac{\omega_n l}{c} = 0. \tag{8.2.2.13}$$

The characteristic determinant Δ with respect to C_1, B_2 and C_2 is:

$$\Delta = \begin{vmatrix} \sin\frac{\omega_n l}{c} & -1 & 0 \\ -GI_o\frac{\omega_n}{c}\cos\frac{\omega_n l}{c} - k\sin\frac{\omega_n l}{c} & Jc^2\left(\frac{\omega_n}{c}\right)^2 & G\frac{I_o}{2}\frac{\omega_n}{c} \\ 0 & -\sin\frac{\omega_n l}{c} & \cos\frac{\omega_n l}{c} \end{vmatrix} = 0. \tag{8.2.2.14}$$

The resulting characteristic equation is

$$-\frac{GI_o\omega_n}{c}+\left(J\omega_n^2-k\right)\cos\frac{\omega_n l}{c}\sin\frac{\omega_n l}{c}+\frac{3GI_o\omega_n}{2c}\sin^2\frac{\omega_n l}{c}=0. \tag{8.2.2.15}$$

Let us first solve this equation in the simplest case required when there is no spring ($k = 0$) and no disc ($J = 0$). Using these values, Equation (8.2.2.15) gives

$$\sin\frac{\omega_n l}{c}=\pm\sqrt{\frac{2}{3}}, \tag{8.2.2.16}$$

whose solutions are

$$\frac{\omega_n l}{c}=0.95532+n\pi,\quad \frac{\omega_n l}{c}=2.18628+n\pi,\quad n=0,1,2,3... \tag{8.2.2.17}$$

The first three natural frequencies are

$$\omega_1=0.95532\frac{c}{l},\quad \omega_2=2.18628\frac{c}{l},\quad \omega_3=4.09691\frac{c}{l}. \tag{8.2.2.18}$$

If there is no disc ($J = 0$), but $k = GI_o/l$, the characteristic equation simplifies to

$$-\frac{\omega_n l}{c}-\cos\frac{\omega_n l}{c}\sin\frac{\omega_n l}{c}+\frac{3\omega_n l}{2c}\sin^2\frac{\omega_n l}{c}=0. \tag{8.2.2.19}$$

The equation was solved numerically to obtain

$$\omega_1=1.18849\frac{c}{l},\quad \omega_2=2.33228\frac{c}{l},\quad \omega_3=4.17389\frac{c}{l}. \tag{8.2.2.20}$$

It is seen that for a given stiffness value, the natural frequencies are slightly higher than in Equation (8.2.2.18).

Finally, the characteristic equation (8.2.2.15) for $k = GI_o/l$, $J = \rho I_o l$ and $\rho = G/c^2$, becomes

$$-\frac{\omega_n l}{c}+\left[\left(\frac{\omega_n l}{c}\right)^2-1\right]\cos\frac{\omega_n l}{c}\sin\frac{\omega_n l}{c}+\frac{3\omega_n l}{2c}\sin^2\frac{\omega_n l}{c}=0. \tag{8.2.2.21}$$

The first three natural frequencies calculated numerically are

$$\omega_1=0.973263\frac{c}{l},\quad \omega_2=1.86722\frac{c}{l},\quad \omega_3=3.4361\frac{c}{l}. \tag{8.2.2.22}$$

Example 8.2.3 The system shown in Figure 8.2.3.1 consists of two shafts: unlike the inner one, the outer one is hollow. They are both attached to a rigid weightless disc at their right ends. The inner shaft is fixed at the left end, while the outer one is attached to a torsional spring of stiffness k. Find the natural frequency for the case when $k = GI_o/l$ as well as for the case when there is no spring.

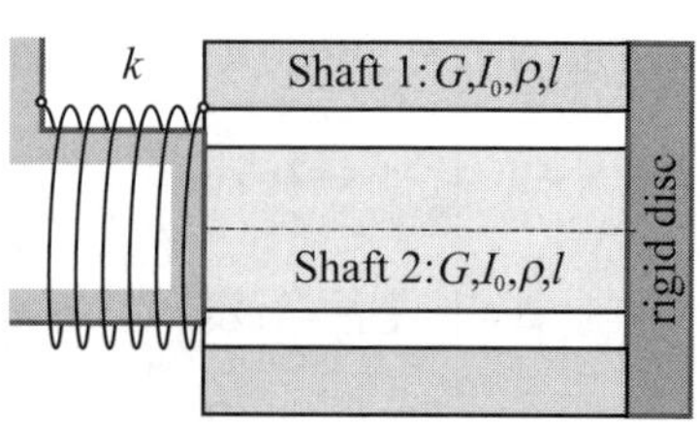

Figure 8.2.3.1

Solution

To define the locations, two coordinates x_1 and x_2 are introduced, and they are related to the outer and inner disc, respectively. The two governing equations are the same as in the previous example, as well as the solutions (8.2.2.3) and (8.2.2.4). The first boundary condition is written for the inner disc, as it is fixed at the left end, which imposes

$$\varphi_2(0,t) = 0 \;\Rightarrow\; X_2(0) = 0, \tag{8.2.3.1}$$

giving $B_2 = 0$.

To analyse the boundary condition for the outer disc at the left end, an infinitesimally small element at this end is considered (Figure 8.2.3.2a).

The equilibrium equation is

$$0 = M_1(0,t) - M_k, \tag{8.2.3.2}$$

where

$$M_1(0,t) = GI_o \frac{\partial \varphi_1(0,t)}{\partial x_1}, \quad M_k = k\varphi_1(0,t), \tag{8.2.3.3}$$

which leads to

$$GI_o X_1'(0) - kX_1(0) = 0, \tag{8.2.3.4}$$

and also to

$$GI_o \frac{\omega_n}{c} C_1 - kB_1 = 0. \tag{8.2.3.5}$$

The fact that the ending sections of Shaft 1 and Shaft 2 are attached to the same rigid disc imposes

$$\varphi_1(l,t) = \varphi_2(l,t) \;\Rightarrow\; X_1(l) = X_2(l), \tag{8.2.3.6}$$

yielding:

$$B_1 \cos\frac{\omega_n l}{c} + C_1 \sin\frac{\omega_n l}{c} - C_2 \sin\frac{\omega_n l}{c} = 0. \tag{8.2.3.7}$$

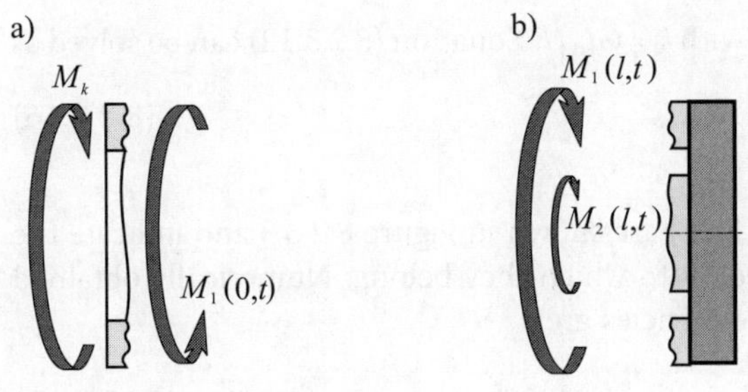

Figure 8.2.3.2

Finally, the equilibrium equation of the torques acting on the weightless disc at the right end (Figure 8.2.3.2b) is

$$-M_1(l,t)-M_2(l,t)=0, \tag{8.2.3.8}$$

which leads to

$$X_1'(l)+X_2'(l)=0. \tag{8.2.3.9}$$

This can be further transformed into

$$-B_1\sin\frac{\omega_n l}{c}+C_1\cos\frac{\omega_n l}{c}+C_2\cos\frac{\omega_n l}{c}=0. \tag{8.2.3.10}$$

The characteristic determinant Δ with respect to B_1, C_1 and C_2 is:

$$\Delta=\begin{vmatrix} -k & GI_o\dfrac{\omega_n}{c} & 0 \\ \cos\dfrac{\omega_n l}{c} & \sin\dfrac{\omega_n l}{c} & -\sin\dfrac{\omega_n l}{c} \\ -\sin\dfrac{\omega_n l}{c} & \cos\dfrac{\omega_n l}{c} & \cos\dfrac{\omega_n l}{c} \end{vmatrix}=0. \tag{8.2.3.11}$$

This can be developed into

$$\frac{\omega_n}{c}GI_o\cos\frac{2\omega_n l}{c}+k\sin\frac{2\omega_n l}{c}=0. \tag{8.2.3.12}$$

If there is no spring, the characteristic equation becomes

$$\cos\frac{2\omega_n l}{c}=0, \tag{8.2.3.13}$$

which gives

$$\omega_n=(2n-1)\frac{\pi c}{4l},\quad n=1,2,3,\ldots, \tag{8.2.3.14}$$

that is,

$$\omega_1=\frac{\pi c}{4l},\quad \omega_2=\frac{3\pi c}{4l},\quad \omega_3=\frac{5\pi c}{4l},\quad \ldots \tag{8.2.3.15}$$

When there is a spring in the system with $k=GI_o/l$, Equation (8.2.3.12) can be solved as

$$\tan 2y=-y. \tag{8.2.3.16}$$

where $y=\omega_n l/c$.

The two functions in Equation (8.2.3.16) are shown in Figure 8.2.3.3 and indicate the values of the solutions, that is, the region to which they belong. Numerically obtained solutions for the first three natural frequencies are

$$\omega_1=1.14446\frac{c}{l},\quad \omega_2=2.54349\frac{c}{l},\quad \omega_3=4.04808\frac{c}{l}. \tag{8.2.3.17}$$

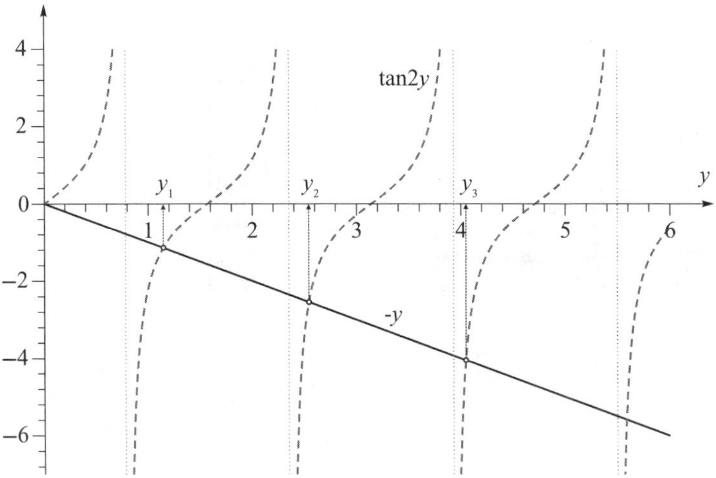

Figure 8.2.3.3

Example 8.2.4 The system shown in Figure 8.2.4.1 consists of two shafts: the upper one is fixed at the left end and carries a disc at the right end; this disc is externally coupled with another disc, carried by a lower shaft, whose right end also carries another disc. There is no slipping between the discs in coupling. For the parameters presented in Figure 8.2.4.1, determine the characteristic determinant and the fundamental frequency, assuming $J = \rho I_o l = GI_o l/c^2$.

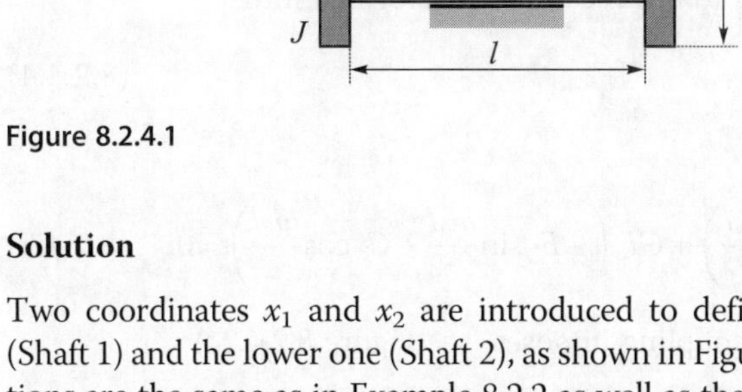

Figure 8.2.4.1

Solution

Two coordinates x_1 and x_2 are introduced to define the locations at the upper shaft (Shaft 1) and the lower one (Shaft 2), as shown in Figure 8.2.4.2a. The two governing equations are the same as in Example 8.2.2 as well as the solutions (8.2.2.3) and (8.2.2.4). The first boundary condition is written for Shaft 1, as its left end is fixed, which imposes

$$\varphi_1(0,t) = 0 \Rightarrow X_1(0) = 0, \tag{8.2.4.1}$$

giving $B_1 = 0$.

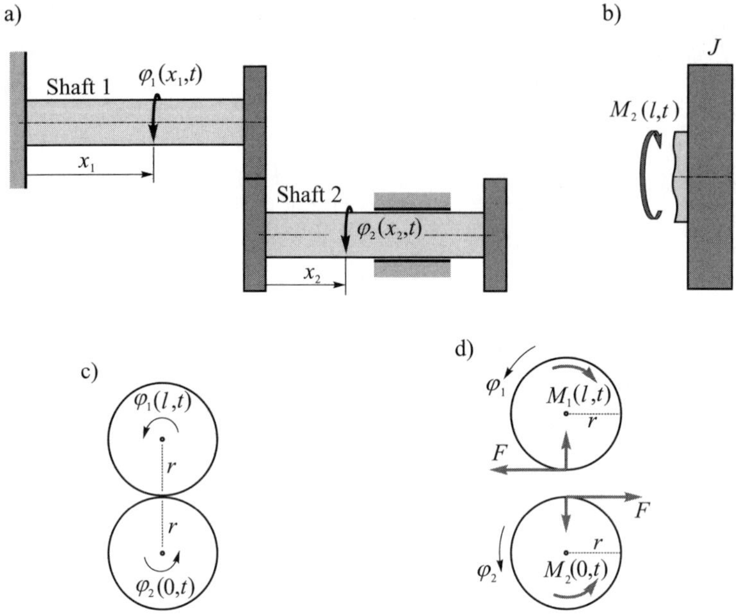

Figure 8.2.4.2

Then, the right disc at Shaft 2 is considered (Figure 8.2.4.2b). The dynamic equation of its rotation has the form:

$$J\frac{\partial^2\varphi_2(l,t)}{\partial t^2} = -M_2(l,t), \tag{8.2.4.2}$$

where

$$M_2(l,t) = GI_o\frac{\partial\varphi_2(l,t)}{\partial x_2}. \tag{8.2.4.3}$$

Using $\ddot{\varphi}_2(l,t) = c^2\varphi_2''(l,t)$, Equation (8.2.4.2) can be transformed into

$$Jc^2X_2''(l) = -GI_oX_2'(l), \tag{8.2.4.4}$$

and further into

$$-Jc^2\frac{\omega_n}{c}\left(B_2\cos\frac{\omega_n l}{c} + C_2\sin\frac{\omega_n l}{c}\right) + GI_o\left(-B_2\sin\frac{\omega_n l}{c} + C_2\cos\frac{\omega_n l}{c}\right) = 0. \tag{8.2.4.5}$$

No slipping between the discs in coupling imposes (see Figure 8.2.4.2c)

$$\varphi_1(l,t) = -\varphi_2(0,t), \tag{8.2.4.6}$$

which can be transformed into

$$C_1\sin\frac{\omega_n l}{c} + B_2 = 0. \tag{8.2.4.7}$$

Considering the dynamics of both discs separately in accordance with their free-body diagrams shown in Figure 8.2.4.2d, one can write the following dynamic equations of their rotations:

$$J\frac{\partial^2\varphi_1(l,t)}{\partial t^2} = -M_1(l,t) - Fr, \tag{8.2.4.8}$$

$$J\frac{\partial^2\varphi_2(0,t)}{\partial t^2} = M_2(0,t) - Fr. \tag{8.2.4.9}$$

When combined and transformed, these give

$$Jc^2\left[X_2''(0) - X_1''(l)\right] = GI_oX_2'(0) + GI_oX_1'(l), \tag{8.2.4.10}$$

that is,

$$-Jc^2\frac{\omega_n}{c}\left(B_2 - C_1\sin\frac{\omega_n l}{c}\right) - GI_o\left(C_2 + C_1\cos\frac{\omega_n l}{c}\right) = 0. \tag{8.2.4.11}$$

The characteristic determinant Δ is found to be:

$$\Delta = \begin{vmatrix} 0 & -\frac{\omega_n l}{c}\cos\frac{\omega_n l}{c} - \sin\frac{\omega_n l}{c} & -\frac{\omega_n l}{c}\sin\frac{\omega_n l}{c} + \cos\frac{\omega_n l}{c} \\ \sin\frac{\omega_n l}{c} & 1 & 0 \\ \frac{\omega_n l}{c}\sin\frac{\omega_n l}{c} - \cos\frac{\omega_n l}{c} & -\frac{\omega_n l}{c} & -1 \end{vmatrix} = 0. \tag{8.2.4.12}$$

The fundamental frequency calculated from this characteristic determinant is:

$$\omega_1 = 0.44812\frac{c}{l}. \tag{8.2.4.13}$$

8.3 Theoretical Introduction: Transversal Vibration of Beams

In this section, transversal (lateral) oscillations of beams are considered. These oscillations cause bending, as shown in Figure 8.3.1a. A uniform beam is assumed to be of length l, cross-sectional area A, mass density ρ and made of Hooke's material whose

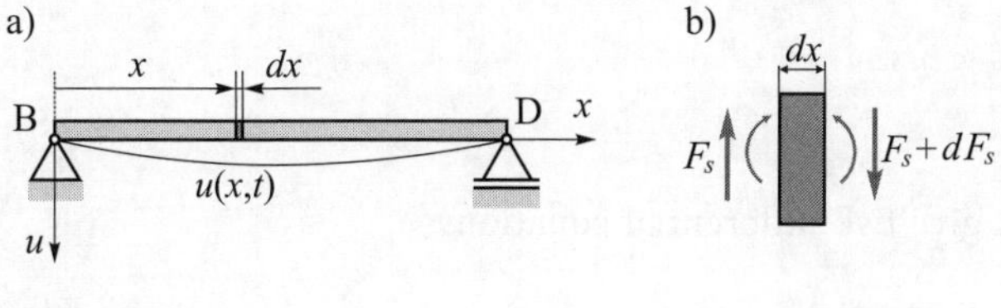

Figure 8.3.1

Young's modulus is E. The displacement u of each section depends on its location x and on time t, that is, $u = u(x,t)$. The free-body diagram for an infinitesimally small element of the beam is shown in Figure 8.3.1b. This element has length dx. It is subjected to internal forces (only the shear forces are shown for brevity) and bending moments. Noting that this element has elementary mass $\rho A dx$, the projection of Newton's second law onto the u-axis is

$$\rho A dx \frac{\partial^2 u(x,t)}{\partial t^2} = -F_s + F_s + dF_s. \tag{8.3.1}$$

To define dF_s, the elementary theory of bending of beams is applied, that is, the known expression relating the shear force and the bending moment is used:

$$F_s = \frac{dM}{dx}, \tag{8.3.2}$$

as well as the expression relating the bending moment and the deflection:

$$M = -EI\frac{\partial^2 u(x,t)}{\partial x^2}, \tag{8.3.3}$$

where I stands for the moment of inertia of the beam cross-section about the neutral axis, which is assumed to be constant here.

Combining the last two expressions gives

$$F_s = -EI\frac{\partial^3 u(x,t)}{\partial x^3} \quad \Rightarrow \quad dF_s = -EI\frac{\partial^4 u(x,t)}{\partial x^4}dx. \tag{8.3.4}$$

Substituting it into Equation (8.3.1), one can derive

$$\frac{\partial^2 u(x,t)}{\partial t^2} + c^2\frac{\partial^4 u(x,t)}{\partial x^4} = 0, \tag{8.3.5}$$

which can be represented as

$$\ddot{u}(x,t) + c^2 u^{\text{iv}}(x,t) = 0, \tag{8.3.6}$$

where $c^2 = EI/(\rho A)$. Note that this constant c is expressed in m^2/s. The governing equation is again a partial differential equation, but with the fourth partial derivative with respect to x.

To find the solution of Equation (8.3.6), the method of separation of variables is used, assuming

$$u(x,t) = X(x)T(t), \tag{8.3.7}$$

which yields

$$-\frac{\ddot{T}}{T} = c^2\frac{X^{\text{iv}}}{X} = \omega_n^2. \tag{8.3.8}$$

The ratios in Equation (8.3.8) give two differential equations:

$$\ddot{T} + \omega_n^2 T = 0, \tag{8.3.9}$$

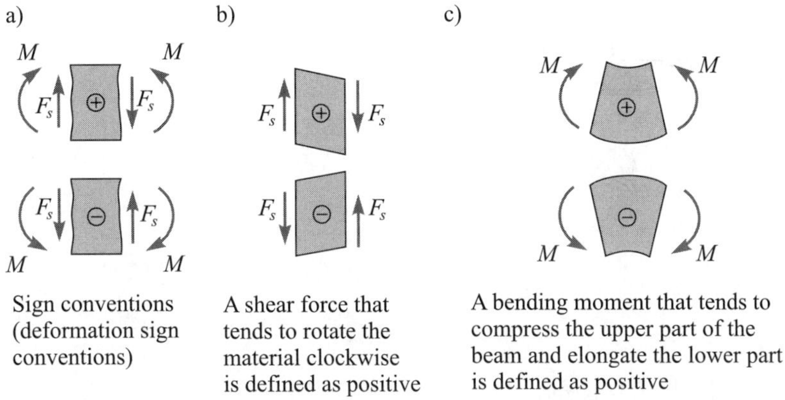

Figure 8.3.2

$$X^{\text{iv}} - \frac{\omega_n^2}{c^2}X = 0. \tag{8.3.10}$$

The solution of Equation (8.3.9) is easily recognised as

$$T = P\cos\omega_n t + Q\sin\omega_n t. \tag{8.3.11}$$

To find the solution of Equation (8.3.10), one can assume it in the form $X = Ce^{\lambda x}$, where C and λ are constants. Substituting it into Equation (8.3.10) and using $\kappa_n^4 = \omega_n^2/c^2$, gives

$$\lambda^4 - \kappa_n^4 = 0, \tag{8.3.12}$$

that is,

$$\lambda_{1,2} = \pm\kappa_n, \quad \lambda_{3,4} = \pm\kappa_n i, \tag{8.3.13}$$

where i is the complex unit. The solution sought is

$$X = Ae^{\kappa_n x} + Be^{-\kappa_n x} + Ce^{i\kappa_n x} + De^{-i\kappa_n x}, \tag{8.3.14}$$

or, it can be expressed as

$$X = A\cos\kappa_n x + B\sin\kappa_n x + C\cosh\kappa_n x + D\sinh\kappa_n x. \tag{8.3.15}$$

The constants A–D are related to the boundary conditions. There will be an infinite number of values of κ_n and an infinite number of natural frequencies ω_n, each of which is associated with one mode of vibration. This is illustrated in the forthcoming examples. Before that, it is worth noting the sign convention (deformation sign convention) shown in Figure 8.3.2. This convention will be consistently used in the examples and it was also used in the derivation of the governing equation above.

Example 8.3.1 Find the natural frequencies and modes of vibration of the simply supported (pinned/hinged) beam shown in Figure 8.3.1.1a. Determine also the characteristic equation when a torsional spring of stiffness k is attached to the beam at the right-hand support (Figure 8.3.1.1b).

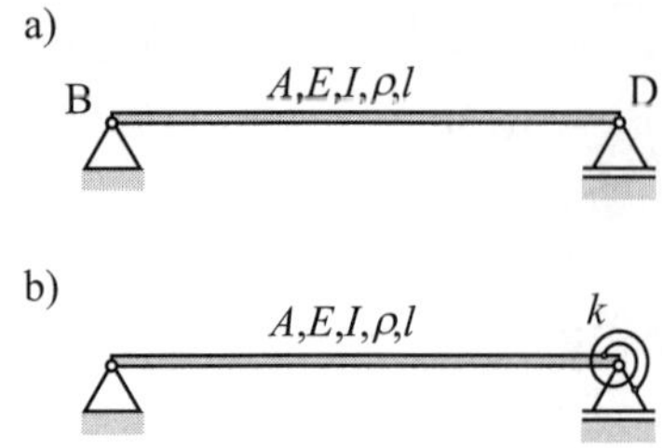

Figure 8.3.1.1

Solution

The displacement at both ends is zero

$$u(0,t)=0,\quad u(l,t)=0, \tag{8.3.1.1}$$

which, given Equation (8.3.7), yields

$$X(0)=0,\quad X(l)=0. \tag{8.3.1.2}$$

Now, Equation (8.3.15) leads to

$$A+C=0, \tag{8.3.1.3}$$
$$A\cos\kappa_n l+B\sin\kappa_n l+C\cosh\kappa_n l+D\sinh\kappa_n l=0. \tag{8.3.1.4}$$

When the beam is only supported at its ends and there is no spring, one can focus on the small elements at its ends, and form the corresponding free-body diagram as shown in Figure 8.3.2.1a,b. It is clearly seen that equating the moment sum about the supports to zero leads to

$$M(0,t)=0,\quad M(l,t)=0. \tag{8.3.1.5}$$

Due to (8.3.3) and (8.3.7), these conditions become:

$$X''(0)=0,\quad X''(l)=0, \tag{8.3.1.6}$$

which can be expressed as

$$-A+C=0, \tag{8.3.1.7}$$
$$-A\cos\kappa_n l-B\sin\kappa_n l+C\cosh\kappa_n l+D\sinh\kappa_n l=0. \tag{8.3.1.8}$$

Equations (8.3.1.3) and (8.3.1.7) give $A=C=0$, while Eqations 8.3.1.4 and 8.3.1.8 simplify to

$$B\sin\kappa_n l+D\sinh\kappa_n l=0, \tag{8.3.1.9}$$

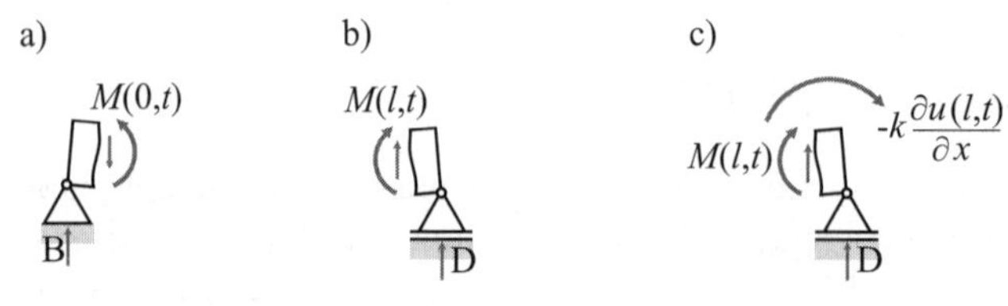

Figure 8.3.1.2

$$-B\sin\kappa_n l + D\sinh\kappa_n l = 0. \tag{8.3.1.10}$$

These equations are linear and homogenous and their non-trivial solution exists provided that the determinant with respect to B and D is equal to zero. This determinant represents the so-called characteristic determinant:

$$\Delta = \begin{vmatrix} \sin\kappa_n l & \sinh\kappa_n l \\ -\sin\kappa_n l & \sinh\kappa_n l \end{vmatrix} = 0, \tag{8.3.1.11}$$

and the following characteristic (frequency) equation is obtained:

$$\sin\kappa_n l \sinh\kappa_n l = 0. \tag{8.3.1.12}$$

This is satisfied for

$$\kappa_n l = n\pi, \quad n = 1,2,3,\ldots \tag{8.3.1.13}$$

or,

$$\omega_n = \kappa_n^2 c = \frac{n^2\pi^2}{l^2} c, \quad n = 1,2,3,\ldots \tag{8.3.1.14}$$

The values given by Equation (8.3.1.13) are shown in Table 8.3.1.1 for the first three modes (this table collects all the common boundary conditions analysed in this section, and the rest of the results presented will be derived in other examples given below).

Table 8.3.1.1

	$\kappa_n l$		
Beam configuration	**First (fundamental)**	**Second**	**Third**
Pinned-pinned	π	2π	3π
Pinned-free	3.92660	7.06858	10.21018
Clamped-free	1.8751	4.69409	7.85476
Clamped-clamped	4.73004	7.8532	10.9956

Recognising the characteristic equation in Equation (8.3.1.9) or (8.3.1.10), one can conclude that $D = 0$. Thus, the solution for X_n is

$$X_n = B_n \sin\kappa_n x = B_n \sin\frac{n\pi x}{l}. \tag{8.1.1.15}$$

The corresponding mode shapes are plotted in Figure (8.3.1.3) for the first three modes.

Let us consider now the case of the beam with a spring at the right end (Figure 8.3.1.1b). Only the second condition in Equation (8.3.1.5) is different due to the existence of the spring. Thus, only Equation (8.3.1.10) will be changed, while $A = C = 0$ will hold. Focusing on the small element at the right-hand end (Figure 8.3.1.2c), one needs to add the resisting torque due to the spring, so that equating to zero the moment sum about the support leads to

$$M(l,t) - k\frac{\partial u(l,t)}{\partial x} = 0, \tag{8.3.1.16}$$

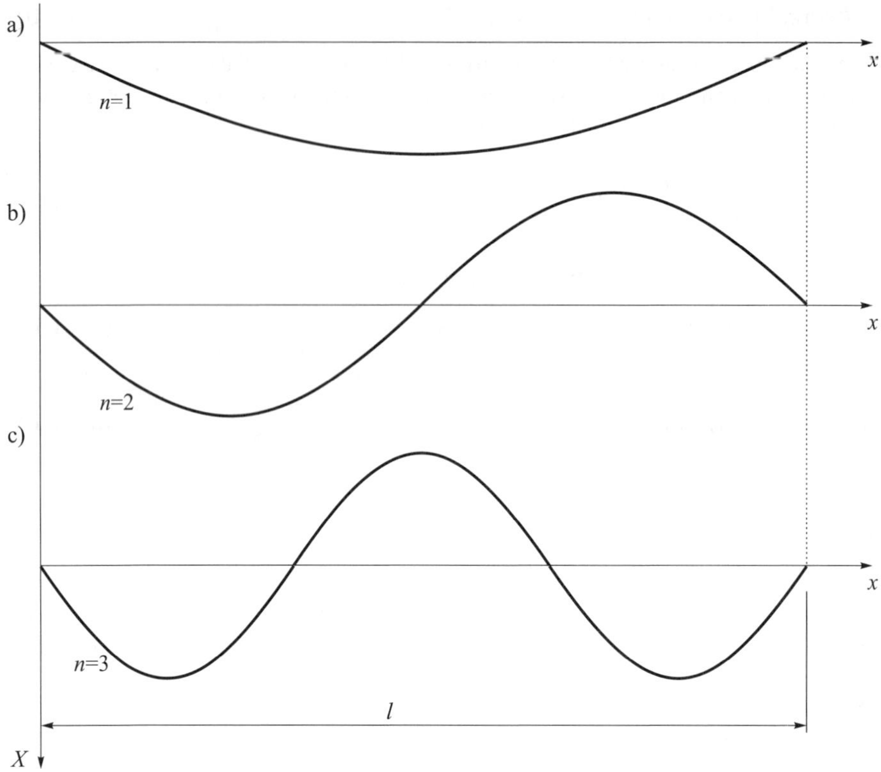

Figure 8.3.1.3

that is,

$$-EI\frac{\partial^2 u(l,t)}{\partial x^2} - k\frac{\partial u(l,t)}{\partial x} = 0. \tag{8.3.1.17}$$

Using (8.3.3) and (8.3.7), this condition gives:

$$EIX''(l) + kX'(l) = 0, \tag{8.3.1.18}$$

which can be expressed as

$$-BEI\kappa_n \sin\kappa_n l + DEI\kappa_n \sinh\kappa_n l + k(B\cos\kappa_n l + D\cosh\kappa_n x) = 0. \tag{8.3.1.19}$$

The characteristic determinant has the form:

$$\Delta = \begin{vmatrix} \sin\kappa_n l & \sinh\kappa_n l \\ -EI\kappa_n \sin\kappa_n l + k\cos\kappa_n l & EI\kappa_n \sinh\kappa_n l + k\cosh\kappa_n x \end{vmatrix} = 0. \tag{8.3.1.20}$$

The resulting characteristic equation is

$$2EI\kappa_n \sin\kappa_n l \sinh\kappa_n l + k(\cosh\kappa_n l \sin\kappa_n l - \cos\kappa_n l \sinh\kappa_n l) = 0. \tag{8.3.1.21}$$

Specifying the system parameters, the values of κ_n can be calculated from this characteristic equation.

Example 8.3.2 A beam is simply supported at its left end and attached to this support via a torsional spring of stiffness k_1 (Figure 8.3.2.1a). The beam carries a mass m at the right end, and this mass is attached to a translational spring of stiffness k_2. Determine the characteristic determinant and then, as a special case, find the natural frequencies and mode shapes of the pinned–free beam (Figure 8.3.2.1b), removing the mass and the translational spring.

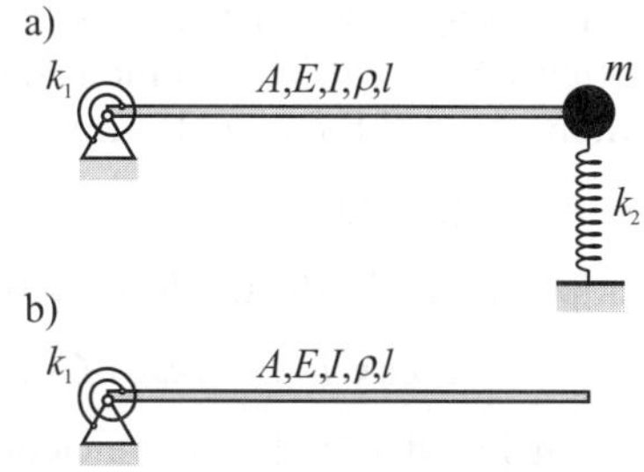

Figure 8.3.2.1

Solution

The left end is fixed, and the displacement at this location is zero:

$$u(0,t) = 0, \tag{8.3.2.1}$$

which given Equation (8.3.7), yields $X(0) = 0$ and further also:

$$A + C = 0. \tag{8.3.2.2}$$

Let us consider now the infinitesimally small element at the left end and its free-body diagram (Figure 8.3.2.2a).

The moment sum about the support involves the internal bending moment and the resistance torque from the spring. The equilibrium equation is

$$M(0,t) + k_1 \frac{\partial u(0,t)}{\partial x} = 0, \tag{8.3.2.3}$$

that is,

$$-EI\frac{\partial^2 u(0,t)}{\partial x^2} + k_1 \frac{\partial u(0,t)}{\partial x} = 0. \tag{8.3.2.4}$$

Using (8.3.3) and (8.3.7), this equation transforms into:

$$-EIX''(0) + k_1 X'(0) = 0, \tag{8.3.2.5}$$

which can be expressed as

$$-EI(-\kappa_n A + \kappa_n C) + k_1(B + D) = 0. \tag{8.3.2.6}$$

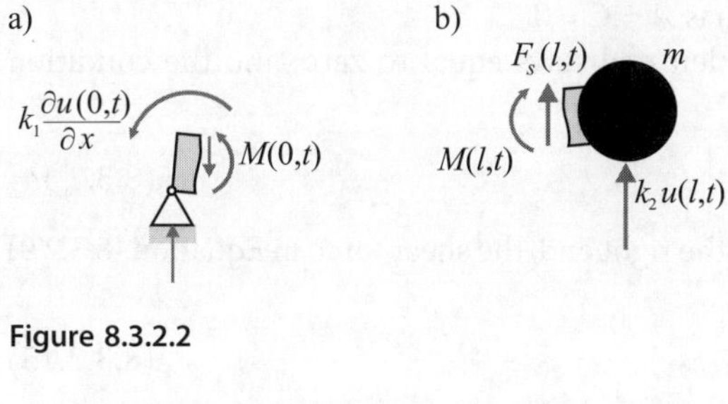

Figure 8.3.2.2

Focusing now on the mass at the right end, its free-body diagram is considered (Figure 8.3.2.2b). The moment sum about the ending point involves the internal bending moment, and the equilibrium condition requires this moment to be equal to zero:

$$M(l,t)=0, \tag{8.3.2.7}$$

implying $X''(l)=0$, and leading to

$$-A\cos\kappa_n l - B\sin\kappa_n l + C\cosh\kappa_n l + D\sinh\kappa_n l = 0. \tag{8.3.2.8}$$

Further, due to the existence of the mass at the left end, its equation of motion in the u-direction needs to be taken into account

$$m\frac{\partial^2 u(l,t)}{\partial t^2} = -F_s(l,t) - k_2 u(l,t). \tag{8.3.2.9}$$

This equation can further be written as

$$-mc^2\frac{\partial^4 u(l,t)}{\partial x^4} = EI\frac{\partial^3 u(l,t)}{\partial x^3} - k_2 u(l,t), \tag{8.3.2.10}$$

or, as:

$$-mc^2\kappa_n^4 X(l) = EIX'''(l) - k_2 X(l). \tag{8.3.2.11}$$

For the solution (8.3.15), one has

$$\begin{aligned}&(k_2 - mc^2\kappa_n^4)(A\cos\kappa_n l + B\sin\kappa_n l + C\cosh\kappa_n l + D\sinh\kappa_n l)\\&-EI\kappa_n^3(A\sin\kappa_n l - B\cos\kappa_n l + C\sinh\kappa_n l + D\cosh\kappa_n l) = 0.\end{aligned} \tag{8.3.2.11}$$

Using Equation (8.3.2.2), one concludes $C = -A$ and the following characteristic determinant is formed

$$\Delta = \begin{vmatrix} 2EI\kappa_n & k_1 & k_1 \\ -\cos\kappa_n l - \cosh\kappa_n l & -\sin\kappa_n l & \sinh\kappa_n l \\ (k_2 - mc^2\kappa_n^4)(\cos\kappa_n l - \cosh\kappa_n l) - & (k_2 - mc^2\kappa_n^4)\sin\kappa_n l + & (k_2 - mc^2\kappa_n^4)\sinh\kappa_n l - \\ EI\kappa_n^3(\sin\kappa_n l - \sinh\kappa_n x) & EI\kappa_n^3\cos\kappa_n l & EI\kappa_n^3\cosh\kappa_n l \end{vmatrix} = 0. \tag{8.3.2.12}$$

Let us now analyse the special case of a pinned–free beam. The first boundary condition (8.3.2.1) remains the same. However, if there is no spring at the left end, the condition (8.3.2.3) simplifies to $M(0,t)=0$, and Equation (8.3.2.6) becomes

$$-A + C = 0. \tag{8.3.2.13}$$

Thus, due to Equation (8.3.2.2), one has $A = C = 0$.

The internal bending moment at the left end stays equal to zero, and the condition (8.3.2.8) simplifies to

$$-B\sin\kappa_n l + D\sinh\kappa_n l = 0. \tag{8.3.2.14}$$

Since there is no mass and no spring at the right end, the shear force in Equation (8.3.2.9) must be equal to zero, leading to

$$-B\cos\kappa_n l + D\cosh\kappa_n l = 0. \tag{8.3.2.15}$$

The following characteristic determinant is now obtained

$$\Delta = \begin{vmatrix} -\sin\kappa_n l & \sinh\kappa_n l \\ -\cos\kappa_n l & \cosh\kappa_n l \end{vmatrix} = 0, \tag{8.3.2.16}$$

resulting in the characteristic equation:

$$-\cosh\kappa_n l \sin\kappa_n l + \cos\kappa_n l \sinh\kappa_n l = 0. \tag{8.3.2.17}$$

The characteristic equation can also be expressed as

$$\tan\kappa_n l - \tanh\kappa_n l = 0. \tag{8.3.2.18}$$

Its solutions can be obtained numerically and the first three of them are:

$$\kappa_1 l = 3.9266, \quad \kappa_2 l = 7.06858, \quad \kappa_3 l = 10.2102. \tag{8.3.2.19}$$

These solutions are also included in Table 8.3.1.1.
Based on Equation (8.3.2.14), one has

$$D = B\frac{\sin\kappa_n l}{\sinh\kappa_n l}, \tag{8.3.2.20}$$

The final solution is

$$X(x) = B\left(\sin\kappa_n x + \frac{\sin\kappa_n l}{\sinh\kappa_n l}\sinh\kappa_n x\right). \tag{8.3.2.21}$$

The corresponding mode shapes are plotted in Figure 8.3.2.3.

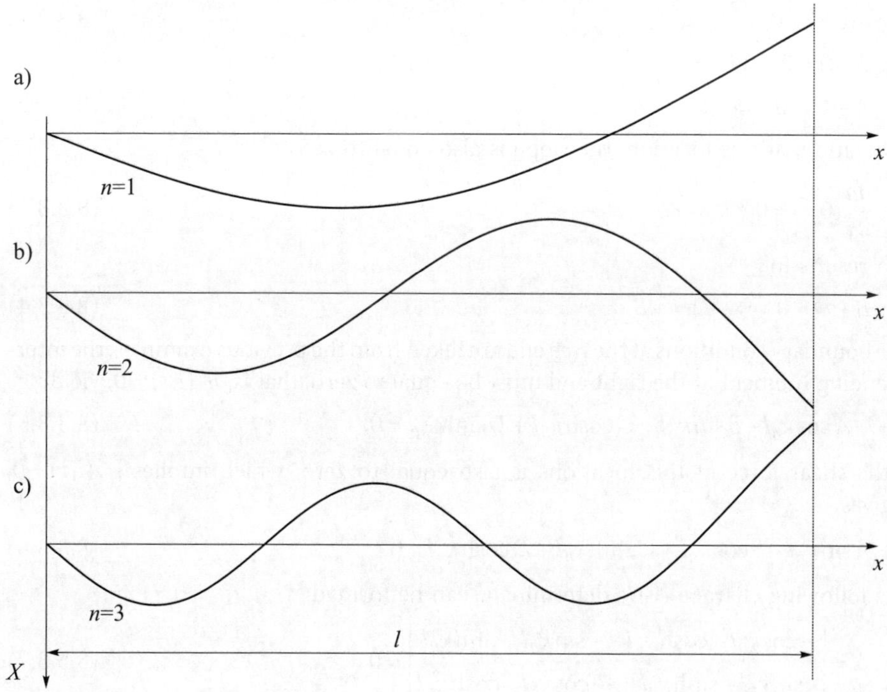

Figure 8.3.2.3

Example 8.3.3 A beam is fixed (clamped) at the left end and free at the right end (Figure 8.3.3.1a). Determine the natural frequencies and plot the mode shapes. Find also the characteristic determinant for a beam that carries a mass m attached to the spring of stiffness k. The constraint on this end allows an unimpeded vertical displacement (Figure 8.3.2.1b).

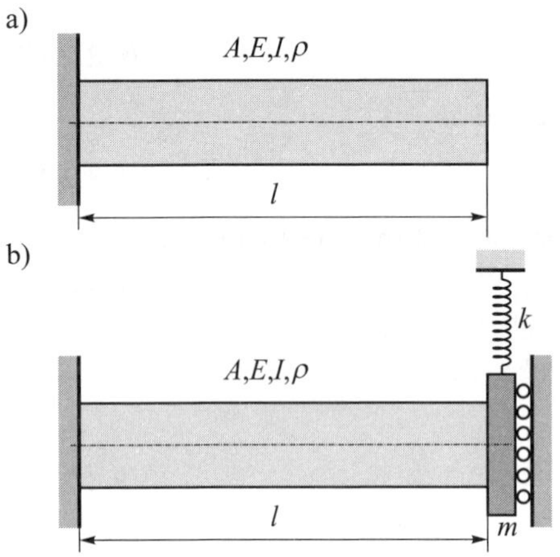

Figure 8.3.3.1

Solution

The fixed left end implies that the displacement is zero when $x = 0$:

$$u(0,t) = 0, \tag{8.3.3.1}$$

giving $X(0) = 0$, and also

$$A + C = 0 \;\Rightarrow\; C = -A. \tag{8.3.3.2}$$

In addition, at this location, the slope is also equal to zero:

$$\frac{\partial u}{\partial x}(0,t) = 0, \tag{8.3.3.3}$$

which results in

$$B + D = 0 \;\Rightarrow\; D = -B. \tag{8.3.3.4}$$

The boundary conditions at the free end are taken from the previous examples: the internal bending moment at the right end must be equal to zero, that is, $u''(l,t) = 0$, yielding

$$-A\cos\kappa_n l - B\sin\kappa_n l + C\cosh\kappa_n l + D\sinh\kappa_n l = 0, \tag{8.3.3.5}$$

and the shear force at this locations is also equal to zero, which implies $u'''(l,t) = 0$, and gives

$$A\sin\kappa_n l - B\cos\kappa_n l + C\sinh\kappa_n l + D\cosh\kappa_n l = 0. \tag{8.3.3.6}$$

The following characteristic determinant can be formed

$$\Delta = \begin{vmatrix} -\cos\kappa_n l - \cosh\kappa_n l & -\sin\kappa_n l - \sinh\kappa_n l \\ \sin\kappa_n l - \sinh\kappa_n l & -\cos\kappa_n l - \cosh\kappa_n l \end{vmatrix} = 0. \tag{8.3.3.7}$$

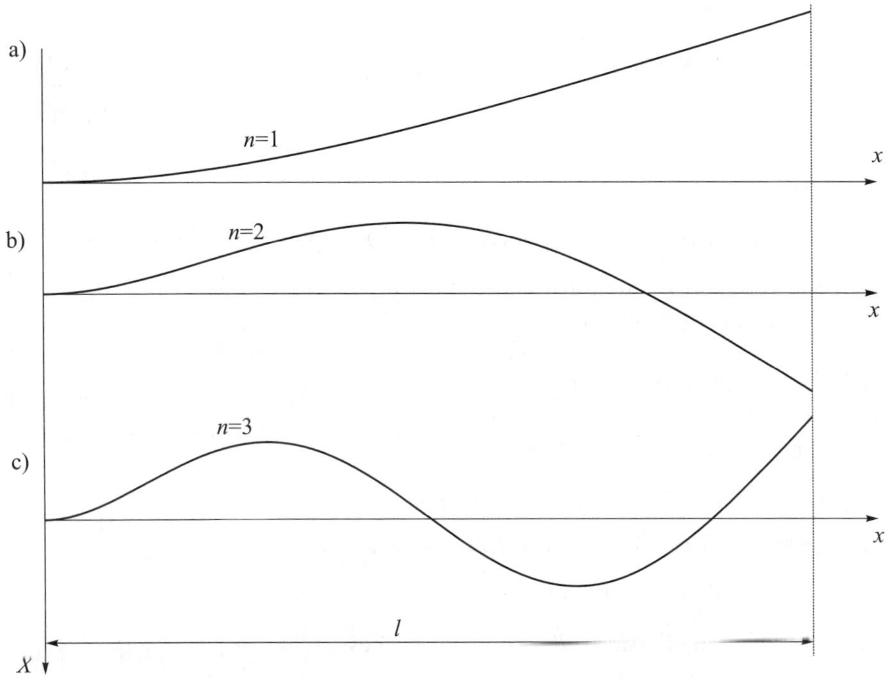

Figure 8.3.3.2

The resulting characteristic equation is

$$\cos\kappa_n l \cosh\kappa_n l + 1 = 0. \tag{8.3.3.8}$$

Its solutions are obtained numerically, and the first three of them are

$$\kappa_1 l = 1.8751, \quad \kappa_2 l = 4.69409, \quad \kappa_3 l = 7.85476 \tag{8.3.3.9}$$

These solutions are also included into Table 8.3.1.1.

Based on the previously derived results (see the second row of the characteristic determinant), one can express:

$$B = A\frac{\sin\kappa_n l - \sinh\kappa_n l}{\cos\kappa_n l + \cosh\kappa_n l}. \tag{8.3.3.10}$$

The final solution is, then,

$$X(x) = A\left[\cos\kappa_n x - \cosh\kappa_n x + \frac{\sin\kappa_n l - \sinh\kappa_n l}{\cos\kappa_n l + \cosh\kappa_n l}(\sin\kappa_n x - \sinh\kappa_n x)\right]. \tag{8.3.3.11}$$

The corresponding mode shapes are plotted in Figure 8.3.3.2.

Let us now consider the system shown in Figure 8.3.3.1b. It is fixed at the left end as for the system presented in Figure 8.3.3.1a. Thus, the first two boundary conditions derived above will be the same:

$$A + C = 0 \;\Rightarrow\; C = -A. \tag{8.3.3.12}$$

$$B + D = 0 \Rightarrow D = -B. \tag{8.3.3.13}$$

The constraint on the right end makes the slope equal to zero:

$$\frac{\partial u}{\partial x}(l,t) = 0, \tag{8.3.3.14}$$

which leads to

$$-A\sin\kappa_n l + B\cos\kappa_n l + C\sinh\kappa_n l + D\cosh\kappa_n l = 0. \tag{8.3.3.15}$$

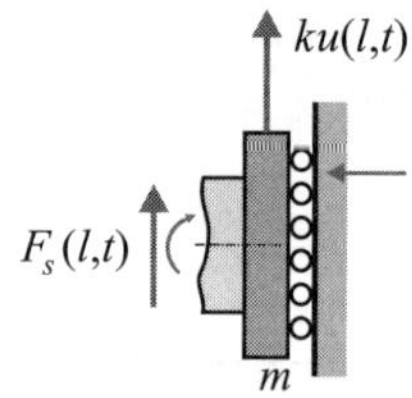

Figure 8.3.3.3

Further, the equation of motion for the mass at the right end (Figure 8.3.3.3) in the u-direction is

$$m\frac{\partial^2 u(l,t)}{\partial t^2} = -F_s(l,t) - ku(l,t), \tag{8.3.3.16}$$

yielding (see the previous example for details)

$$\begin{aligned}&(k - mc^2\kappa_n^4)(A\cos\kappa_n l + B\sin\kappa_n l + C\cosh\kappa_n l + D\sinh\kappa_n l) - \\ &EI\kappa_n^3(A\sin\kappa_n l - B\cos\kappa_n l + C\sinh\kappa_n l + D\cosh\kappa_n l) = 0.\end{aligned} \tag{8.3.3.17}$$

The characteristic determinant is found to be

$$\Delta = \begin{vmatrix} -\sin\kappa_n l - \sinh\kappa_n l & \cos\kappa_n l - \cosh\kappa_n l \\ (k - mc^2\kappa_n^4)(\cos\kappa_n l - \cosh\kappa_n l) - & (k - mc^2\kappa_n^4)(\sin\kappa_n l - \sinh\kappa_n l) + \\ EI\kappa_n^3(\sin\kappa_n l - \sinh\kappa_n l) & EI\kappa_n^3(B\cos\kappa_n l + \cosh\kappa_n l) \end{vmatrix} = 0. \tag{8.3.3.18}$$

Example 8.3.4 A beam is clamped at both ends and attached to a translational spring in the middle as shown in Figure 8.3.4.1a. Define the number of boundary conditions and their expressions. Remove the spring (Figure 8.3.4.1b), and find the natural frequency and mode shapes of the resulting clamped–clamped beam.

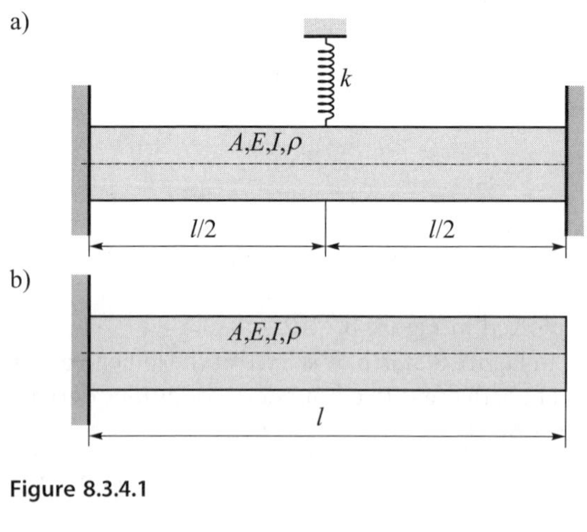

Figure 8.3.4.1

Solution

Due to the existence of the spring, the beam will be divided into Part 1 and Part 2, as shown in Figure 8.3.4.2a.

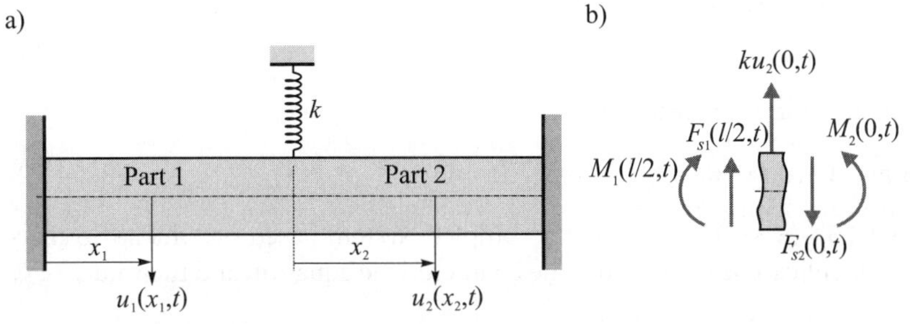

Figure 8.3.4.2

Consequently, the two governing equations are

$$\ddot{u}_1(x_1,t) + c^2 u_1{}^{\text{iv}}(x_1,t) = 0, \tag{8.3.4.1}$$

$$\ddot{u}_2(x_2,t) + c^2 u_2{}^{\text{iv}}(x_2,t) = 0. \tag{8.3.4.2}$$

For each one them, one has $u_{1,2}(x_{1,2},t) = X_{1,2}(x_{1,2})T(t)$, where

$$X_1 = A_1 \cos\kappa_n x_1 + B_1 \sin\kappa_n x_1 + C_1 \cosh\kappa_n x_1 + D_1 \sinh\kappa_n x_1, \tag{8.3.4.3}$$

$$X_2 = A_2 \cos\kappa_n x_2 + B_2 \sin\kappa_n x_2 + C_2 \cosh\kappa_n x_2 + D_2 \sinh\kappa_n x_2. \tag{8.3.4.4}$$

Thus, there are eight constants and eight boundary conditions are needed. The first two of them hold for the left end of Part 1 as the displacement and the slope is zero at $x_1 = 0$:

$$u_1(0,t) = 0 \;\Rightarrow\; X_1(0) = 0, \tag{8.3.4.5}$$

$$\frac{\partial u_1}{\partial x_1}(0,t) = 0 \;\Rightarrow\; X_1'(0) = 0. \tag{8.3.4.6}$$

The same conditions hold for the right end of Part 2:

$$u_2(l/2,t) = 0 \;\Rightarrow\; X_2(l/2) = 0, \tag{8.3.4.7}$$

$$\frac{\partial u_2}{\partial x_2}(l/2,t) = 0 \;\Rightarrow\; X_2'(l/2) = 0. \tag{8.3.4.8}$$

Further, the continuity conditions at the end of Part 1 and at the beginning of Part 2 impose:

$$u_1(l/2,t) = u_2(0,t) \;\Rightarrow\; X_1(l/2) = X_2(0), \tag{8.3.4.9}$$

$$\frac{\partial u_1}{\partial x_1}(l/2,t) = \frac{\partial u_2}{\partial x_2}(0,t) \;\Rightarrow\; X_1'(l/2) = X_2'(0). \tag{8.3.4.10}$$

Focusing on the infinitesimally small element in the middle and the corresponding free-body diagram (Figure 8.3.4.2b), and balancing the internal shear forces and the spring force, gives:

$$-F_{s1}(l/2,t)-ku_2(0,t)+F_{s2}(0,t)=0 \;\Rightarrow\; EIX_1'''(l/2)-kX_2(0)-EIX_2'''(0)=0. \tag{8.3.4.11}$$

Balancing the bending moments gives:

$$M_1(l/2,t)=M_2(0,t) \;\Rightarrow\; X_1''(l/2)-X_2''(0)=0. \tag{8.3.4.12}$$

Equations (8.3.4.5)–(8.3.4.12) represent a complete system based on which the characteristic determinant can be formed and the characteristic equation and the values of κ_n derived.

Let us focus now on the simplified case, that is, a clamped–clamped beam from Figure 8.3.4.1b. The four boundary conditions are:

$$u(0,t)=0,\;\; \frac{\partial u}{\partial x}(0,t)=0,\;\; u(l,t)=0,\;\; \frac{\partial u}{\partial x}(l,t), \tag{8.3.4.13}$$

which yield the following system of algebraic equations:

$$A+C=0, \tag{8.3.4.14}$$

$$B+D=0, \tag{8.3.4.15}$$

$$A\cos\kappa_n l+B\sin\kappa_n l+C\cosh\kappa_n l+D\sinh\kappa_n l=0, \tag{8.3.4.16}$$

$$-A\sin\kappa_n l+B\cos\kappa_n l+C\sinh\kappa_n l+D\cosh\kappa_n l. \tag{8.3.4.17}$$

Forming the characteristic determinant and solving it, one can derive the following characteristic equation:

$$\cos\kappa_n l\cosh\kappa_n l-1=0. \tag{8.3.4.18}$$

Its solutions are obtained numerically, and the first three of them are

$$\kappa_1 l=4.73004,\;\; \kappa_2 l=7.8532,\;\; \kappa_3 l=10.9956. \tag{8.3.4.19}$$

These solutions are presented in Table 8.3.1.1. Thus, the lowest natural frequency (fundamental frequency) is:

$$\omega_1=\frac{22.3733}{l^2}\sqrt{\frac{EI}{\rho A}}. \tag{8.3.4.20}$$

Going back to the system (8.3.4.14)–(8.3.4.17), the final solution can be written down in the form

$$X(x)=A\left[\cos\kappa_n x-\cosh\kappa_n x+\frac{\sin\kappa_n l+\sinh\kappa_n l}{\cos\kappa_n l-\cosh\kappa_n l}(\sin\kappa_n x-\sinh\kappa_n x)\right]. \tag{8.3.3.21}$$

The first three mode shapes are plotted in Figure 8.3.4.3.

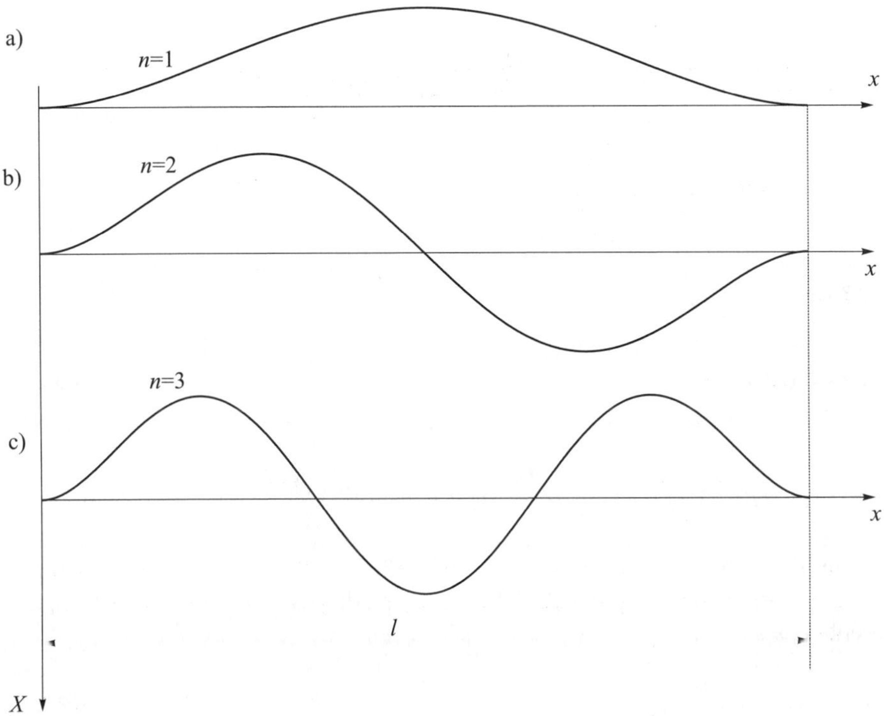

Figure 8.3.4.3

Example 8.3.5 A uniform cantilever beam of length l has a rigid disc of mass m_d and radius $r = l/4$ attached to its free end (Figure 8.3.5.1). Define how the first three natural frequencies change with the ratio between the mass of the beam m and the disc.

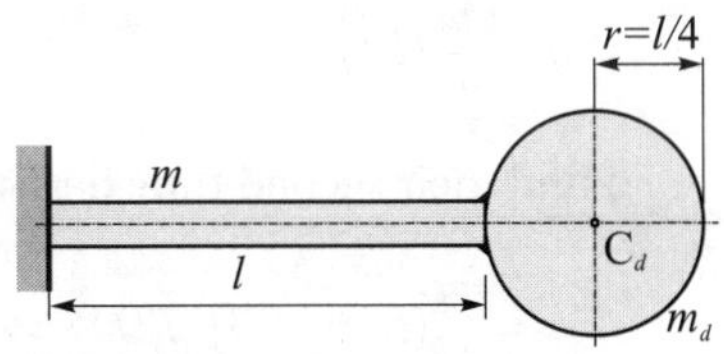

Figure 8.3.5.1

Solution

The displacement and the slope at the left end are zero: $u(0,t) = 0$, $\partial u(0,t)/\partial x = 0$. As a result, the following algebraic equations between the coefficients in the solution (8.3.15) hold:

$$A + C = 0, \quad (8.3.5.1)$$

$$B + D = 0. \quad (8.3.5.2)$$

Two additional equations can be found by considering the free-body diagram for the disc (Figure 8.3.5.2.a), which is subjected to a shear force and a bending moment. The disc executes general plane motion and the corresponding dynamic equations of motion are:

$$m_d \ddot{u}_{C_d} = -F_s(l,t), \quad (8.3.5.3)$$

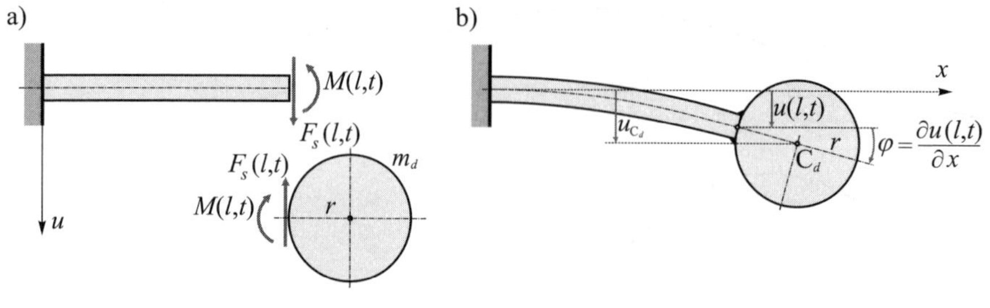

Figure 8.3.5.2

$$J\ddot{\varphi} = F_s(l,t)r + M(l,t), \tag{8.3.5.4}$$

where

$$F_s(l,t) = -EI\frac{\partial^3 u(l,t)}{\partial x^3}, \quad J = \frac{1}{2}m_d r^2, \quad M(l,t) = -EI\frac{\partial^2 u(l,t)}{\partial x^2}. \tag{8.3.5.5}$$

To obtain $\ddot{u}_{C_d}$ and $\ddot{\varphi}$, one can present the system in the displaced position, bearing in mind that the disc is rigid (Figure 8.3.5.2.b). Thus, the displacement of its centre C_d can be presented as

$$u_{C_d} = u(l,t) + r\frac{\partial u}{\partial x}(l,t). \tag{8.3.5.6}$$

The angle of rotation φ corresponds to the slope at $x = l$, that is,

$$\varphi = \frac{\partial u}{\partial x}(l,t). \tag{8.3.5.7}$$

To find their second time derivatives needed for (8.3.5.3) and (8.3.5.4), one can derive:

$$\varphi = \frac{\partial u}{\partial x}(l,t) = X'(l)T(t) \;\Rightarrow\; \ddot{\varphi} = X'(l)\ddot{T}(t) = -X'(l)\omega_n^2 T(t) = -\omega_n^2\frac{\partial u}{\partial x}(l,t). \tag{8.3.5.8}$$

Similarly, the following can be obtained:

$$\ddot{u}_{C_d} = -\omega_n^2\left[u(l,t) + r\frac{\partial u}{\partial x}(l,t)\right]. \tag{8.3.5.9}$$

The corresponding characteristic equation is derived in the form

$$\begin{aligned} &-y^4 - 32\mu^2 + \cosh y[(y^2 - 32\mu^2)\cos y + y(32 + 3y^2)\mu\sin y] \\ &+ \mu y\sinh y[(3y^2 - 32)\cos y + 16y\sin y] = 0. \end{aligned} \tag{8.3.5.10}$$

where $y = \kappa_n l$ and the parameter μ is introduced as the ratio between the mass of the beam m and disc $m_d (\mu = m/m_d)$. This equation is solved numerically for a range of values of the mass ratio μ, and the values of the product $y = \kappa_n l$ that give the first, second and third natural frequencies, are calculated. The way this product changes with μ in these three cases is plotted in Figure 8.3.5.3.

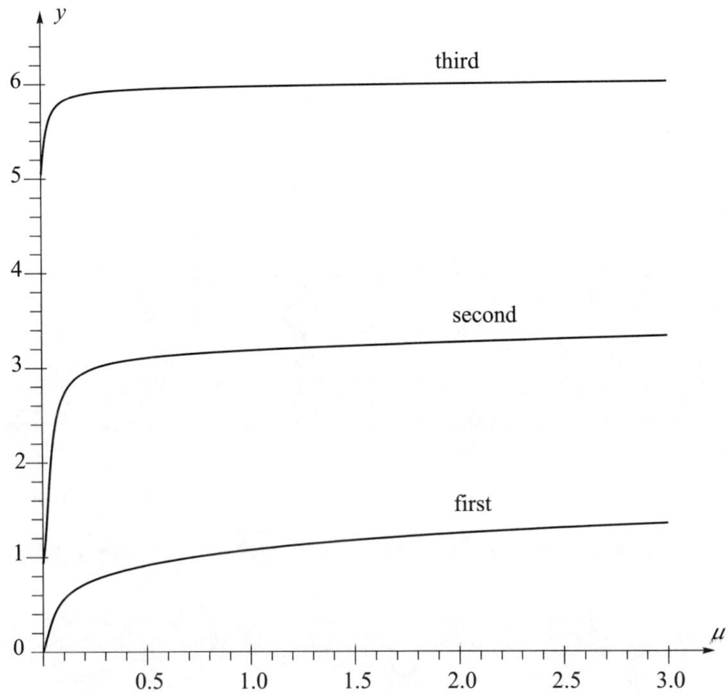

Figure 8.3.5.3

9

Additional Topics

Chapter Outline

This chapter starts with the concept of equivalent stiffness and its simple illustration. This concept is then extended and used in more complicated cases, such as in-plane concurrent springs and nonlinear springs. In addition to springs, the concept of equivalency is also demonstrated on linear viscous dampers. In order to deal with certain nonlinear problems, approximate expressions for the deflection and potential energy of nonlinear springs are derived in terms of generalised coordinates.

Chapter Objectives

- To present the concept of equivalent stiffness
- To show how a system of concurrent springs and dampers can be replaced by the equivalent, but simpler, system consisting of two orthogonal springs and two orthogonal dampers
- To illustrate these replacements with a few examples
- To define the cases when the equivalent spring can be obtained for two nonlinear springs in series

9.1 Theoretical Introduction

Engineering structures often contain stiffness and damping elements combined appropriately to achieve a specific aim or to exhibit a desired behaviour. For the sake of easier analysis, there is a tendency to simplify these systems by using the concept of equivalent stiffness and equivalent damping or to adjust them to have certain properties. One of the simplest cases is the one-dimensional problem of springs connected serially shown in Figure 9.1.1a, where the displacements of points A and B are defined by x and $x+y$, respectively. The first spring deforms for $\Delta l_1 = x$ and the second one for $\Delta l_2 = x + y - x = y$. The equivalent spring (Figure 9.1.1b) deforms for $\Delta l_{eq} = x + y \equiv z$.

Mechanical Vibration: Fundamentals with Solved Examples, First Edition.
Ivana Kovačić and Dragi Radomirović.

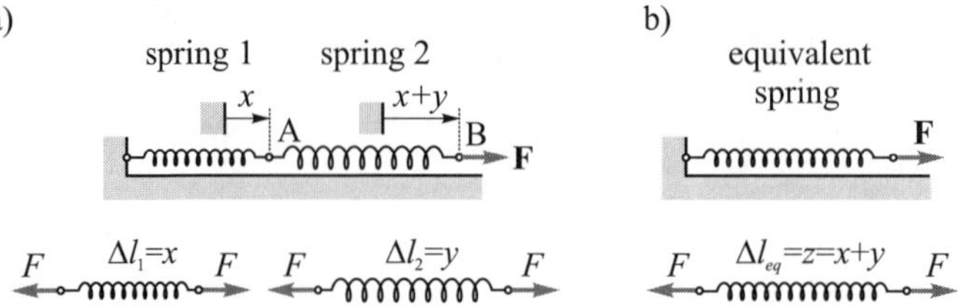

Figure 9.1.1

In accordance with Newton's third law, the magnitude of the force in springs in series is equal – this magnitude is labelled by F (Figure 9.1.1 a and b). If the springs are linear with the spring/stiffness constants k_1 and k_2, then $F = k_1x$ and $F = k_2y$. Expressing the deflections and summing them up, one derives:

$$F = \frac{1}{\dfrac{1}{k_1} + \dfrac{1}{k_2}} z. \tag{9.1.1}$$

This implies that the equivalent spring is also linear, satisfying $F = k_e z$, and yielding

$$k_e = \frac{1}{\dfrac{1}{k_1} + \dfrac{1}{k_2}} = \frac{k_1 k_2}{k_1 + k_2}, \tag{9.1.2}$$

which gives the expression for the equivalent spring constant.

The aim of this chapter is to show how the concept of equivalent stiffness can be extended and used in more complicated cases, such as in-plane concurrent springs (Section 9.2) and nonlinear springs (Sections 9.3 and 9.5). In addition to springs, the concept of equivalency will be demonstrated on linear viscous dampers as well (Section 9.2). In order to deal with certain nonlinear problems in Section 9.5, approximate expressions for the deflection and potential energy of nonlinear springs are derived in terms of generalised coordinates in Section 9.4.

9.2 Equivalent Two-Element System for Concurrent Springs and Dampers

Let us consider a particle of mass m, which is connected with n arbitrarily positioned linear springs and N arbitrarily positioned linear viscous dampers, as shown in Figure 9.2.1a. Let us also assume that a constant static force **F** acts on the particle. Its position with respect to the horizontal axis is defined by the angle α. The springs are, thus, prestressed (predeformed) in the stable static equilibrium M_0 around which the particle performs small in-plane oscillations. The resulting static deformations of the springs are labelled by δ_i ($i = 1,\ldots, n$), and the stiffness of each of them is k_i; the corresponding lengths in the static equilibrium position are l_i and the angle between each of them and the horizontal is labelled by φ_i. The dampers are characterised by the damping coefficient c_j ($j = 1,\ldots, N$) and their positions with respect to the horizontal axis are

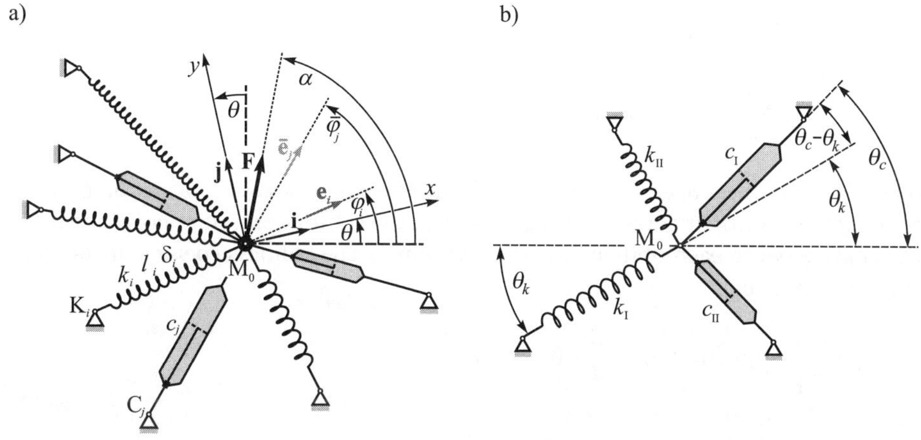

Figure 9.2.1

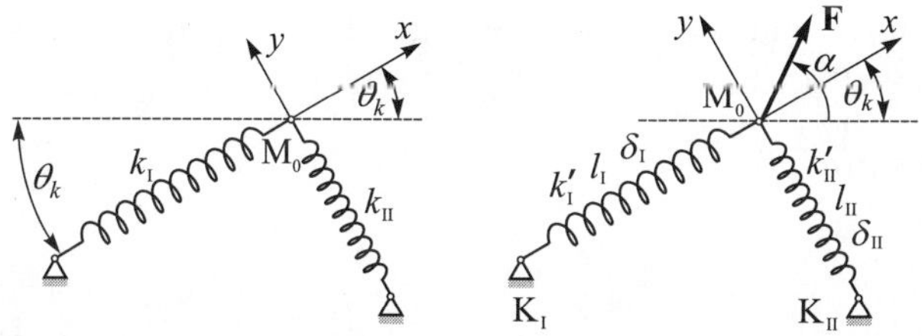

Figure 9.2.2

defined by the angle $\bar{\varphi}_j$. The rectangular coordinate system x-y is introduced in the point M_0, where the angle θ defines the position of the x-axis with respect to the horizontal.

Of interest here is the replacement of this concurrent system of springs and dampers by the equivalent, but simpler system that consists of two orthogonal springs and two orthogonal dampers, as shown in Figure 9.2.1b. The task is to find their positions, defined by the angles θ_k and θ_c, respectively, as well as the stiffness coefficients k_{I} and k_{II}, and the damping coefficients c_{I} and c_{II}.

9.2.1 Concurrent Springs

Let us consider now the system of concurrent springs described above and shown in Figure 9.2.1a with a view to finding the equivalent system of two orthogonal springs shown in Figure 9.2.2a. The following theorems and corollaries define this equivalent system.

Theorem 1 The system of concurrent prestressed linear springs performing small in-plane oscillations can be replaced by the equivalent system of two orthogonal springs on which the original constant static force does not act and which do not have the corresponding static deformations. The position of one of them with respect to the horizontal is given by the angle θ_k, which satisfies the following equation

$$\tan 2\theta_k = \frac{\sum_{i=1}^{n} k_i \left(1 - \frac{\delta_i}{l_i}\right) \sin 2\varphi_i}{\sum_{i=1}^{n} k_i \left(1 - \frac{\delta_i}{l_i}\right) \cos 2\varphi_i}, \tag{9.2.1}$$

where k_i is the stiffness of each of the original springs, δ_i is the static deformation for each spring, l_i stands for their corresponding lengths in the static equilibrium position and φ_i is the angle between each of them and the horizontal. The stiffness coefficients of these springs are given by

$$k_{\mathrm{I}} = \sum_{i=1}^{n} k_i \left[\cos^2(\varphi_i - \theta_k) + \frac{\delta_i}{l_i} \sin^2(\varphi_i - \theta_k)\right],$$
$$k_{\mathrm{II}} = \sum_{i=1}^{n} k_i \left[\sin^2(\varphi_i - \theta_k) + \frac{\delta_i}{l_i} \cos^2(\varphi_i - \theta_k)\right]. \tag{9.2.2a, b}$$

Proof: The equivalence of the systems will be achieved by considering their potential energies V in an arbitrary position. The one corresponding to the original system consists of the potential energy of the force and the potential energy of each spring V_i. The former is

$$V(\mathbf{F}) = -F\cos(\alpha - \theta)x - F\sin(\alpha - \theta)y, \tag{9.2.3}$$

while the latter takes the form

$$V_i = \frac{1}{2} k_i \left(\delta_i + u_i + \frac{1}{2l_i} v_i^2\right)^2, \tag{9.2.4}$$

where u_i is the deformation in the direction collinear with the spring in the static equilibrium position and v_i is the deformation in the orthogonal direction. The deformation u_i can be calculated by using $u_i = \mathbf{r} \cdot \mathbf{e}_i$, which results in

$$u_i = (x\mathbf{i} + y\mathbf{j}) \cdot (\cos(\varphi_i - \theta)\mathbf{i} + \sin(\varphi_i - \theta)\mathbf{j}) = x\cos(\varphi_i - \theta) + y\sin(\varphi_i - \theta), \tag{9.2.5}$$

while the deformation v_i is

$$v_i^2 = |\mathbf{r}|^2 - u_i^2 = x^2 + y^2 - u_i^2. \tag{9.2.6}$$

Thus, the term in the brackets in Equation (9.2.4) transforms into

$$\left(\delta_i + u_i + \frac{1}{2l_i} v_i^2\right)^2 = \delta_i^2 + u_i^2 + 2\delta_i u_i + \frac{\delta_i}{l_i}(x^2 + y^2 - u_i^2) + \ldots$$
$$= \delta_i^2 + 2\delta_i[x\cos(\varphi_i - \theta) + y\sin(\varphi_i - \theta)] + \left[\left(1 - \frac{\delta_i}{l_i}\right)\sin(2\varphi_i - 2\theta)\right]xy$$
$$+ \left[\cos^2(\varphi_i - \theta) + \frac{\delta_i}{l_i}\sin^2(\varphi_i - \theta)\right]x^2 + \left[\sin^2(\varphi_i - \theta) + \frac{\delta_i}{l_i}\cos^2(\varphi_i - \theta)\right]y^2 + \ldots \tag{9.2.7}$$

The system is in the static equilibrium position, so the following conditions should be satisfied:

$$F\cos(\alpha-\theta)=\sum_{i=1}^{n}k_i\delta_i\cos(\varphi_i-\theta),\quad F\sin(\alpha-\theta)=\sum_{i=1}^{n}k_i\delta_i\sin(\varphi_i-\theta). \tag{9.2.8a,b}$$

The overall potential energy becomes

$$\begin{aligned}V=&\frac{1}{2}\left\{\sum_{i=1}^{n}k_i\left[\cos^2(\varphi_i-\theta)+\frac{\delta_i}{l_i}\sin^2(\varphi_i-\theta)\right]\right\}x^2\\&+\frac{1}{2}\left\{\sum_{i=1}^{n}k_i\left[\sin^2(\varphi_i-\theta)+\frac{\delta_i}{l_i}\cos^2(\varphi_i-\theta)\right]\right\}y^2\\&+\frac{1}{2}\left\{\sum_{i=1}^{n}k_i\left(1-\frac{\delta_i}{l_i}\right)\sin(2\varphi_i-2\theta)\right\}xy.\end{aligned} \tag{9.2.9}$$

Note that as the system performs small oscillations around the stable equilibrium position, the potential energy has a quadratic form in the generalised coordinates.

The angle θ_k, which defines the position of one of the spring from the equivalent system (Figure 9.2.2a) is obtained by equating the coefficient in front of the mixed term xy with zero, which yields

$$\sum_{i=1}^{n}k_i\left(1-\frac{\delta_i}{l_i}\right)\sin(2\varphi_i-2\theta_k)=0. \tag{9.2.10}$$

The solution of this equation is equal to that given by Equation (9.2.1). The expressions in brackets in front of x^2 and y^2 in Equation (9.2.9) define the stiffness coefficients k_{I} and k_{II} given by (9.2.2a,b), which completes the proof of Theorem 1.

Corollary 1.1 A system of concurrent linear springs that are not prestressed in the stable static equilibrium position and that perform small oscillations can be replaced by an equivalent system of two orthogonal springs, where the position of one of them with respect to the horizontal is given by the angle θ_k, and whose stiffness coefficients are

$$\tan 2\theta_k=\frac{\sum_{i=1}^{n}k_i\sin 2\varphi_i}{\sum_{i=1}^{n}k_i\cos 2\varphi_i},\quad k_{\mathrm{I}}=\sum_{i=1}^{n}k_i\cos^2(\varphi_i-\theta_k),\quad k_{\mathrm{II}}=\sum_{i=1}^{n}k_i\sin^2(\varphi_i-\theta_k). \tag{9.2.11a–c}$$

Proof: These expressions follow directly from Theorem 1 for the case when $\delta_i=0$.

Corollary 1.2 The sum of the stiffness coefficients of the springs from the equivalent systems is given by

$$k_{\mathrm{I}}+k_{\mathrm{II}}=\sum_{i=1}^{n}\left(k_i+\frac{k_i\delta_i}{l_i}\right). \tag{9.2.12}$$

When the springs from the original system are not prestressed in the stable static equilibrium position ($\delta_i = 0$), the sum of their stiffness coefficients is equal to that from the equivalent system:

$$k_{\mathrm{I}} + k_{\mathrm{II}} = \sum_{i=1}^{n} k_i, \tag{9.2.13}$$

that is, the sum of the stiffness coefficients is an *invariant* of these systems.

Proof: It is sufficient to observe that the sum of Equations (9.2.2a,b) gives Equation (9.2.12) and that for $\delta_i = 0$, this sum yields Equation (9.2.13).

Let us now consider the system of concurrent springs described above and shown in Figure 9.2.1a with the aim of defining the equivalent system of two orthogonal springs shown in Figure 9.2.2b.

Theorem 2 A system of concurrent prestressed linear springs performing small in-plane oscillations can be replaced by an equivalent system of two orthogonal linear springs on which the original constant static force also acts and which are, consequently, prestressed. The position of one of them with respect to the horizontal is given by the angle θ_k, satisfying

$$\tan 2\theta_k = \frac{\sum_{i=1}^{n} k_i \left(1 - \frac{\delta_i}{l_i}\right) \sin 2\varphi_i}{\sum_{i=1}^{n} k_i \left(1 - \frac{\delta_i}{l_i}\right) \cos 2\varphi_i}, \tag{9.2.14}$$

where k_i is the stiffness of each of the original springs, δ_i is their static deformations, l_i stands for their corresponding lengths in the static equilibrium position and φ_i is the angle between each spring and the horizontal. The stiffness coefficients k_{I}' and k_{II}' of these springs are given by

$$k_{\mathrm{I}}' = k_{\mathrm{I}} - \frac{F \sin(\alpha - \theta_k)}{l_{\mathrm{II}}}, \quad k_{\mathrm{II}}' = k_{\mathrm{II}} - \frac{F \cos(\alpha - \theta_k)}{l_{\mathrm{I}}}, \tag{9.2.15a,b}$$

where

$$\begin{aligned} k_{\mathrm{I}} &= \sum_{i=1}^{n} k_i \left[\cos^2(\varphi_i - \theta_k) + \frac{\delta_i}{l_i} \sin^2(\varphi_i - \theta_k)\right], \\ k_{\mathrm{II}} &= \sum_{i=1}^{n} k_i \left[\sin^2(\varphi_i - \theta_k) + \frac{\delta_i}{l_i} \cos^2(\varphi_i - \theta_k)\right], \end{aligned} \tag{9.2.16a,b}$$

and where l_{I} and l_{II} stand for the lengths of two new springs in the static equilibrium position, which need to be prescribed.

Proof: The part of the proof related to the angle θ_k follows from Equation (9.2.10), as in the case of Theorem 1. However, here the new springs (Figure 9.2.2b) are prestressed and their static deformations can be obtained from the static equilibrium equations

$$\delta_{\rm I} k_{\rm I}' = F\cos(\alpha - \theta_k), \quad \delta_{\rm II} k_{\rm II}' = F\sin(\alpha - \theta_k). \tag{9.2.17a,b}$$

Equating the potential energies of the springs, one can derive

$$k_{\rm I} = k_{\rm I}' + \frac{\delta_{\rm II}\, k_{\rm II}'}{l_{\rm II}}, \quad k_{\rm II} = k_{\rm II}' + \frac{\delta_{\rm I}\, k_{\rm I}'}{l_{\rm I}}. \tag{9.2.18a,b}$$

Combining (9.2.17a,b) and (9.2.18a,b), one can obtain (9.2.15a,b) as well as the static deformations (9.2.17a,b).

Remark: In this approach, one needs to define *a priori* the lengths of the new springs in the static equilibrium position, $l_{\rm I}$ and $l_{\rm II}$. However, there are other possibilities as well, because the equivalent system contains six unknown quantities $k_{\rm I}'$, $k_{\rm II}'$, $l_{\rm I}$, $l_{\rm II}$, $\delta_{\rm I}$ and $\delta_{\rm II}$ that are mutually dependent by means of four equations (9.2.17a,b) and (9.2.18a,b). Thus, one needs to define two of them *a priori*, and then calculate the rest of them from (9.2.17a,b) and (9.2.18a,b).

Corollary 2.1 The sum of the stiffness coefficients of two types of equivalent systems of springs is given by the following relationship

$$k_{\rm I} + k_{\rm II} = k_{\rm I}' + k_{\rm II}' + \frac{k_{\rm I}'\delta_{\rm I}}{l_{\rm I}} + \frac{k_{\rm II}'\delta_{\rm II}}{l_{\rm II}}. \tag{9.2.19}$$

Proof: This follows directly from (9.2.18a,b).

9.2.2 Concurrent Dampers

Let us consider now the system of concurrent dampers from Figure 9.2.1a and define the equivalent system of two orthogonal dampers shown in Figure 9.2.3.

Theorem 3 A system of N concurrent in-plane linear viscous dampers can be replaced by an equivalent system of two mutually orthogonal linear viscous dampers, where the position of one of them with respect to the horizontal is given by the angle θ_c that satisfies

$$\tan 2\theta_c = \frac{\sum_{j=1}^{N} c_j \sin 2\bar{\varphi}_j}{\sum_{j=1}^{N} c_j \cos 2\bar{\varphi}_j}, \tag{9.2.20}$$

where c_j is the damping coefficient of each of the original dampers and $\bar{\varphi}_j$ is the angle between each of them and the horizontal. The damping coefficients of the new dampers are

$$c_{\rm I} = \sum_{j=1}^{N} c_j \cos^2\left(\bar{\varphi}_j - \theta_c\right), \quad c_{\rm II} = \sum_{j=1}^{N} c_j \sin^2\left(\bar{\varphi}_j - \theta_c\right). \tag{9.2.21a,b}$$

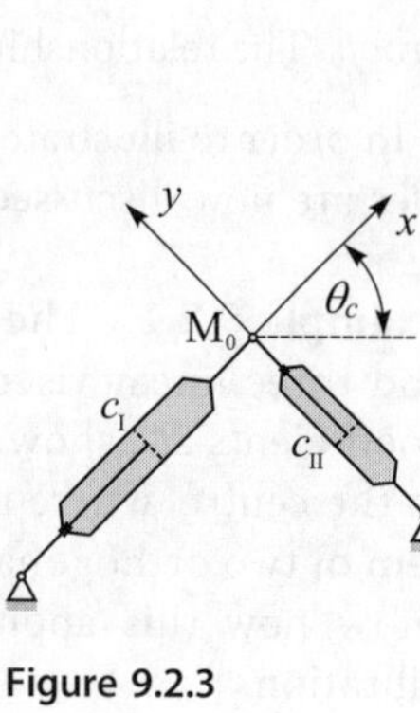

Figure 9.2.3

Proof: The equivalence of the systems is based on the dissipative function. For each of the original dampers, this depends on the relative velocity of the piston with respect to the cylinder $(v_{p/c})_j$, which is found as follows

$$(v_{p/c})_j = \mathbf{v}\cdot\bar{\mathbf{e}}_j = (\dot{x}\mathbf{i}+\dot{y}\mathbf{j})\cdot\left(\cos\left(\bar{\varphi}_j-\theta\right)\mathbf{i}+\sin\left(\bar{\varphi}_j-\theta\right)\mathbf{j}\right) = \dot{x}\cos\left(\bar{\varphi}_j-\theta\right)+\dot{y}\sin\left(\bar{\varphi}_j-\theta\right). \tag{9.2.22}$$

Thus, the dissipative function corresponding to one damper becomes

$$\begin{aligned} D_j &= \frac{1}{2}c_j(v_{p/c})_j^2 = \frac{1}{2}c_j\left[\dot{x}\cos\left(\bar{\varphi}_j-\theta\right)+\dot{y}\sin\left(\bar{\varphi}_j-\theta\right)\right]^2 \\ &= \frac{1}{2}c_j\left[\dot{x}^2\cos^2\left(\bar{\varphi}_j-\theta\right)+\dot{y}^2\sin^2\left(\bar{\varphi}_j-\theta\right)+\dot{x}\dot{y}\sin\left(2\bar{\varphi}_j-2\theta\right)\right]. \end{aligned} \tag{9.2.23}$$

The overall dissipative function is

$$\begin{aligned} D &= \sum_{j=1}^{N} D_j \\ &= \frac{1}{2}\left[\sum_{j=1}^{N} c_j\cos^2\left(\bar{\varphi}_j-\theta\right)\right]\dot{x}^2+\frac{1}{2}\left[\sum_{j=1}^{N} c_j\sin^2\left(\bar{\varphi}_j-\theta\right)\right]\dot{y}^2+\frac{1}{2}\left[\sum_{j=1}^{N} c_j\sin\left(2\bar{\varphi}_j-2\theta\right)\right]\dot{x}\dot{y}. \end{aligned} \tag{9.2.24}$$

The angle θ_c is obtained from the requirement that the coefficient in front of the mixed term $\dot{x}\dot{y}$ is equal to zero, that is,

$$\sum_{j=1}^{N} c_j\sin\left(2\bar{\varphi}_j-2\theta_c\right) = 0, \tag{9.2.25}$$

which after some transformations leads to Equation (9.2.20).

Corollary 3.1 The sum of the damping coefficients of the dampers from the equivalent systems is given by

$$c_{\mathrm{I}}+c_{\mathrm{II}} = \sum_{j=1}^{N} c_j. \tag{9.2.26}$$

Proof: The relationship (9.2.26) is easily obtained by summing up (9.2.21a) and (9.2.21b).

In order to illustrate the use and benefits of the approach presented above, two examples are now discussed.

Example 9.2.1 The system under consideration consists of three linear springs and three linear viscous dampers whose positions as well as stiffness and damping coefficients are shown next to them in Figure 9.2.4a. All the elements are connected in the centre, where the particle of mass m is located. Let us find the equivalent system of two orthogonal springs and two orthogonal dampers. In addition, we will also show how this approach easily gives the natural frequencies of free undamped vibration.

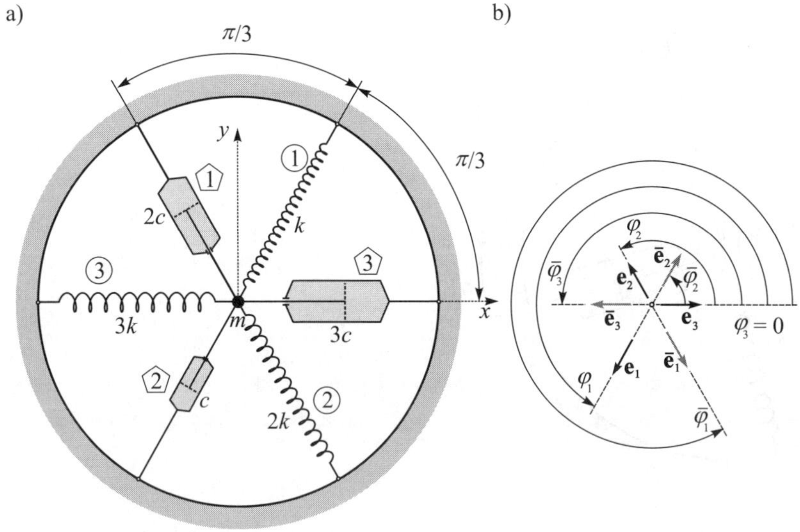

Figure 9.2.4

Solution

Following the notation from Figure 9.2.1a and using explanations given in Figure 9.2.4b, the angles between the axes of the springs and dampers are identified as follows: $\varphi_1 = 4\pi/3,\ \varphi_2 = 2\pi/3,\ \varphi_3 = 0,\ \bar{\varphi}_1 = 5\pi/3,\ \bar{\varphi}_2 = \pi/3,\ \bar{\varphi}_3 = \pi$.

The equivalent two-element system of dampers is obtained by calculating first the position of one of the dampers, that is, the angle θ_c (see Equation (9.2.20))

$$\tan 2\theta_c = \frac{\sum_{j=1}^{3} c_j \sin 2\bar{\varphi}_j}{\sum_{j=1}^{3} c_j \cos 2\bar{\varphi}_j} = \frac{2c\sin(10\pi/3) + c\sin(2\pi/3) + 3c\sin(2\pi)}{2c\cos(10\pi/3) + c\cos(2\pi/3) + 3c\cos(2\pi)} = -\frac{\sqrt{3}}{3}, \tag{9.2.27}$$

the solution of which is

$$\theta_c = -\frac{\pi}{12} = -15^{\circ}. \tag{9.2.28}$$

Equations (9.2.21a,b) yield the values of two damping coefficients:

$$c_{\mathrm{I}} = \sum_{j=1}^{3} c_j \cos^2\left(\bar{\varphi}_j - \theta_c\right) = 2c\cos^2(5\pi/3 + \pi/12) + c\cos^2(\pi/3 + \pi/12)$$
$$+ 3c\cos^2(\pi + \pi/12) = 3.866c,\quad c_{\mathrm{II}} = \sum_{j=1}^{3} c_j \sin^2\left(\bar{\varphi}_j - \theta_c\right) = 2c\sin^2(5\pi/3 + \pi/12)$$
$$+ c\sin^2(\pi/3 + \pi/12) + 3c\sin^2(\pi + \pi/12) = 2.134c. \tag{9.2.29a, b}$$

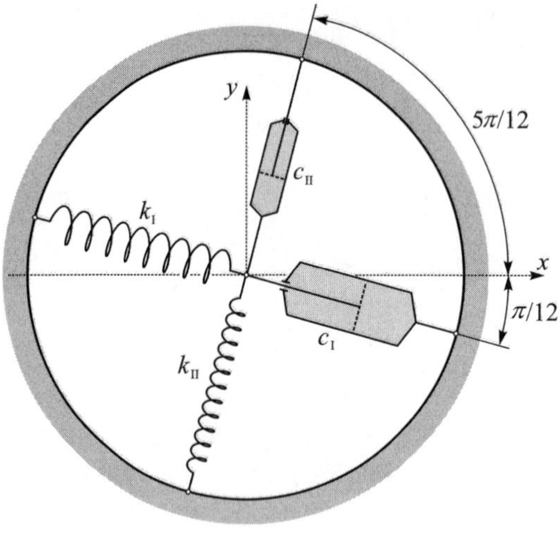

Figure 9.2.5

These two dampers are shown in Figure 9.2.5.

Let us find now the system of equivalent springs. The position of one of the spring follows from Equation (9.2.11a):

$$\tan 2\theta_k = \frac{\sum_{i=1}^{3} k_i \sin 2\varphi_i}{\sum_{i=1}^{3} k_i \cos 2\varphi_i} = \frac{k\sin(8\pi/3) + 2k\sin(4\pi/3) + 3k\sin(0)}{k\cos(8\pi/3) + 2k\cos(4\pi/3) + 3k\cos(0)} = -\frac{\sqrt{3}}{3}, \tag{9.2.30}$$

which gives

$$\theta_k = -\frac{\pi}{12} = -15^\circ. \tag{9.2.31}$$

Equations (9.2.11a) and (9.2.11b) are then used to calculate the stiffness coefficients of the equivalent springs:

$$k_{\text{I}} = \sum_{i=1}^{3} k_i \cos^2(\varphi_i - \theta_k) = k\cos^2(4\pi/3 + \pi/12) + 2k\cos^2(2\pi/3 + \pi/12)$$
$$+ 3k\cos^2(0 + \pi/12) = 3.866k, \quad k_{\text{II}} = \sum_{i=1}^{3} k_i \sin^2(\varphi_i - \theta_k) = k\sin^2(4\pi/3 + \pi/12)$$
$$+ 2k\sin^2(2\pi/3 + \pi/12) + 3k\sin^2(0 + \pi/12) = 2.134k. \tag{9.2.32a, b}$$

These springs are shown in Figure 9.2.5. Note that $k_{\text{I}} + k_{\text{II}} = 6k$, as stated in Corollary 1.2 and Equation (9.2.13).

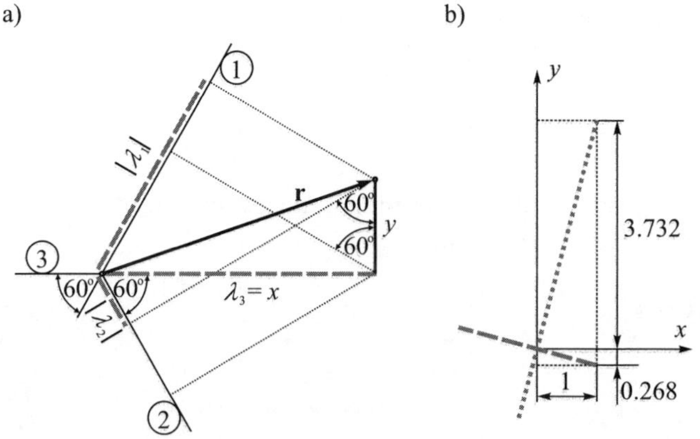

Figure 9.2.6

If there is no damping, the natural frequencies of free vibration are easily calculated as

$$\omega_1 = \sqrt{\frac{k_{\mathrm{II}}}{m}} \approx 1.461\sqrt{\frac{k}{m}}, \quad \omega_2 = \sqrt{\frac{k_{\mathrm{I}}}{m}} \approx 1.966\sqrt{\frac{k}{m}}. \tag{9.2.33a,b}$$

In order to emphasise some of the benefits of the approach presented above, let us show an alternative way of finding the natural frequencies. The springs are underformed in the stable static equilibrium position around which the particle performs small oscillations and, consequently, one needs to determine the deflection in the direction that is collinear with the direction of the spring in the static equilibrium position. Based on Figure 9.2.6a, these deflections are found to be:

$$\lambda_1 = -\left(x/2 + y\sqrt{3}/2\right), \quad \lambda_2 = -\left(x/2 - y\sqrt{3}/2\right), \quad \lambda_3 = x, \tag{9.2.34a–c}$$

and the potential energy is

$$V = \frac{1}{2}k_1\lambda_1{}^2 + \frac{1}{2}k_2\lambda_2{}^2 + \frac{1}{2}k_3\lambda_3{}^2 = \frac{15k}{8}x^2 - \frac{\sqrt{3}k}{4}xy + \frac{9k}{8}y^2. \tag{9.2.35}$$

The kinetic energy of the particle is $T = m(\dot{x}^2 + \dot{y}^2)/2$, so that Lagrange's equations of the second kind for small oscillations,

$$\frac{d}{dt}\left(\frac{\partial T}{\partial \dot{q}}\right) + \frac{\partial V}{\partial q} = 0, \ q \in \{x, y\}, \tag{9.2.36}$$

yield the following differential equations:

$$4m\ddot{x} + 15kx - \sqrt{3}ky = 0, \quad 4m\ddot{y} - \sqrt{3}kx + 9ky = 0. \tag{9.2.37a,b}$$

Assuming the solution for motion in the form $x = A\cos(\omega t + \gamma)$, $y = B\cos(\omega t + \gamma)$, where A and B are constant, the following system of algebraic equations is obtained:

$$\left(15k - 4m\omega^2\right)A + \left(-\sqrt{3}k\right)B = 0, \quad \left(-\sqrt{3}k\right)A + \left(9k - 4m\omega^2\right)B = 0. \qquad (9.2.38\text{a,b})$$

For a non-trivial solution, the corresponding determinant must vanish, leading to the following characteristic equation (see Chapter 6):

$$4m^2\omega^4 - 24mk\omega^2 + 33k^2 = 0, \qquad (9.2.39)$$

which gives:

$$\omega_1 = \sqrt{3 - \frac{\sqrt{3}}{2}}\sqrt{\frac{k}{m}} = 1.461\sqrt{\frac{k}{m}}, \quad \omega_2 = \sqrt{3 + \frac{\sqrt{3}}{2}}\sqrt{\frac{k}{m}} = 1.966\sqrt{\frac{k}{m}}. \qquad (9.2.40\text{a,b})$$

Comparing the expressions for the frequencies from (9.2.40a,b) with the one given by (9.2.33a,b), it is seen that they coincide.

Now, Equations (9.2.38a and b) can be used to calculate the mode shapes ratios:

$$\left.\frac{y}{x}\right|_{\omega=\omega_1} = \left.\frac{B}{A}\right|_{\omega=\omega_1} = \frac{15k - 4m\omega_1{}^2}{\sqrt{3}k} = 2 + \sqrt{3} = 3.732,$$

$$\left.\frac{y}{x}\right|_{\omega=\omega_2} = \left.\frac{B}{A}\right|_{\omega=\omega_2} = \frac{15k - 4m\omega_2{}^2}{\sqrt{3}k} = -2 + \sqrt{3} = -0.268. \qquad (9.2.41\text{a,b})$$

The corresponding mode shapes are given in Figure 9.2.6b, where the one corresponding to ω_1 is plotted as a dotted line and the one corresponding to ω_2 as a dashed line, and the value of A is taken arbitrarily to be equal to unity. It is important to note that the angles between these two directions and the horizontal are 75° and 15°, which agrees with the position of a two-element equivalent system of springs shown in Figure 9.2.5. However, unlike this procedure related to modes of vibration, which involves lengthy calculations, the one related to the equivalent system of springs performed previously is straightforward and mathematically tractable.

Example 9.2.2 The second example is shown and defined in Figure 9.2.7a. The system consists of four springs, whose stiffness coefficients are: $k_1 = 4000$ N/m, $k_2 = 2000$ N/m, $k_3 = 2500$ N/m and $k_4 = 1184.6$ N/m. They are joined in a point M_0, where a particle of mass m ($mg = 900$ N) is located. In the static equilibrium position shown, the first and the third spring are deformed in this position, where: $\delta_1 = 0.18028$ m and $\delta_3 = 0.2$ m (this implies that $\delta_2 = 0$ and $\delta_4 = 0$ m). Note that it is easy to check that the system of all forces acting in M_0 is balanced. In addition, based on Figure 9.2.7a, one can find the lengths of the springs in the static equilibrium position: $l_1 = \sqrt{13}$ m, $l_2 = 3$ m, $l_3 = 5$ m and $l_4 = 6$ m.

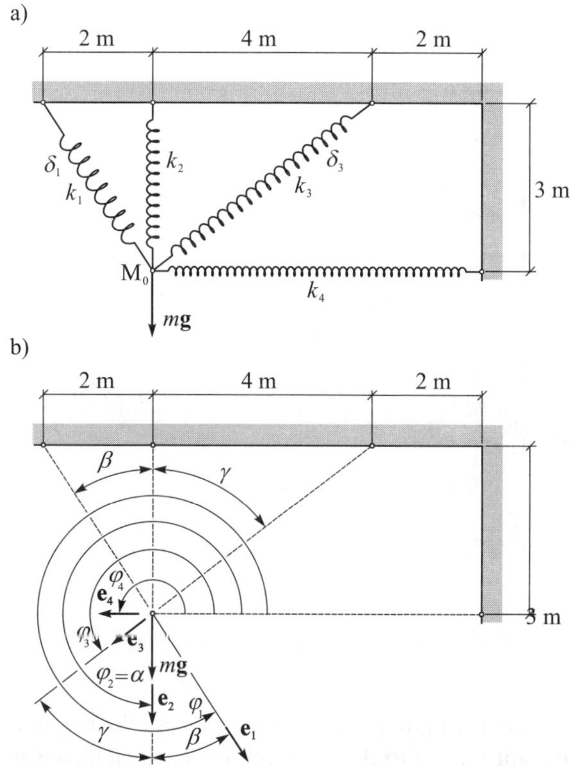

Figure 9.2.7

Solution

The position of each of the springs is determined based on Figure 9.2.7a,b as: $\varphi_1 = 3\pi/2 + \beta$, $\varphi_2 = 3\pi/2$, $\varphi_3 = 3\pi/2 - \gamma$, $\varphi_4 = \pi$, where $\beta = \arctan(2/3) = 0.588$, and $\gamma = \arctan(4/3) = 0.927$.

The angle θ_k is defined by

$$\tan 2\theta_k = \frac{\sum_{i=1}^{4} k_i\left(1-\frac{\delta_i}{l_i}\right)\sin 2\varphi_i}{\sum_{i=1}^{4} k_i\left(1-\frac{\delta_i}{l_i}\right)\cos 2\varphi_i} = 0.75, \tag{9.2.42}$$

which gives $\theta_k = 0.322 = 18.435°$.

Equations (9.2.2a,b) yield the stiffness coefficients of the equivalent system plotted in Figure 9.2.8a:

$$\begin{aligned} k_{\mathrm{I}} &= \sum_{i=1}^{4} k_i\left[\cos^2(\varphi_i-\theta_k) + \frac{\delta_i}{l_i}\sin^2(\varphi_i-\theta_k)\right] = 3989.2\ \mathrm{N/m}, \\ k_{\mathrm{II}} &= \sum_{i=1}^{4} k_i\left[\sin^2(\varphi_i-\theta_k) + \frac{\delta_i}{l_i}\cos^2(\varphi_i-\theta_k)\right] = 5995.4\ \mathrm{N/m}. \end{aligned} \tag{9.2.43}$$

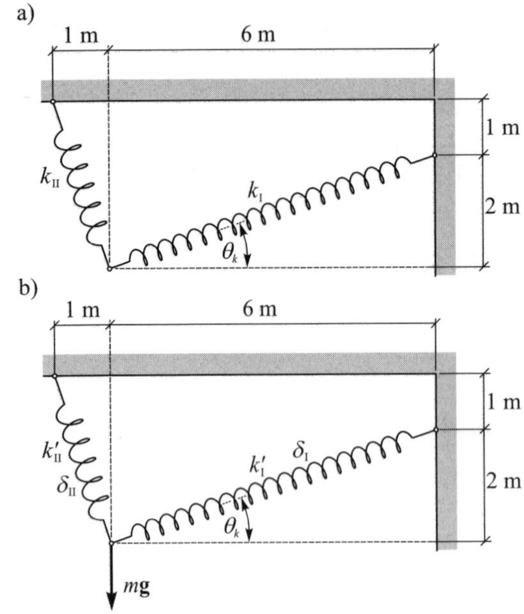

Figure 9.2.8

Note that this equivalent system does not contain the static force $m\mathbf{g}$ and the two new springs are not prestressed. The invariant related to the new stiffness coefficients given by Equation (9.2.12) is satisfied.

If one would like to establish the equivalent system with two springs in which the force $m\mathbf{g}$ is retained (Figure 9.2.8b), the facts expressed in Theorem 2 are used. The angle θ_k is the same as the one defined by Equation (9.2.42), while Equations (9.2.15a and b) for the assumed values of $l_I = \sqrt{40}$ m, $l_{II} = \sqrt{10}$ m give:

$$k'_I = k_I - \frac{mg\cos\theta_k}{l_{II}} = 3719.2 \text{ N/m}, \; k'_{II} = k_{II} - \frac{mg\sin\theta_k}{l_I} = 5950.4 \text{ N/m}. \tag{9.2.44a,b}$$

The corresponding static deflections of these springs are:

$$\delta_I = \frac{mg\sin\theta_k}{k'_I} = 0.0765 \text{ m}, \quad \delta_{II} = \frac{mg\cos\theta_k}{k'_{II}} = 0.143 \text{ m}. \tag{9.2.45a,b}$$

It is easy to verify that the invariant from Corollary 2.1 is satisfied:

$$k_I + k_{II} = 9984.6 \text{ N/m}, \quad k'_I + k'_{II} + \frac{k'_I\delta_I}{l_I} + \frac{k'_{II}\delta_{II}}{l_{II}} = 9984.6 \text{ N/m}. \tag{9.2.46}$$

9.3 Nonlinear Springs in Series

Unlike the case of two linear springs in series discussed in Section 9.1, which is straightforward and easy to solve, the cases of two nonlinear springs in series or one linear and one nonlinear spring have not been recognised as such. These cases have attracted considerable attention from researchers and many approximate analytical approaches have been derived to perform qualitative and mainly quantitative analyses of such systems.

This section demonstrates that there are cases where the equivalent spring can be obtained exactly.

9.3.1 Purely Nonlinear Springs in Series

Here we assume that both springs from Figure 9.1.1a are purely nonlinear, where the power of nonlinearity is $2n+1$, $n = 1, 2, 3\ldots$. The spring constants are K_1 and K_2. Thus, the following are satisfied: $F = K_1 x^{2n+1}$ and $F = K_2 y^{2n+1}$. Expressing x and y and summing them up and by using $z = x + y$, one can derive

$$F = \frac{1}{\left(\frac{1}{\sqrt[2n+1]{K_1}} + \frac{1}{\sqrt[2n+1]{K_2}}\right)^{2n+1}} z^{2n+1}. \tag{9.3.1}$$

Thus, the equivalent spring should be of the same type of nonlinearity, satisfying $F = K_e z^{2n+1}$, which gives its coefficient

$$\frac{1}{\sqrt[2n+1]{K_e}} = \frac{1}{\sqrt[2n+1]{K_1}} + \frac{1}{\sqrt[2n+1]{K_2}}. \tag{9.3.2}$$

Consequently, if the springs are purely cubic ($n = 1$), the equivalent spring is also cubic with the coefficient

$$\frac{1}{\sqrt[3]{K_e}} = \frac{1}{\sqrt[3]{K_1}} + \frac{1}{\sqrt[3]{K_2}}, \tag{9.3.3}$$

or

$$K_e = \frac{K_1 K_2}{\left(\sqrt[3]{K_1} + \sqrt[3]{K_2}\right)^3}. \tag{9.3.4}$$

9.3.2 Equal Duffing Springs in Series

Let us consider the case when two equal springs with linear coefficients k and cubic coefficient K (so-called *Duffing* springs) are attached to each other in series. The fact that the magnitude of the force in springs in series is equal imposes the following equalities:

$$F = kx + Kx^3 = ky + Ky^3, \tag{9.3.5}$$

$$F = kx + Kx^3 = k_e(x+y) + K_e(x+y)^3, \tag{9.3.6}$$

where it is assumed that the equivalent spring is also of Duffing type with the coefficients k_e and K_e. Solving (9.3.5) and (9.3.6), that is, using $z = x + y = 2x = 2y$, gives that the linear and cubic coefficients of the equivalent springs are, respectively:

$$k_e = \frac{k}{2}, \quad K_e = \frac{K}{8}. \tag{9.3.7a,b}$$

This result can also be obtained by considering the potential energies V of the original and the equivalent system $V_1 + V_2 = 2V_1 = V_{eq}$ and by using $y = x$ and $z = 2x$ to derive:

$$2\left(\frac{k}{2}x^2 + \frac{K}{4}x^4\right) = \frac{k_e}{2}(2x)^2 + \frac{K_e}{4}(2x)^4. \tag{9.3.8}$$

The solution of Equation (9.3.8) obviously agrees with Equations (9.3.7a and b).

This consideration of the potential energies can be extended to N equal Duffing springs in series as $N V_1 = V_{eq}$, leading to

$$N\left(\frac{k}{2}x^2 + \frac{K}{4}x^4\right) = \frac{k_e}{2}(Nx)^2 + \frac{K_e}{4}(Nx)^4. \tag{9.3.9}$$

This implies that the coefficients of the equivalent Duffing spring are

$$k_e = \frac{k}{N}, \quad K_e = \frac{K}{N^3}. \tag{9.3.10}$$

9.3.3 Two Different Nonlinear Springs

Finally, let us investigate the case when two springs in series are of the same power-type of nonlinearity, but with different coefficients $k_1,\ K_1$ and $k_2,\ K_2$. The force in the first one is

$$F = k_1 x + K_1 x^{2n+1}, \tag{9.3.11}$$

and in the second one

$$F = k_2 y + K_2 y^{2n+1}. \tag{9.3.12}$$

The force in the system with the equivalent spring is assumed to satisfy

$$F = k_e(x + y) + K_e(x + y)^{2n+1} \tag{9.3.13}$$

If one compares the first, linear terms in (9.3.11)–(9.3.13) and then the nonlinear terms in the same equations, one can conclude that there might be a possibility for these terms and the force itself to be equal if there is a linear relationship between x and y. Thus, one can assume $y = \kappa x$, where κ is to be determined, substitute it into Equation (9.3.12) and making it equal with Equation (9.3.11), deriving

$$k_1 = \kappa\, k_2, \quad K_1 = \kappa^{2n+1} K_2. \tag{9.3.14a,b}$$

These two expressions yield

$$\kappa = \frac{k_1}{k_2} = \sqrt[2n+1]{\frac{K_1}{K_2}}. \tag{9.3.15}$$

So, if the condition given by Equation (9.3.15) is satisfied, the resulting equivalent spring will be of the same type as the original ones. Its coefficients can be derived by substituting again $y = \kappa x$, with κ being defined by Equation (9.3.15), into Equation (9.3.13) and setting it equal to Equation (9.3.11), obtaining

$$k_e = \frac{k_1 k_2}{k_1 + k_2}, \quad K_e = \frac{K_1 K_2}{\left(\sqrt[2n+1]{K_1} + \sqrt[2n+1]{K_2}\right)^{2n+1}}. \tag{9.3.16a,b}$$

It should be noted that in the case of nonlinear springs in series considered, neither the coefficient k_e nor the coefficient K_e can be made equal to zero. Furthermore, as k_1 and k_2 are positive, it follows that due to Equation (9.3.15), the coefficients K_1 and K_2 should be of the same sign. This implies that if both original springs are hardening, the equivalent spring will also be hardening. Analogously, if both original springs are softening, the equivalent spring will also be softening.

For two different Duffing springs ($n = 1$) in series, the condition (9.3.15) becomes

$$\left(\frac{k_1}{k_2}\right)^3 = \frac{K_1}{K_2}. \tag{9.3.17}$$

So, the equivalent spring is also of the Duffing type and its linear coefficient is given by (9.3.16a), which is the same as in the case of two linear springs (9.3.2). The coefficient in front of the cubic term corresponds to

$$K_e = \frac{K_1 K_2}{\left(\sqrt[3]{K_1} + \sqrt[3]{K_2}\right)^3}, \tag{9.3.18}$$

which coincides with Equation (9.3.4) obtained for purely cubic springs.

Further, when Duffing springs in series are equal, one has $k_1 = k_2 = k$ and $K_1 = K_2 = K$, and the conditions (9.3.16a) and (9.3.18) coincide with (9.3.7a,b) derived in the previous subsection.

It should be noted that although the coefficients (9.3.16a,b) have been derived for springs containing only two terms in their expressions (9.3.11), (9.3.12), the results can easily be extended to an arbitrary number of terms. For example, if the springs in series are such that the force F in each of them is given by

$$F = k_1 x + K_1 x^3 + D_1 x^5 + E_1 x^7, \quad F = k_2 y + K_2 y^3 + D_2 y^5 + E_2 y^7, \tag{9.3.19a,b}$$

and the following condition is satisfied

$$\kappa = \frac{k_1}{k_2} = \sqrt[3]{\frac{K_1}{K_2}} = \sqrt[5]{\frac{D_1}{D_2}} = \sqrt[7]{\frac{E_1}{E_2}}, \tag{9.3.20}$$

then the equivalent spring will have the same type of terms, that is,

$$F = k_e z + K_e z^3 + D_e z^5 + E_e z^7, \tag{9.3.21}$$

where

$$k_e = \frac{k_1 k_2}{k_1 + k_2}, \quad K_e = \frac{K_1 K_2}{\left(\sqrt[3]{K_1} + \sqrt[3]{K_2}\right)^3} \quad D_e = \frac{D_1 D_2}{\left(\sqrt[5]{D_1} + \sqrt[5]{D_2}\right)^5} \quad E_e = \frac{E_1 E_2}{\left(\sqrt[7]{E_1} + \sqrt[7]{E_2}\right)^7}. \tag{9.3.22a–d}$$

9.4 On the Deflection and Potential Energy of Nonlinear Springs: Approximate Expressions

The procedure for obtaining the spring deflection by using the expression $\Delta l = l - l_0$, (where l is the length of the spring in an arbitrary position and l_0 is the length of the underformed spring), or $\Delta l = f_{st} + \lambda$ as a sum of a static spring deflection f_{st} and the deflection measured from the static equilibrium position λ is easy and straightforward if the spring deforms axially, that is, collinearly with the direction of the spring in the static equilibrium position. However, in many applications and experiments this is not the case and springs exhibit plane deformation. Then, for oscillatory systems that perform small (linear) oscillations, the alternative approach presented in Chapter 2, Section 2.3.2 can be used. Here, the extension of this approach to springs whose restoring force contains both a linear term and a positive/negative cubic term, that is, to the so-called Duffing-type hardening/softening spring is presented.

9.4.1 Duffing-Type Spring Deformed in the Static Equilibrium Position

The potential energy of the Duffing-type spring is:

$$V = \frac{1}{2}k(\Delta l)^2 \pm \frac{1}{4}K(\Delta l)^4, \tag{9.4.1}$$

where k and K are constants, and the upper sign corresponds to hardening springs and the lower one to softening springs.

By truncating the expansion for the deflection λ given by Equation (2.3.21) up to the terms of fourth order, one obtains

$$\lambda \approx u + \frac{1}{2l_{st}}v^2 - \frac{1}{2(l_{st})^2}uv^2 + \frac{1}{2(l_{st})^3}u^2v^2 - \frac{1}{8(l_{st})^3}v^4. \tag{9.4.2}$$

When expressed in terms of the generalised coordinate, this deflection is expected to have the form

$$\lambda \approx Aq + Bq^2 + Cq^3 + Dq^4, \tag{9.4.3}$$

where A, B, C and D are constants. This implies that the extension/compression in the u-direction should comprise the terms up to fourth order (if they exist), while all those of higher orders can be neglected, that is,

$$u \approx A_1q + B_1q^2 + C_1q^3 + D_1q^4. \tag{9.4.4}$$

On the other hand, the truncation for v should be as follows

$$v \approx A_2q + B_2q^2 + C_2q^3. \tag{9.4.5}$$

9.4.2 Duffing-Type Spring Undeformed in the Static Equilibrium Position

If the Duffing-type spring is undeformed in the static equilibrium position ($f_{st} = 0$), its potential energy is given by

$$V = \frac{1}{2}k\lambda^2 \pm \frac{1}{4}K\lambda^4. \tag{9.4.6}$$

Table 9.4.1

Spring	Linear $f_{st}=0$	Linear $f_{st}\neq 0$	Nonlinear spring of the Duffing type $f_{st}=0$	Nonlinear spring of the Duffing type $f_{st}\neq 0$
Potential energy V	$\frac{1}{2}k(\Delta l)^2$	$\frac{1}{2}k(\Delta l)^2$	$\frac{1}{2}k(\Delta l)^2 \pm \frac{1}{4}K(\Delta l)^4$	$\frac{1}{2}k(\Delta l)^2 \pm \frac{1}{4}K(\Delta l)^4$
Total deflection Δl	$\lambda(q)$	$f_{st}+\lambda(q)$	$\lambda(q)$	$f_{st}+\lambda(q)$
Required truncation for $\lambda(u,v)$	u	$u+\frac{1}{2l_{st}}v^2$	$u+\frac{1}{2l_0}v^2-\frac{1}{2(l_0)^2}uv^2$	$u+\frac{1}{2l_{st}}v^2-\frac{1}{2(l_{st})^2}uv^2 + \frac{1}{2(l_{st})^3}u^2v^2-\frac{1}{8(l_{st})^3}v^4$
Required truncation for $u(q)$	A_1q	$A_1q+B_1q^2$	$A_1q+B_1q^2+C_1q^3$	$A_1q+B_1q^2+C_1q^3+D_1q^4$
Required truncation for $v(q)$	none	A_2q	$A_2q+B_2q^2$	$A_2q+B_2q^2+C_2q^3$

The required form for the deflection λ is now

$$\lambda \approx Aq+Bq^2+Cq^3. \tag{9.4.7}$$

Based on its series expansion (2.3.21) of the same order, one has

$$\lambda \approx u+\frac{1}{2l_0}v^2-\frac{1}{2(l_0)^2}uv^2, \tag{9.4.8}$$

where the fact that $l_0 \equiv l_{st}$ has been used.

Comparing Equations (9.4.7) and (9.4.8), one concludes that the deflection in the u-direction should be truncated to cubic terms

$$u \approx A_1q+B_1q^2+C_1q^3, \tag{9.4.9}$$

and the deflection in the v-direction to quadratic terms

$$v \approx A_2q+B_2q^2. \tag{9.4.10}$$

For the sake of the reader, all of the expressions derived in Section 2.4.2 and from this section are collected and presented in Table 9.4.1.

To enable the reader to gain a deeper understanding of the approximations derived for nonlinear springs, an example that illustrates their use follows.

Example 9.4.1 Let us consider the example defined in Chapter 6, and Figure 6.2.1, with two equal sliders each of which has a mass m. They can slide along two smooth circular wires of radius R, so that the system performs small oscillations around the equilibrium position shown in Figure 6.2.1 when the spring is horizontal. It is assumed now that the spring is of a hardening Duffing type. The task is to find the expression for the potential energy by using the approximations derived previously.

Solution

When the system in placed in a vertical plane, the equilibrium of one of the sliders is achieved if Equation (6.2.1) is satisfied, but with the restoring force containing the cubic term

$$kf_{st} + K(f_{st})^3 - \frac{\sqrt{3}}{3}mg = 0, \tag{9.4.11}$$

from which f_{st} can be calculated. As the spring is nonlinear and deformed in the equilibrium position, the deformations in the characteristic deflections given u and v (6.2.2), (6.2.3) should be truncated as defined by (9.4.4) and (9.4.5), that is, to the fourth and third order, respectively:

$$u = \frac{R\sqrt{3}}{2}(\varphi+\psi) + \frac{R}{4}(\varphi^2+\psi^2) - \frac{R\sqrt{3}}{12}(\varphi^3+\psi^3) - \frac{R}{48}(\varphi^4+\psi^4),$$
$$v = \frac{R}{2}(\psi-\varphi) - \frac{R\sqrt{3}}{4}(\psi^2-\varphi^2) - \frac{R}{12}(\psi^3-\varphi^3). \tag{9.4.12a,b}$$

Therefore, the deflection (9.4.2) for $l_{st} = R$ is

$$\lambda = \frac{R\sqrt{3}}{2}(\varphi+\psi) + \frac{3R}{8}(\varphi^2+\psi^2) - \frac{R}{4}\varphi\psi - \frac{13\sqrt{3}}{48}R(\varphi^3+\psi^3) + \frac{3\sqrt{3}}{16}R(\varphi^2\psi+\varphi\psi^2)$$
$$+ \frac{35R}{128}(\varphi^4+\psi^4) + \frac{13}{96}R(\varphi^3\psi+\varphi\psi^3) - \frac{55}{64}R\varphi^2\psi^2. \tag{9.4.13}$$

The static deflection defined by (9.4.11) and the additional deflection given by (9.4.13) define the potential energy of the spring. It should be noted that as this potential energy contains fourth order terms, this implies that, for consistency reasons, the potential energy of the sliders should also be truncated to the same order (if needed for forming the equations of motion).

Note also that if the system is placed in a horizontal plane (assuming that it is a part of a bigger system that enables it to oscillate around a stable equilibrium), the spring is undeformed in the static equilibrium position. Based on Equations (9.4.9) and (9.4.10), the deflection in the u- and v-direction should be truncated to cubic and quadratic terms, respectively:

$$u = \frac{R\sqrt{3}}{2}(\varphi+\psi) + \frac{R}{4}(\varphi^2+\psi^2) - \frac{R\sqrt{3}}{12}(\varphi^3+\psi^3),$$
$$v = \frac{R}{2}(\psi-\varphi) - \frac{R\sqrt{3}}{4}(\psi^2-\varphi^2). \tag{9.4.14a,b}$$

When substituted in Equation (9.4.8), these expressions give λ, which then defines the potential energy (9.4.6).

9.5 Corrections of Stiffness Properties of Certain Oscillatory Systems

In this section, as in Section 9.1, the energy approach is used to examine certain one- and two-degree-of-freedom systems. One-degree-of-freedom systems are considered first, and include a main spring and an additional spring. The questions of interest are:

(1) What are the stiffness properties of the equivalent system; (2) How are they influenced by the characteristics of the additional part; and (3) Is the resulting motion oscillatory and if so, what is the corresponding period of vibration? To answer these questions, several combinations related to the springs in the main and additional system are analysed. Then, two-degree-of-freedom systems comprising two pairs of orthogonal or symmetrically attached springs are examined. Equivalent stiffness properties are obtained and the possibilities for their variation and corrections are also discussed.

9.5.1 One-Degree-of-Freedom Systems

We consider a system containing a main part and an additional part (Figure 9.5.1a). The main part comprises a mass that moves along a smooth horizontal surface and is attached to a horizontal spring, performing oscillations around the equilibrium $x = 0$. This spring can be linear with stiffness coefficient k, or of a Duffing type with stiffness coefficients k and K, that is, the related restoring force is $F = kx \pm Kx^3$ (the plus sign corresponds to a hardening spring and the minus sign to a softening spring). The additional part also includes a spring attached to the mass, but this spring is vertical when $x = 0$ and is deformed for the length δ in this position (its length in this position is $l_{st} = l_0 \pm \delta$, where l_0 is its length in an underformed position, with the plus sign corresponding to the extended and the minus sign to the compressed spring). This spring can be linear with the stiffness coefficient k_a or of a Duffing type with the stiffness coefficients k_a and K_a. The objective is to determine the stiffness properties of the equivalent system (Figure 9.5.1b) for several combinations of the main and additional spring.

9.5.1.1 Linear–Linear System

Let us first investigate the case when the spring in the main system is linear and the one added is also linear but prestressed (predeformed). The overall potential energy of the system performing small (linear) oscillations around the position $x = 0$ is (see Table 9.4.1):

$$V(x) = \frac{k}{2}x^2 + \frac{k_a}{2}\left(\pm\delta + \frac{1}{2l_{st}}x^2\right)^2. \tag{9.5.1}$$

By calculating $\mathbf{F} = -\frac{\partial V(x)}{\partial x}\mathbf{i} = -\left(k \pm \frac{\delta}{l_{st}}k_a\right)x\mathbf{i}$, we conclude that the spring in the equivalent system can be approximately substituted with a linear spring whose stiffness coefficient is

$$k_e = k \pm \frac{\delta}{l_{st}}k_a. \tag{9.5.2}$$

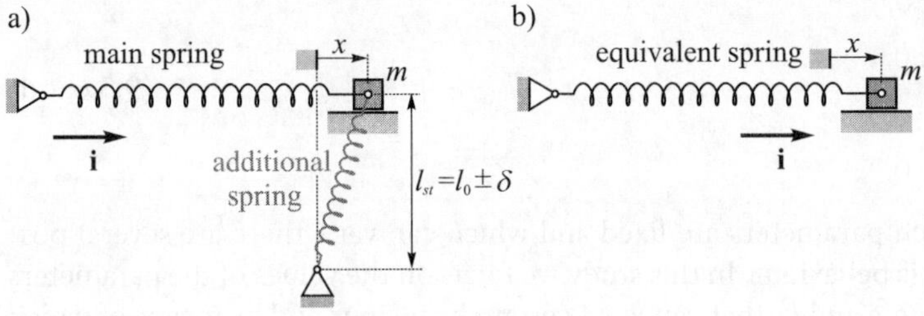

Figure 9.5.1

Note that the results (9.5.2) can also be derived if one would consider the exact solution for the potential energy $V(x) = kx^2/2 + k_a\left(\sqrt{(l_0 \pm \delta)^2 + x^2} - l_0\right)^2/2$, expanding it and truncating it to the terms quadratic with respect to x. However, this approach includes multiple steps, while the one based on Equation (9.5.1) is straightforward and easy to perform.

Depending on the characteristics of the additional system, one can distinguish three values $k_e < 0$, $k_e = 0$ and $k_e > 0$ in Equation (9.5.2), which are, respectively, associated with the cases when the position $x = 0$ is unstable, neutrally stable or stable. The first one obtains when the additional spring is compressed when $x = 0$ for the value $\delta > kl_{st}/k_a$, and the second one when $\delta = kl_{st}/k_a$. The system performs small oscillations around the stable equilibrium position $x = 0$ when $k_e > 0$, that is, when the additional spring is compressed for $\delta < kl_{st}/k_a$ or when this spring is extended for any value. In these cases, the resulting period and angular frequency of vibration are:

$$T = \frac{2\pi}{\omega_e}, \quad \omega_e = \sqrt{\frac{k_e}{m}}. \tag{9.5.3a,b}$$

Thus, the length δ for which the main spring is predeformed in the position $x = 0$ gives the possibility of achieving the desired stiffness in the horizontal direction by changing the stiffness coefficient of the main spring as defined by Equation (9.5.2).

9.5.1.2 Duffing–Linear System

If the main spring is of a Duffing type and the additional spring is linear and predeformed, the overall potential energy is (see Table 9.4.1):

$$V(x) = \frac{k}{2}x^2 \pm \frac{K}{4}x^4 + \frac{k_a}{2}\left(\pm\delta + \frac{1}{2l_{st}}x^2 - \frac{1}{8l_{st}^3}x^4\right)^2. \tag{9.5.4}$$

By extending this expression and grouping the terms next to x^2 and x^4, neglecting those of higher order (assuming that they are negligible), one can conclude that the spring in the equivalent system is of a Duffing type with the following characteristics:

$$k_e = k \pm \frac{\delta}{l_{st}}k_a, \quad K_e = \pm K + \frac{k_a}{2l_{st}^2}\left(1 \mp \frac{\delta}{l_{st}}\right). \tag{9.5.5a,b}$$

The coefficient k_e can be changeable as discussed above, but also the values of K_e can vary. Thus, several scenarios are possible. First, there is no oscillatory motion when:

$$\begin{aligned} &k_e < 0 \ \wedge \ K_e < 0, \\ &k_e = 0 \ \wedge \ K_e < 0, \\ &k_e = 0 \ \wedge \ K_e = 0, \\ &k_e < 0 \ \wedge \ K_e = 0. \end{aligned} \tag{9.5.6a–d}$$

Depending on which parameters are fixed and which can vary, there are several possibilities to achieve this behaviour. In this study we focus on the values of the parameters of the additional spring δ and k_a that can yield certain behaviour and by means of which the stiffness property in the horizontal direction can be changed. So, for the behaviour satisfying Equations (9.5.6a–d) to occur ($k_e \leq 0$, $K_e \leq 0$), the main spring should be softening and the additional spring should have linear stiffness coefficient $k_a \leq K l_{st}^2$ and be compressed for $\delta \geq k l_{st}/k_a$.

The equivalent systems can also behave as a softening or hardening Duffing oscillator if

$$k_e > 0 \ \wedge \ K_e < 0,$$
$$k_e > 0 \ \wedge \ K_e > 0. \tag{9.5.7a,b}$$

First, if the additional spring is extended, the condition $k_e > 0$ is always satisfied, while $K_e < 0$ is attainable if the main spring is softening and if the additional spring is such that $k_a < 2 l_{st}^2 K/(1 - \delta/l_{st})$ (analogously, the condition $K_e > 0$ is attainable if $k_a > 2 l_{st}^2 K/(1 - \delta/l_{st})$). Second, if the additional spring is compressed, the condition $k_e > 0$ is satisfied for $\delta < k l_{st}/k_a$, while $K_e < 0$ ($K_e > 0$) can be achieved with a softening main spring and the additional spring satisfying $k_a < 2 l_{st}^2 K/(1 + \delta/l_{st})$ or ($k_a > 2 l_{st}^2 K/(1 + \delta/l_{st})$) in the respective case.

When the conditions (9.5.7a,b) are fulfilled, the resulting equation of motion can be solved exactly in terms of the Jacobi elliptic function. The period of vibration for the softening (9.5.7a) and hardening case (9.5.7b) are given by

$$T = \frac{4\text{EllipticK}\left(\sqrt{\dfrac{|K_e|A^2}{2K_e - |K_e|A^2}}\right)}{\sqrt{k_e - \dfrac{|K_e|A^2}{2}}}, \quad T = \frac{4\text{EllipticK}\left(\sqrt{\dfrac{|K_e|A^2}{2(K_e - |K_e|A^2)}}\right)}{\sqrt{k_e - K_e A^2}}, \tag{9.5.8a,b}$$

where EllipticK is the elliptic integral of the first kind, and A is a constant that can be calculated from the initial conditions.

The equivalent system can behave as a bistable Duffing oscillator when

$$k_e < 0 \ \wedge \ k_e > 0. \tag{9.5.9}$$

This can be the consequence of the additional spring compressed for $\delta > k l_{st}/k_a$. If the main spring is hardening, the second condition $K_e > 0$ is always satisfied; if the main spring is softening, the condition $K_e > 0$ is satisfied for $k_a > 2 l_{st}^2 K/(1 + \delta/l_{st})$. The corresponding period of vibration is

$$T = \frac{4\text{EllipticK}\left(\sqrt{\dfrac{K_e A^2}{2(K_e A^2 - |K_e|)}}\right)}{\sqrt{K_e A^2 |k_e|}}. \tag{9.5.10}$$

Note that this period corresponds to larger-amplitude oscillations, surrounding the unstable equilibrium $x_{eq1} = 0$ and two stable ones $x_{eq2,3} = \pm\sqrt{|k_e|/K_e}$. These oscillations will be realised provided that the initial total mechanical energy in the equivalent system is higher than zero.

The equivalent system can also behave as a pure cubic oscillator when

$$k_e = 0 \ \wedge \ K_e > 0. \tag{9.5.11}$$

The first condition can be achieved with the additional spring compressed for $\delta > kl_{st}/k_a$. If the main spring is hardening, the second condition $K_e > 0$ is always satisfied; if the main spring is softening, the condition $K_e > 0$ is satisfied for $k_a > 2l_{st}^2 K - k$. The corresponding period of vibration is

$$T = \frac{4\text{EllipticK}\left(\sqrt{2}/2\right)}{A\sqrt{c_e}}. \tag{9.5.12}$$

Finally, the equivalent system can behave as a linear harmonic oscillator when

$$k_e > 0 \quad K_e = 0, \tag{9.5.13}$$

and its period and angular frequency of vibration are given by (9.5.3a,b).

9.5.1.3 Duffing–Duffing System

If both the main and the additional spring are of a Duffing type, the overall potential energy reads as:

$$V(x) = \frac{k}{2}x^2 \pm \frac{K}{4}x^4 + \frac{k_a}{2}\left(\pm\delta + \frac{1}{2l_{st}}x^2 - \frac{1}{8\,l_{st}^3}x^4\right)^2 \pm \frac{K_a}{4}\left(\pm\delta + \frac{1}{2l_{st}}x^2 - \frac{1}{8\,l_{st}^3}x^4\right)^2. \tag{9.5.14}$$

By truncating this expression to the terms of the fourth order, one can find that the coefficients of x^2 and x^4, that is, the stiffness coefficients in the equivalent system, are:

$$k_e = k \pm \frac{\delta}{l_{st}}\left(k_a \pm K_a\delta^2\right), \quad K_e = \pm K\frac{k_a}{2\,l_{st}^2}\left(1 \mp \frac{\delta}{l_{st}}\right) \pm \frac{K_a\delta^2}{2\,l_{st}^2}\left(3 \mp \frac{\delta}{l_{st}}\right). \tag{9.5.15a, b}$$

These expressions indicate that there are a number of outcomes related to the sign and values of k_e and K_e, as discussed in the previous subsection. So, it is possible to select δ, k_a and K_a to influence the sign and the values of k_e and K_e and to achieve a desirable stiffness property in the horizontal direction, as well as a desired motion.

Table 9.5.1 summarises all the cases considered regarding the combinations of the main and additional springs and the resulting stiffness expressions.

9.5.2 Two-Degree-of-Freedom Systems

9.5.2.1 System with Two Pairs of Orthogonal Duffing Springs

In this section we consider a Duffing oscillator with two degrees of freedom (Figure 9.5.2a): a point M of mass m is attached to two horizontal Duffing springs with a linear stiffness coefficient k_1 and a cubic stiffness coefficient K_1, while the length of the undeformed springs is l_{01}. The are two more orthogonal Duffing springs, whose analogous properties are k_2, K_2 and l_{02} (this notation will be used through the whole section).

The overall potential energy of this system is:

$$V = \frac{k_1}{2}(\lambda_{11})^2 + \frac{K_1}{4}(\lambda_{11})^4 + \frac{k_1}{2}(\lambda_{12})^2 + \frac{K_1}{4}(\lambda_{12})^4 + \frac{k_2}{2}(\lambda_{21})^2 + \frac{K_2}{4}(\lambda_{21})^4 + \frac{k_2}{2}(\lambda_{22})^2 + \frac{K_2}{4}(\lambda_{22})^4, \tag{9.5.16}$$

Table 9.5.1

Equivalent stiffness properties		Additional Spring	
		Linear	Duffing
Main spring	Linear	$k_e = k \pm \frac{\delta}{l_{st}} k_a$	$k_e = k \pm \frac{\delta}{l_{st}}(k_a \pm K_a\delta^2)$ $K_e = \frac{k_a}{2\,l_{st}^2}\left(1 \mp \frac{\delta}{l_{st}}\right) \pm \frac{K_a\delta^2}{2\,l_{st}^2}\left(3 \mp \frac{\delta}{l_{st}}\right)$
	Duffing	$k_e = k \pm \frac{\delta}{l_{st}} k_a$ $K_e = \pm K + \frac{k_a}{2\,l_{st}^2}\left(1 \mp \frac{\delta}{l_{st}}\right)$	$k_e = k \pm \frac{\delta}{l_{st}}(k_a \pm c_a\delta^2)$ $K_e = \pm K + \frac{k_a}{2\,l_{st}^2}\left(1 \mp \frac{\delta}{l_{st}}\right) \pm \frac{K_a\delta^2}{2\,l_{st}^2}\left(3 \mp \frac{\delta}{l_{st}}\right)$

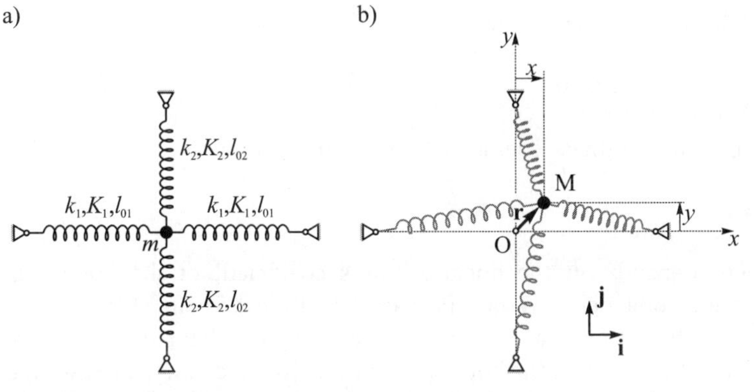

Figure 9.5.2

where λ_{11}, λ_{12}, λ_{21} and λ_{22} are the deflections of, respectively, the left, right, lower and upper spring. By using the expression:

$$\lambda \approx u + \frac{1}{2l_{st}}v^2 - \frac{1}{2\,l_{st}^2}uv^2 \tag{9.5.17}$$

where u and v stand for the characteristic deformation in the spring direction and the one orthogonal to it, and the length in the static position l_{st} is l_{01} and l_{02}. This yields:

$$\lambda_{11} = x + \frac{1}{2l_{01}}y^2 - \frac{1}{2\,l_{01}^2}xy^2,\quad \lambda_{12} = -x + \frac{1}{2l_{01}}y^2 + \frac{1}{2\,l_{01}^2}xy^2,$$

$$\lambda_{21} = y + \frac{1}{2l_{02}}x^2 - \frac{1}{2\,l_{02}^2}yx^2,\quad \lambda_{22} = -y + \frac{1}{2l_{02}}x^2 + \frac{1}{2\,l_{02}^2}yx^2, \tag{9.5.18a–d}$$

where x represents the displacement of the mass in the horizontal direction and y in the vertical direction. By substituting (9.5.18a–d) into (9.5.16), one obtains

$$V = k_1 x^2 + k_2 y^2 + \frac{1}{4}\left(2K_1 + \frac{k_2}{l_{02}^2}\right)x^4 + \frac{1}{4}\left(2K_2 + \frac{k_1}{l_{01}^2}\right)y^4 - \left(\frac{k_1}{l_{01}^2} + \frac{k_2}{l_{02}^2}\right)x^2 y^2. \tag{9.5.19}$$

The kinetic energy is $T = m(\dot{x}^2 + \dot{y}^2)/2$. The Lagrange equations of the second kind for a system performing small (linear) oscillations around a stable equilibrium $\frac{d}{dt}\frac{\partial T}{\partial \dot{q}} - \frac{\partial V}{\partial q} = 0$, where $q \in \{x, y\}$, take the following form:

$$m\ddot{x} + 2k_1 x + \left(2K_1 + \frac{k_2}{l_{02}^2}\right)x^3 - 2\left(\frac{k_1}{l_{01}^2} + \frac{k_2}{l_{02}^2}\right)xy^2 = 0,$$
$$m\ddot{x} + 2k_2 y + \left(2K_2 + \frac{k_1}{l_{01}^2}\right)y^3 - 2\left(\frac{k_1}{l_{01}^2} + \frac{k_2}{l_{02}^2}\right)yx^2 = 0. \tag{9.5.20a, b}$$

In a general case, the oscillators are coupled and motion is exhibited in two directions. However, this system of equations indicates that if, for example, the initial conditions are such $y(0) = 0$, $\dot{y}(0) = 0$, while $x(0)$ or/and $\dot{x}(0)$ are different from zero, then there are only oscillations in the x-direction, governed by

$$m\ddot{x} + 2k_1 x + \left(2K_1 + \frac{k_2}{l_{02}^2}\right)x^3 = 0. \tag{9.5.21}$$

Thus, the linear and cubic stiffness coefficients in this direction are

$$k_x = 2k_1, \quad K_x = 2K_1 + \frac{k_2}{l_{02}^2}. \tag{9.5.22a, b}$$

The first coefficient depends on the linear stiffness coefficient of the horizontal springs, while the cubic one depends on their cubic stiffness coefficient and the properties k_2 and l_{02} of the orthogonal springs. The resulting oscillator has a positive linear stiffness coefficient as $k_x > 0$. If the horizontal springs are hardening, the resulting oscillator (9.5.21) is also hardening as $K_x > 0$. However, if the horizontal springs are softening, the coefficient K_x (9.5.22b) can be negative, zero or positive, resulting, respectively, in a softening Duffing oscillator, linear harmonic oscillator or hardening Duffing oscillator.

Given the fact that this system is conservative, the energy conservation law can be written down in the form

$$\frac{m}{2}(\dot{x}^2 + \dot{y}^2) + k_1 x^2 + k_2 y^2 + \frac{1}{4}\left(2K_1 + \frac{k_2}{l_{02}^2}\right)x^4 + \frac{1}{4}\left(2K_2 + \frac{k_1}{l_{02}^2}\right)y^4 - \left(\frac{k_1}{l_{01}^2} + \frac{k_2}{l_{02}^2}\right)x^2 y^2 = E = const. \tag{9.5.23}$$

where E stands for the initial total mechanical energy. The function

$$f(x, y) = k_1 x^2 + k_2 y^2 + \frac{1}{4}\left(2K_1 + \frac{k_2}{l_{02}^2}\right)x^4 + \frac{1}{4}\left(2K_2 + \frac{k_1}{l_{02}^2}\right)y^4 - \left(\frac{k_1}{l_{01}^2} + \frac{k_2}{l_{02}^2}\right)x^2 y^2 - E = 0, \tag{9.5.24}$$

represents a closed symmetric curve. During the motion, the point M is always on the surface defined by $V(x, y) \le E$, which is shown in grey in Figure 9.5.3a. The case

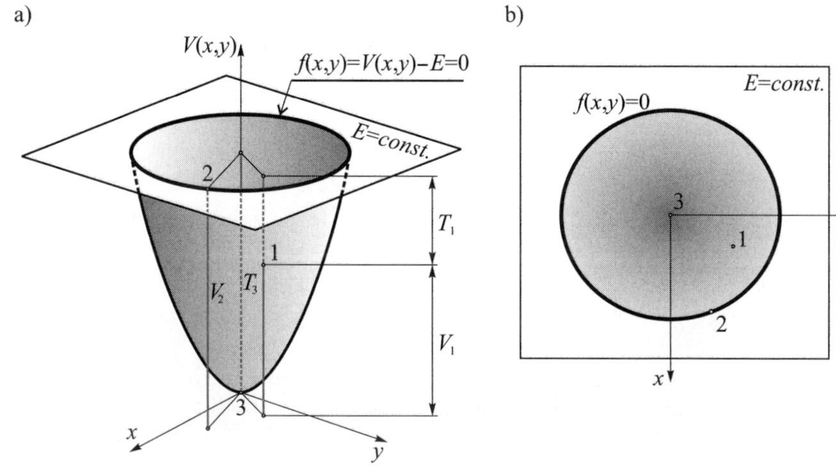

Figure 9.5.3

corresponding to an arbitrary position – Point 1 – is also depicted in this figure, where the distance from the xy-plane to this point represents the potential energy V_1 and the remaining distance to the plane $E = const.$ is the kinetic energy T_1 of the point M in this position. When the point M is on the curve $f(x,y) = 0$, which is depicted by Point 2 in Figure 9.5.3a, its kinetic energy is zero and its potential energy has a maximum value $V_2 \equiv E$. Another special case is when the potential energy is zero (this is depicted by Point 3 in Figure 9.5.3a) and then $T_3 \equiv E$. The projection of these three cases on the plane $E = const.$ and their location with respect to the curve $f(x,y) = 0$ are plotted in Figure 9.5.3b.

Equivalent Systems with Four Duffing Springs

Let us now focus on two systems of four Duffing springs shown in Figure 9.5.4a and 9.5.4b, with a view to answering the question of when these two systems are equivalent, that is, when they result in the same motion of the mass m.

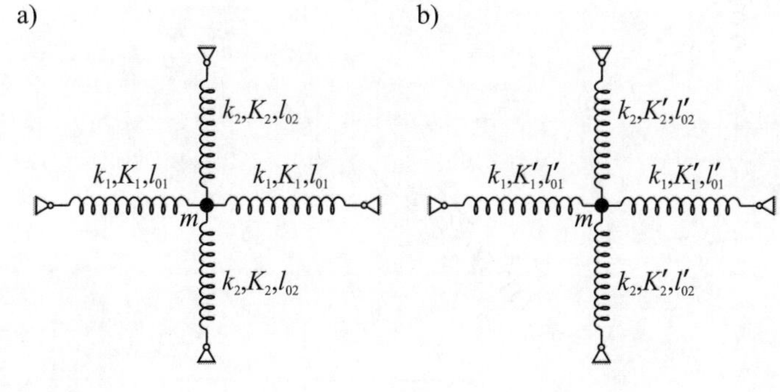

Figure 9.5.4

To that end, Equations (9.5.20a and b) are used for the parameters indicated in Figure 9.5.4a and 9.5.4b, yielding the conclusion that the following equalities should be satisfied:

$$2K_1{}' + \frac{k_2}{(l_{02}{}')^2} = 2K_2 + \frac{k_2}{l_{01}^2},$$
$$2K_2{}' + \frac{k_1}{(l_{01}{}')^2} = 2K_2 + \frac{k_1}{l_{01}^2}, \qquad (9.5.25\text{a–c})$$
$$\frac{k_1}{(l_{01}{}')^2} + \frac{k_2}{(l_{02}{}')^2} = \frac{k_1}{l_{01}^2} + \frac{k_2}{l_{01}^2}.$$

By summing up (9.5.25a,b) and using (9.5.25c), one concludes that these two systems should be such that the sum of their linear stiffness coefficients should be equal, that is,

$$K_1{}' + K_2{}' = K_1 + K_2. \qquad (9.5.26)$$

The rest of the parameters can be chosen in non-unique way, satisfying Equations (9.5.25a–c).

9.5.2.2 System with Two Pairs of Equal and Orthogonal Duffing Springs

Let us now examine the system consisting of four equal Duffing spring shown in Figure 9.5.5a. This system has two pairs of symmetry axes: x-y and ξ-η.

Based on the conclusions reached in Section 9.5.3.1 and Equations (9.5.22a and b), one finds that for certain initial conditions, the system can perform oscillations in the x-direction only (Figure 9.5.5b) with the stiffness characteristics

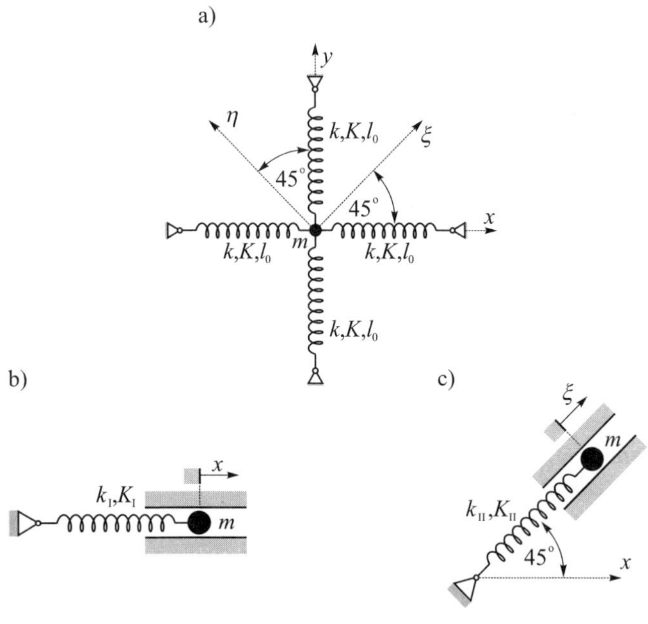

Figure 9.5.5

$$k_x = k_{\mathrm{I}} = 2k,\ \ K_x = K_{\mathrm{I}} = 2\left(K + \frac{k}{2\,l_0^2}\right). \tag{9.5.27a, b}$$

The type of the resulting oscillations is the same as discussed in the paragraph following Equations (9.5.22a and b).

To obtain the governing equations of motion in terms of ξ and η, the following transformations are introduced:

$$x = (\xi - \eta)\frac{\sqrt{2}}{2},\ \ y = (\xi + \eta)\frac{\sqrt{2}}{2}. \tag{9.5.28a, b}$$

The expressions for the kinetic and potential energy are now

$$T = \frac{m}{2}\left(\dot{\xi}^2 - \dot{\eta}^2\right),\ \ V = k(\xi^2 - \eta^2) + \frac{1}{8}\left(2K - 2\frac{k}{l_0^2}\right)(\xi^4 - \eta^4) + \frac{1}{4}\left(6K - 7\frac{k}{l_0^2}\right)\xi^4\eta^2. \tag{9.5.29a, b}$$

The corresponding Lagrange equations of the second kind have the following form

$$\begin{aligned} m\ddot{\xi} + 2k\xi + \frac{1}{2}\left(2K - 3\frac{k}{l_0^2}\right)\xi^3 + \frac{1}{2}\left(6K - 7\frac{k}{l_0^2}\right)\xi\eta^2 = 0, \\ m\ddot{\eta} + 2k\eta + \frac{1}{2}\left(2K - 3\frac{k}{l_0^2}\right)\eta^3 + \frac{1}{2}\left(6K - 7\frac{k}{l_0^2}\right)\eta\xi^2 = 0. \end{aligned} \tag{9.5.30a, b}$$

So, the mass performs coupled oscillations in the ξ and η directions. However, if the initial condition with respect to one of these coordinates are zero, the system performs oscillations in the orthogonal direction with the stiffness characteristics (Figure 9.5.5c)

$$k_{\mathrm{II}} = 2k,\ \ K_{\mathrm{II}} = \frac{1}{2}\left(2K - 3\frac{k}{l_0^2}\right). \tag{9.5.31a, b}$$

The corresponding oscillator has a positive linear stiffness coefficient. If the horizontal springs are softening, the oscillator (9.5.31) is also softening. If the horizontal springs are hardening, the coefficient K_{II} (9.5.31) can be negative ($K < 3k/2l_0^2$), zero ($K < 3k/2l_0^2$) or positive ($K > 3k/2l_0^2$), giving, respectively, a softening Duffing oscillator, linear harmonic oscillator or hardening Duffing oscillator.

9.5.2.3 System with Two Pairs of Equal and Symmetrically Attached Duffing Springs

The system of interest is shown in Figure 9.5.6a. It consists of four equal Duffing springs attached to the mass so that their angle with respect to the x-axes is α. The deflection of these springs can be obtained from the expressions

$$\lambda_i \approx u_i + \frac{1}{2l}v_i - \frac{1}{2\,l_0^2}u_i v_i^2,\ i = 1, 2, 3, 4, \tag{9.5.32}$$

where the deformations u_{i} and v_{i} are (see Figure 9.5.5.b and c):

$$\begin{aligned} u_1 = -u_4 = x\cos\alpha + y\sin\alpha,\ \ v_1 = v_4 = x\sin\alpha - y\cos\alpha, \\ u_2 = -u_3 = x\cos\alpha - y\sin\alpha,\ \ v_2 = v_3 = x\sin\alpha + y\cos\alpha. \end{aligned} \tag{9.5.33a–d}$$

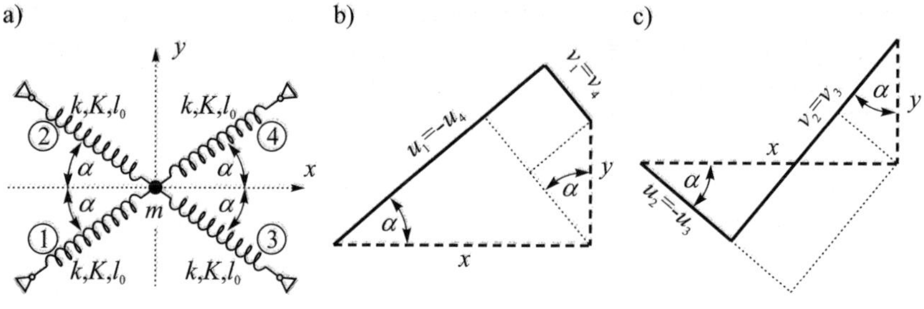

Figure 9.5.6

The overall potential energy $V = \frac{k}{2}\sum_{i=1}^{4}(\lambda_i)^2 + \frac{K}{4}\sum_{i=1}^{4}(\lambda_i)^4$ becomes

$$V = 2k\left(x^2\cos^2\alpha + y^2\sin^2\alpha\right) + \frac{1}{2}x^4\left(2K\cos^4\alpha + \frac{k}{l_0^2}\sin^4\alpha - \frac{k}{l_0^2}\sin^2\alpha\cos^2\alpha\right)$$
$$+\frac{1}{2}y^4\left(2K\sin^4\alpha + \frac{k}{l_0^2}\cos^4\alpha - 4\frac{k}{l_0^2}\sin^2\alpha\cos^2\alpha\right) - x^2y^2\frac{1}{l_0^2}\left(2k - \left(15k + 6Kl_0^2\right)\sin^2\alpha\cos^2\alpha\right), \tag{9.5.34}$$

and the equations of motion are

$$m\ddot{x} + 4kx\cos^2\alpha + 2x^3\left(2K\cos^4\alpha + \frac{k}{l_0^2}\sin^4\alpha - \frac{4k}{l_0^2}\sin^2\alpha\cos^2\alpha\right)$$
$$-\frac{2xy^2}{l_0^2}\left(2k - \left(15k + 6Kl_0^2\right)\sin^2\alpha\cos^2\alpha\right) = 0,$$
$$m\ddot{y} + 4ky\sin^2\alpha + 2y^3\left(2K\sin^4\alpha + \frac{k}{l_0^2}\cos^4\alpha - \frac{4k}{l_0^2}\sin^2\alpha\cos^2\alpha\right)$$
$$-\frac{2yx^2}{l_0^2}\left(2k - \left(15k + 6Kl_0^2\right)\sin^2\alpha\cos^2\alpha\right) = 0. \tag{9.5.35a,b}$$

For the case when, for example, $y(0) = 0$, $\dot{y}(0) = 0$, with $x(0)$ and/or $\dot{x}(0)$ being non-zero, the mass performs oscillations in the x-direction only, and the corresponding stiffness properties are

$$k_x = 4k\cos^2\alpha,\ K_x = 2\left(2K\cos^4\alpha + \frac{k}{l_0^2}\sin^4\alpha - 4\frac{k}{l_0^2}\sin^2\alpha\cos^2\alpha\right). \tag{9.5.36}$$

In this case, the sign and the value of the coefficient K_x can also be influenced by the value of α, which gives the user an additional possibility for influencing the behaviour of the resulting oscillator.

Appendix

Mathematical Topics

A.1 Geometry

Angle

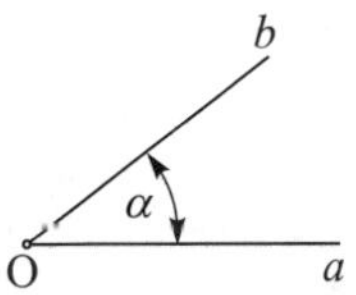

Acute: $\alpha < 90^\circ$
Right: $\alpha = 90^\circ$
Obtuse: $90^\circ < \alpha < 180^\circ$
Straight: $\alpha = 180^\circ$
Reflex: $\alpha > 180^\circ$

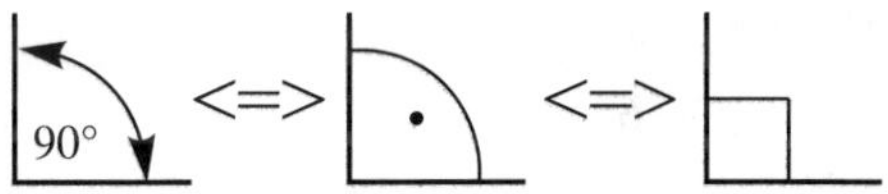

Complementary angles
$\alpha + \beta = 90^\circ$

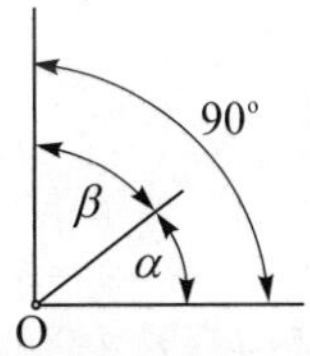

Supplementary angles
$\alpha + \beta = 180^\circ$

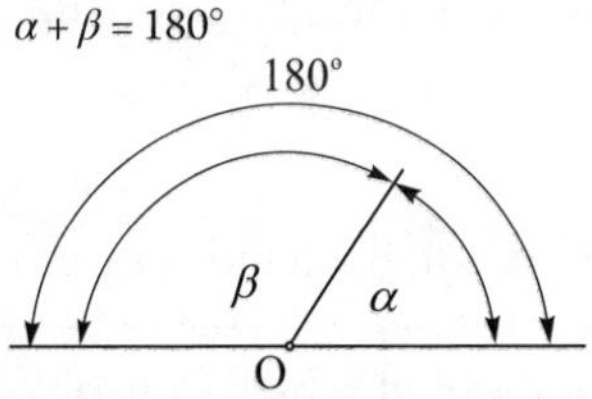

Equal (congruent) angles $\alpha = \beta$:

Opposite angles

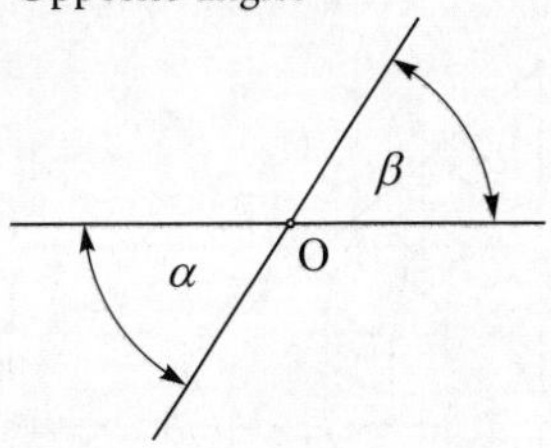

Transversal angles

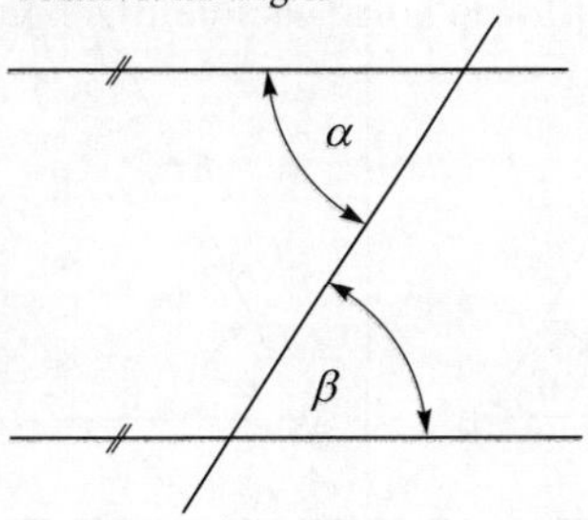

Mechanical Vibration: Fundamentals with Solved Examples, First Edition.
Ivana Kovačić and Dragi Radomirović.

Angles with parallel rays	Angles with orthogonal rays
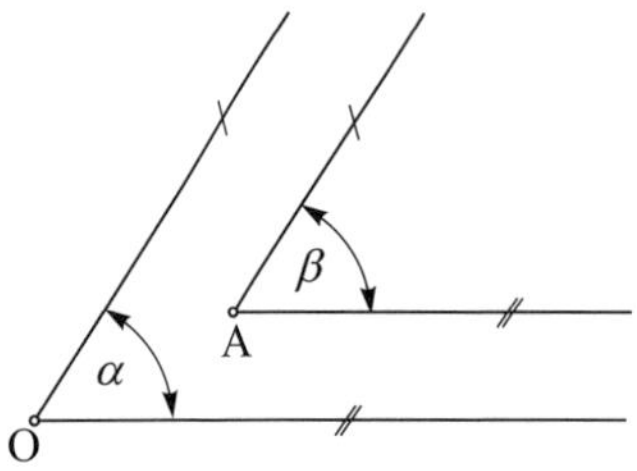	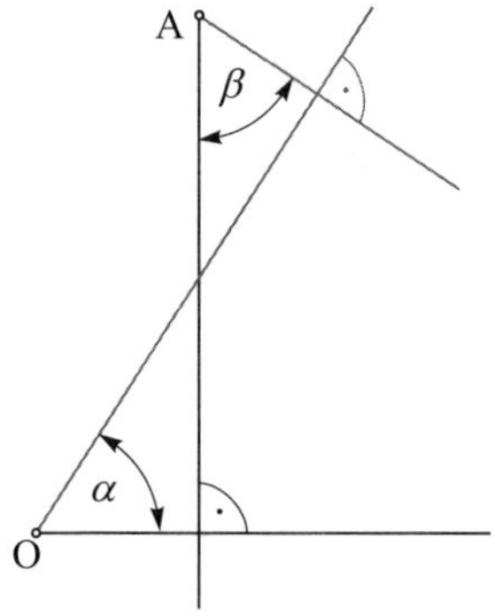

Triangle

$\alpha + \beta + \gamma = 180°$

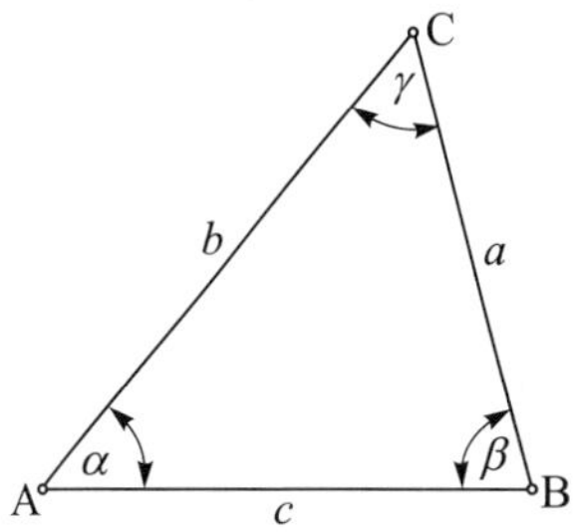

Isosceles triangle	Equilateral triangle	Right triangle $a^2 + b^2 = c^2$ c – hypotenuse, a, b – sides
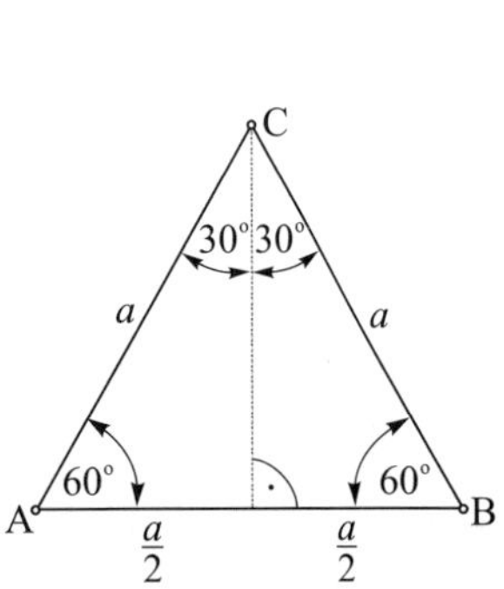	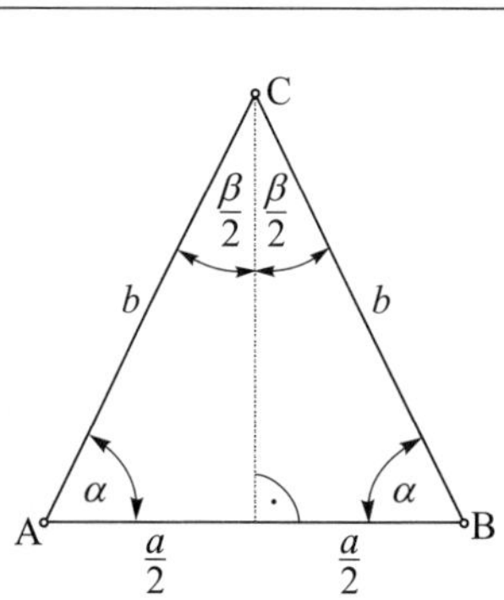	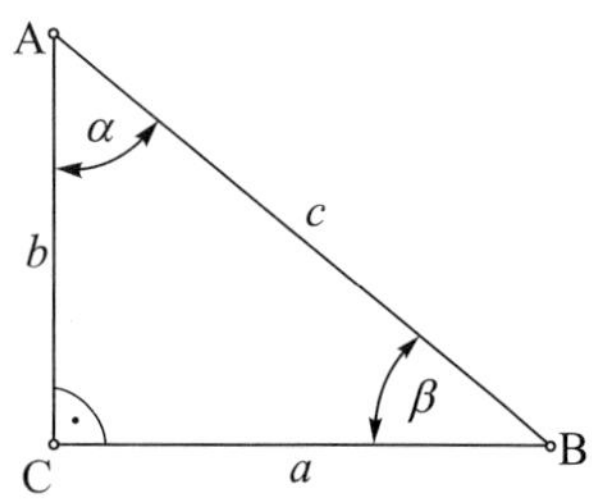

A.2 Trigonometry

Trigonometric functions:

$$\sin\alpha = \frac{\text{opposite side}}{\text{hypotenuse}} \qquad \cos\alpha = \frac{\text{adjucent side}}{\text{hypotenuse}}$$

$$\tan\alpha = \frac{\text{opposite side}}{\text{adjucent side}} = \frac{\sin\alpha}{\cos\alpha} \qquad \cot\alpha = \frac{\text{adjucent side}}{\text{opposite side}} = \frac{\cos\alpha}{\sin\alpha}$$

If α and β are complementary, then: $\sin\alpha = \cos\beta$, $\cos\alpha = \sin\beta$.

Trigonometric table

α	0°	30°	45°	60°	90°	180°	270°
$\sin\alpha$	0	1/2	$\sqrt{2}/2$	$\sqrt{3}/2$	1	0	−1
$\cos\alpha$	1	$\sqrt{3}/2$	$\sqrt{2}/2$	1/2	0	−1	0
$\tan\alpha$	0	$\sqrt{3}/3$	1	$\sqrt{3}$	$\pm\infty$	0	$\pm\infty$
$\cot\alpha$	$\pm\infty$	$\sqrt{3}$	1	$\sqrt{3}/3$	0	$\pm\infty$	0

Miscellaneous relations: –

$$\begin{cases} \sin^2\theta + \cos^2\theta = 1 \\ \sin(\theta \pm \varphi) = \sin\theta\cos\varphi \pm \cos\theta\sin\varphi \\ \cos(\theta \pm \varphi) = \cos\theta\cos\varphi \mp \sin\theta\sin\varphi \\ \sin 2\theta = 2\sin\theta\cos\theta \\ \cos 2\theta = \cos^2\theta - \sin^2\theta \end{cases}$$

Law of Sines:

$$\frac{a}{\sin\alpha} = \frac{b}{\sin\beta} = \frac{c}{\sin\gamma}$$

Law of Cosines:

$$a^2 = b^2 + c^2 - 2bc\cos\alpha$$

$$b^2 = a^2 + c^2 - 2ac\cos\beta$$

$$c^2 = a^2 + b^2 - 2ab\cos\gamma$$

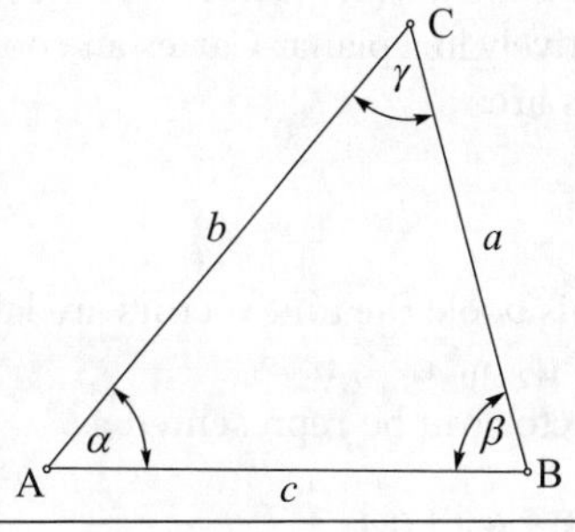

A.3 Algebra

Quadratic equation

$$ax^2 + bx + c = 0$$

$$x = \frac{-b \pm \sqrt{b^2 - 4ac}}{2a}$$

$$b^2 \geq 4ac \text{ for real roots}$$

Determinants

2nd order	3rd order
$\begin{vmatrix} a & b \\ c & d \end{vmatrix} = ad - bc$	$\begin{vmatrix} a & b & c \\ d & e & f \\ g & h & i \end{vmatrix} = \begin{vmatrix} a & b & c \\ d & e & f \\ g & h & i \end{vmatrix} \begin{matrix} a & b \\ d & e \\ g & h \end{matrix} = aei + bfg + cdh - gec - hfa - idb$

A.4 Vectors

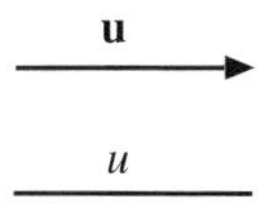

Scalar quantitites have only magnitude. Vector quantities have both magnitude and direction. For example, scalar quantities in mechanics are: mass, energy, time, work, and so on. Some vector quantities in mechanics are: force, position vector, velocity and acceleration.

Vectors are labelled in this book as upright bold. Thus, the vector **u** has a scalar magnitude *u*. In some other publications and longhand work, vectors are indicated by an arrow: $\vec{u}$.

Unit vectors

Unit vectors have a magnitude equal to unity. They are mainly used to define certain directions, such as, for example, the unit vectors **i** and **j**, which define the direction of the *x*-axis and *y*-axis respectively in a planar Cartesian coordinate system. Their magnitudes are:

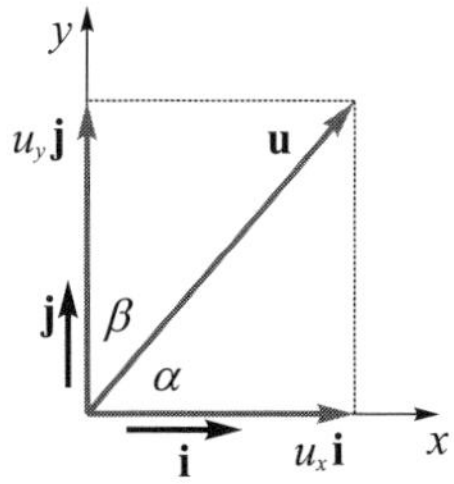

$$i = j = 1$$

In this book, the unit vectors are labelled by $\mathbf{i}$, $\mathbf{j}$, $\mathbf{i}_1$, $\mathbf{j}_1$, $\mathbf{i}_2$, $\mathbf{j}_2$, $\mathbf{e}_1$, $\mathbf{e}_2$, $\boldsymbol{\mu}_1$, $\boldsymbol{\mu}_2$, $\mathbf{r}_0$, $\mathbf{c}_0$, $\mathbf{t}$, $\mathbf{n}$.

A vector can be represented as:

$$\mathbf{u} = u_x\mathbf{i} + u_y\mathbf{j}$$

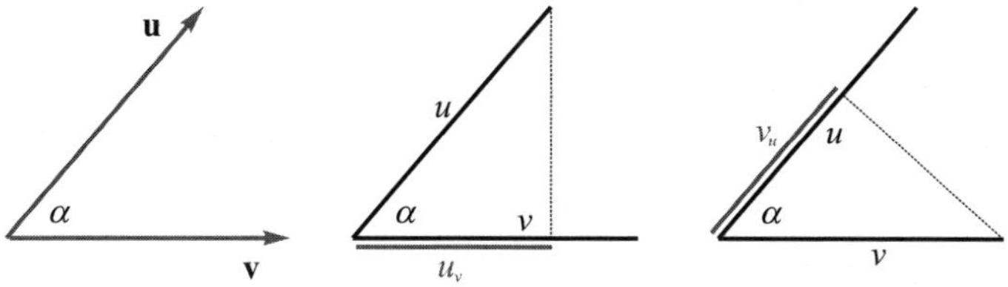

Scalar product

$$\mathbf{u} \cdot \mathbf{v} = w$$

$$w = uv\cos\alpha = vu\cos\alpha$$

$$w = vu_v = uv_u$$

A commutative law holds: $\mathbf{u} \cdot \mathbf{v} = \mathbf{v} \cdot \mathbf{u}$
Thus, the scalar products of the unit vectors are:

$$\mathbf{i} \cdot \mathbf{i} = \mathbf{j} \cdot \mathbf{j} = 1$$

$$\mathbf{i} \cdot \mathbf{j} = \mathbf{j} \cdot \mathbf{i} = 0$$

The scalar products of two vectors can be obtained as:

$$\mathbf{u} \cdot \mathbf{v} = \left(u_x\mathbf{i} + u_y\mathbf{j}\right) \cdot \left(v_x\mathbf{i} + v_y\mathbf{j}\right) = u_x v_x + u_y v_y$$

$$\mathbf{u} \cdot \mathbf{u} = u^2 = \left(u_x\mathbf{i} + u_y\mathbf{j}\right) \cdot \left(v_x\mathbf{i} + v_y\mathbf{j}\right) = u_x^2 + u_y^2$$

Projections on the axes are:

$$\mathbf{u} \cdot \mathbf{i} = u_x = u\cos\alpha$$

$$\mathbf{u} \cdot \mathbf{j} = u_y = u\cos\beta = u\sin\alpha$$

A.5 Derivatives (Notation: $\dot{f} \equiv \frac{df}{dt}$)

$$\frac{d}{dt}(t^n) = nt^{n-1}$$

$$\frac{d}{dt}(e^t) = e^t$$

$$\frac{d}{dt}(\sin t) = \cos t$$

$$\frac{d}{dt}(\cos t) = -\sin t$$

$$\frac{d}{dt}(\cosh t) = \frac{d}{dt}\left(\frac{e^t + e^{-t}}{2}\right) = \sinh t$$

$$\frac{d}{dt}(\sinh t) = \frac{d}{dt}\left(\frac{e^t - e^{-t}}{2}\right) = \cosh t$$

Rules:

Rule	Function	Derivative
Multiplication by constant	Cf	$C\dot{f}$
Sum rule	$u \pm v$	$\dot{u} \pm \dot{v}$
Product rule	uv	$\dot{u}v + u\dot{v}$
Quotient rule	$\frac{u}{v}$	$\frac{\dot{u}v - u\dot{v}}{v^2}$
Chain rule	$\frac{d}{dt}(f(g(t)))$	$\frac{df}{dg}\frac{dg}{dt}$

A.6 Variation (Virtual Displacements)

$$f = f(q) \Rightarrow \delta f = \frac{df}{dq}\delta q = f'\delta q$$

$$f = f(x,y) \Rightarrow \delta f = \frac{\partial f}{\partial x}\delta x + \frac{\partial f}{\partial y}\delta y$$

A.7 Series

A Maclaurin series is a series expansion of a function about 0, and is defined by

$$f(q) \approx f(0) + f'(0)q + \frac{f''(0)}{2!}q^2 + \frac{f'''(0)}{3!}q^3 + \ldots + \frac{f^{(n)}(0)}{n!}q^n + \ldots$$

Maclaurin series for some functions are:

$$\sin q \approx q - \frac{q^3}{6} + \frac{q^5}{120} + \ldots$$

$$\cos q \approx 1 - \frac{q^2}{2} + \frac{q^4}{24} + \ldots$$

$$e^q \approx 1 + q + \frac{q^2}{2} + \ldots$$

A Maclaurin series may also be generalised to functions of more than one variable. This generalisation is given here for functions of two variables:

$$f(x,y) \approx f(0,0) + f_x(0,0)x + f_y(0,0)y + \frac{1}{2}\left[f_{xx}(0,0)x^2 + 2f_{xy}(0,0)xy + f_{yy}(0,0)y^2\right] + \ldots$$

Here the subscripts denote the respective partial derivatives.

Index

Mechanical Vibration: Fundamentals with Solved Examples, First Edition.
Ivana Kovačić and Dragi Radomirović.